Nanotechnologies and Nanomaterials Applied to Chemical Sensors and Biosensors

Series Editor
Jean-Charles Pomerol

Nanotechnologies and Nanomaterials Applied to Chemical Sensors and Biosensors

Applications to the Environment, Medicine and Health

Pierre Camille Lacaze
Benoît Piro
Jean-Christophe Lacroix

iSTE

WILEY

First published 2024 in Great Britain and the United States by ISTE Ltd and John Wiley & Sons, Inc.

ISTE Ltd
27-37 St George's Road
London SW19 4EU
UK

www.iste.co.uk

John Wiley & Sons, Inc.
111 River Street
Hoboken, NJ 07030
USA

www.wiley.com

Library of Congress Control Number: 2024941764

British Library Cataloguing-in-Publication Data
A CIP record for this book is available from the British Library
ISBN 978-1-78630-660-9

Contents

Introduction

The environment, health and the agri-food sector are areas that are subject to very strict regulatory standards with an increasing use of detection and analysis systems (sensors) to carry out the numerous controls and verifications. A sensor is thus a device that measures, continuously where possible, either a physical quantity (temperature, pressure, light intensity, object position, acceleration, etc.) or detects and accurately measures the concentration of a chemical or biological compound present in the surrounding environment.

Long before nanotechnologies emerged, the design of all kinds of chemical sensors was a hot topic in analytical chemistry. The operation of a chemical or biochemical sensor involves a cause and effect relationship between three entities: the compound to be analyzed (analyte)[1], the receptor and a transduction device that transforms the interaction between the analyte and the receptor into a measurable signal, which, depending on the case, can be electrical, electrochemical, optical, thermal or gravimetric (Figure I.1).

One of the oldest commercialized systems is probably the electrochemical detection of glucose in the blood of the human body, initially proposed by Clark and Lyons in 1962 (Clark and Lyons 1962), which carries the greatest number of associated research studies and has seen the greatest technological evolution. This is how various enzymatic or non-enzymatic detection systems subsequently emerged, based on nanomaterials, and were miniaturized and designed as portable systems, making it possible to carry out precise continuous measurements of glucose in the blood (Wang 2008), with a semi-invasive approach (e.g. by insertion of a transcutaneous needle) or by new methods aiming toward non-invasive systems (Chen et al. 2017).

1 Analyte refers to the chemical entity that we wish to identify and/or quantify.

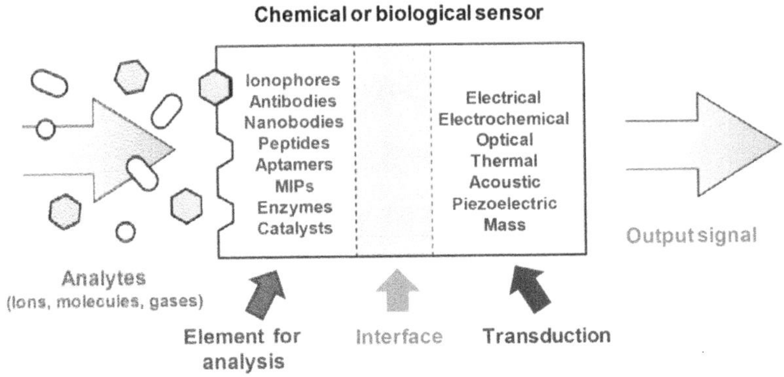

Figure I.1. *Working principle of a chemical sensor or biosensor Adapted from Chambers et al. (2008). For a color version of this figure, see www.iste.co.uk/lacaze/nanotechnologies.zip*

Over the last few decades, numerous increasingly efficient chemical and biological sensors have been manufactured, including receptors made of carbon nanomaterials (carbon nanotubes [CNTs], graphene and carbon nanodots) as well as inorganic nanomaterials (various shaped metal nanoparticles, nanocrystals, nanowires, semiconductor quantum dots [QDs]). The advantages provided by these nanomaterials are significant, particularly when they are involved in the design of systems involving electrical or optical transduction. In the case of graphene and CNTs, their very high conductivity as well as their very high charge mobilities, which are susceptible to significant variations as soon as slight surface modifications occur, make them extremely sensitive receptors. Furthermore, the very high specific surface areas of these materials (~2,600 m^2/g for graphene) mean that, in the form of an atomic monolayer, they are capable of adsorbing significant quantities of the compound to be measured and, therefore, of producing a high intensity transduction signal, much better than that obtained with other micron-sized carbonaceous materials.

In parallel with this research, significant efforts have been devoted to the "portability" of very high sensitivity sensors equipped with various transduction systems, with a signal amplification effect based on the development of the sensor in the form of a field-effect transistor (FET). The ultimate design objective is that these sensors have total autonomy in terms of operation based on the capture and transformation of surrounding energy into electrical energy (e.g. self-charging power systems) (Kausar et al. 2014; Pu et al. 2018a).

This book at a glance

Our objective is to analyze the different challenges faced in the design of a sensor and to show what methods are used to, on the one hand, separate the analytes from a mixture and, on the other hand, choose a physical method that gives an intense transduction signal leading to selective detection. This book therefore has two parts: the first is a description of the most common nanomaterials used in the making of sensors and the physical methods to identify and separate these nanomaterials; the second is more applied and describes the different sensors used in the environmental and biomedical fields.

Part 1: nanomaterials, separation, amplification, recognition and transduction

Chapter 1 describes the properties of some of the most important organic and inorganic nanomaterials. We highlight the main contribution of new carbon nanomaterials such as CNTs, graphene, carbon nanodots, conductive polymers in thin layers, but also various forms of inorganic nanomaterials (nanowires, nanotubes and nanosheets), which have allowed enormous progress in the design of portable detection and analysis systems (Kim et al. 2019).

Chapter 2 is devoted to separation methods of compound mixtures and concerns the different chromatography and electrophoresis techniques. Furthermore, DNA amplification techniques necessary for element identification and sequencing, often present only in trace amounts, are also briefly described.

Chapter 3 develops the main analyte–receptor recognition systems, which generally involve pairs of self-assembling molecules, with very high affinities for each other, as observed with the protein pairs avidin–biotin, antibody–antigen, as well as associations between complementary DNA strands.

Chapter 4 describes the characteristics of the main physical methods used to produce a transduction signal, based essentially on the use of electronic, electrochemical, piezoelectric or optoelectronic devices, illustrated using some examples.

Part 2: environmental and biological sensors

Chapter 5 is devoted to chemical sensors specific for environmental issues. It mainly concerns the analysis of mineral salts and gases in terms of toxicology. This is a field that has seen significant developments in ion-selective (membrane and ionophore) electrodes (ISE), subsequently adapted to the production of all-solid portable devices with, in the latter case, a strong involvement of nanomaterials. The same evolution occurred for the analysis of gases, applied mainly to a few toxic gases.

Chapter 6 focuses primarily on biomedical analysis problems. Innovative solutions are proposed with the contribution of nanomaterials (Lin et al. 2018) being decisive. The real-time control of the primary molecules secreted by the body is a main focal point that we describe in the case of glucose, urea and cholesterol, three key chronic disease-causing molecules. Biomarkers of certain diseases (cancers, heart diseases, etc.), food pathogens, as well as viruses and bacteria are also areas covered, for which portable analysis devices are proving decisive for their effective prevention.

PART 1

Nanomaterials, Amplification, Separation, Recognition and Transduction

1

Nanomaterials

This chapter[1] is not an exhaustive review of nanomaterials. Its objective is to describe some of the carbon-based and inorganic nanomaterials, frequently used in the design of sensors. Some nanomaterials with opto-electronic properties that have not really been used in the design of sensors but also have specific properties that have led to their application in the biomedical field, in particular in biological imaging, will also be described.

1.1. Carbon nanomaterials

Fullerenes, carbon nanotubes (CNTs), nano-diamonds (NDs) and carbon quantum particles or carbon quantum dots (CQDs) were all discovered before the 2000s, followed by graphene in 2005. Due to their remarkable physicochemical properties, fullerenes, CNTs and graphene (G) have given rise to innumerable applications and their exploitation has remained continuous, but with a much quicker development of G compared to fullerenes and CNTs.

1.1.1. *Fullerenes*

In 1985, Kroto et al. (1985) discovered Buckminsterfullerene C_{60} (hereinafter referred to as fullerene C_{60}) and revealed its balloon-shaped structure, consisting of an assembly of sp^2 carbon atoms, organized in the form of pentagons (12) and hexagons (20). Fullerene chemistry then developed and gave rise to a multitude of applications, in fields as diverse as electronics, energy, biology and medicine. Along with C_{60} and C_{70}, the two most stable forms (Figure 1.1), new fullerenes were subsequently synthesized, with more carbon atoms.

1 This chapter is inspired by one of our previous books entitled *Nanotechnology and Nanomaterials for Energy* (Lacaze and Lacroix 2021). We have reused some illustrations.

1.1.1.1. *Electrophilic and antioxidant properties of fullerenes*

The electrophilic nature of fullerenes allows the formation of anions with six negative charges, corresponding to the acceptance of six electrons by lower unoccupied molecular orbitals (LUMOs). The progressive electrochemical reduction in six steps of C_{60} and C_{70} shows that the six reduction states of fullerene can be obtained with good stability (Xie et al. 1992) (Figure 1.1).

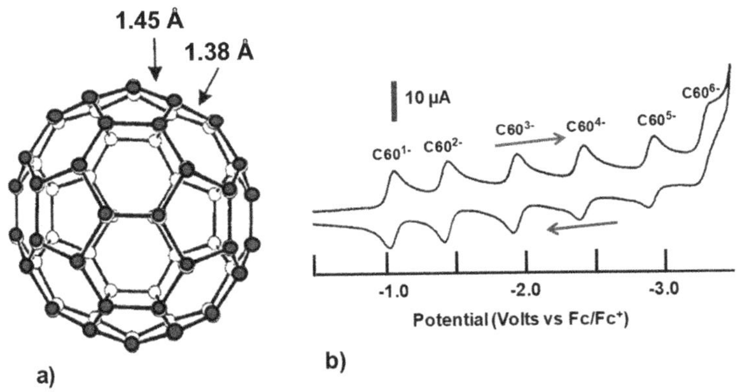

Figure 1.1. *(a) Image of fullerene C_{60}. Adapted from Balch et al. (1998). (b) Voltametric redox curves of C_{60} in solution in an acetonitrile/toluene mixture at low temperature (–10°C). Adapted from Xie et al. (1992). For a color version of this figure, see www.iste.co.uk/lacaze/nanotechnologies.zip*

COMMENT ON FIGURE 1.1.– *The red and white symbols (a) represent the C atoms in front and behind the projection plane. The reduction curve obtained by a potential sweep from –0.5 to –3.5 V (blue arrow) clearly shows the six reduction peaks, each corresponding to a gain of one electron. The reverse oxidation curve from –3.5 to –0.5 V (red arrow) shows six oxidation peaks, corresponding to the six intermediate and successive anionic states, thus proving the stability and reversibility of these six redox states. Scan rate: 100 mV/s.*

This electronic affinity has been exploited in the fields of medicine and biology. C_{60} and its derivatives, functionalized by various hydrophilic groups, have been used in the fight against cancer and AIDS (antiretroviral therapy against HIV-1). Functionalized by carbohydrate chains, they have also proven to be good antibacterial agents. Their high affinity for radicals, a property resulting from radical addition to the many fullerene double bonds, means that they are considered as "radical sponges". They are powerful antioxidants, used in biology to neutralize radical oxygen species (ROS), such as the superoxide ion $O_2^{-\bullet}$, hydroxyls $HO^{\bullet}$ or hydrogen peroxide H_2O_2, particularly impressive regarding DNA and certain proteins. All these properties used for medical purposes have been described in review articles (Bakry et al. 2007; Lalwani and Sitharaman 2013; Acquah et al. 2017; Castro et al. 2017).

1.1.1.2. *Chemical reactivity and exofunctionalization*

Given their cage-shaped carbon structure, involving only covalent bonds between sp^2 carbons, fullerenes are naturally hydrophobic and therefore insoluble in aqueous environments. Their use for biological purposes requires transformations to make them hydrophilic, which is achieved by the grafting of hydroxyl groups (COOH or OH), corresponding to exofunctionalization reactions, typically carried out with C_{60}, and some commercially produced fullerenes (C_{70}, C_{80}, etc.), however at high prices, because of the difficulties in producing them in large quantities (Taylor and Walton 1993; Georgakylas et al. 2015).

Among all the products derived from fullerenes, fullerenols and carboxyfullerenes, obtained by grafting hydroxyl or carboxylic groups onto the surface of fullerenes, have the advantage over simple fullerenes in that they are soluble in aqueous and biological media. This is the case of hexacarboxylated fullerenes (carboxyfullerenes), which are fullerenes with three pairs of carboxylic acids, composed of a mixture of two stereoisomers, C3-C_{60} and D3-C_{60} (Figure 1.2)[2].

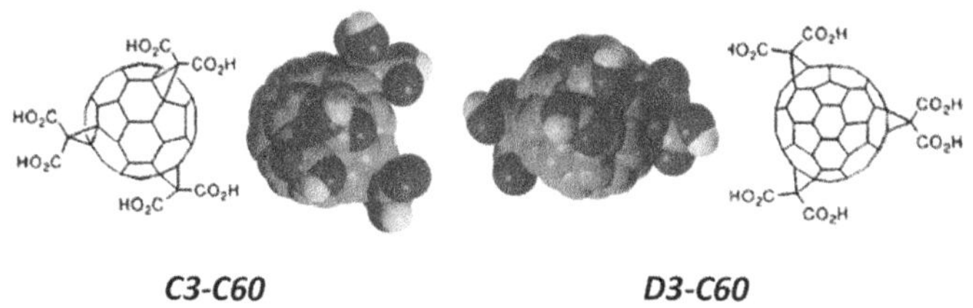

Figure 1.2. *Structure of the two stereoisomeric carboxyfullerenes C3-C_{60} and D3-C_{60} as per Dugan et al. (1997). For a color version of this figure, see www.iste.co.uk/lacaze/nanotechnologies.zip*

Like simple fullerenes, but with the added advantage of solubility in aqueous media, they have strong antioxidant properties, including the ability to destroy the peroxide ion $O_2^{\bullet -}$, a toxic by-product of cell metabolism. This property makes them very good neuroprotective agents, and research in this area is still ongoing (Dugan et al. 1997; Ali et al. 2004; Gharbi et al. 2005; Ye et al. 2015).

---

2 Both isomers, C3-C_{60} and D3-C_{60}, obtained by reacting ethyl 2-bromomalonate with C_{60} in the presence of NaH, belong to different symmetry groups (C3 and D3).

1.1.1.3. *Endometallofullerenes (EMFs)*

The presence of a cavity allows the encapsulation of chemical species, which is of interest for the design of various markers, used for example in medical imaging. As early as 1985, Smalley et al. identified with mass spectrometry the first fullerene with 60 carbon atoms containing a lanthanum atom (Heath et al. 1985). A few years later, they isolated several metallofullerenes with 60, 70, 74 and 82 carbon atoms, the latter La@C_{82}[3] being the only one that is stable in contact with air (Chai et al. 1991).

What is remarkable is that for all these new compounds there is no metal release when the compound is placed in a biological medium, which is a considerable improvement compared to metal chelates. This stability is likely due to the transfer of electrons between the La atom and the fullerene, which leads to the ion pair La^{3+}@C_{82}^{3-}, and to the fact that fullerene meshes are small enough that they prevent diffusion of the ion La^{3+} outward.

This discovery, which opened promising prospects in the medical field for diagnostic and therapeutic applications, triggered significant research into the synthesis of new EMFs, with some having applications in cancer therapy. Thus, gadolinium fullerenol Gd@$C_{82}(OH)_{22}$, initially used as a contrast agent in nuclear magnetic resonance imaging (MRI), also turned out to have a strong anti-cancer activity, different from that of simple fullerene, with low cytotoxicity (Kang et al. 2014)[4].

1.1.2. *Carbon nanodiamonds (NDs)*

Although known since the 1960s[5], they only began to be exploited at the end of the 1990s when they were produced by detonation of a mixture of explosive

3 The notation La@C_{60} indicates that the lanthanum atom is inside the cage formed by the fullerene.

4 The gadolinium ion (Gd^{3+}), used as a contrast agent in NMR imaging, has the disadvantage of being very toxic. This toxicity can be greatly reduced when Gd is encapsulated in a hydrophilic fullerene, soluble in biological media, after exo-functionalization by OH or COOH groups of the fullerene molecule.

5 First discovered by Danilenko et al. in 1963 and overlooked by the scientific community, they were synthesized a second time in 1982 by another Russian team, during research into the synthesis of diamonds using the detonation technique in a pressurized chamber (Danilenko 2004). Obtained in the form of carbonaceous soot and considered uninteresting by-products, they only really began to interest scientists when they realized their functionalization capacity and areas of exploitation in the biomedical field.

products (trinitrotoluene [TNT] and trinitrobenzene [TNB]) (Greiner et al. 1988). These nanomaterials are now produced commercially. They are used for medical, diagnostic and therapeutic purposes, or to improve the mechanical properties of plastics.

From a mechanical point of view, NDs inherited properties from pure diamonds. They are characterized by their very high hardness and a very high Young's elasticity modulus, making them ideal for use in polishing hard surfaces (ceramics). Their very high chemical stability also allows them to be used in very aggressive environments. Two other particularly interesting properties are their fluorescence and biocompatibility, which make them suitable for biomedical applications (diagnosis and treatment) due to their easy surface functionalization.

Fluorescent nanodiamonds (FNDs) are a new family of nanomaterials, with sizes between 35 and 100 nm (Hsiao et al. 2016) and characterized by the presence of a structural defect inside the crystal (Figure 1.3(a)). This defect, called the nitrogen-vacancy (NV) defect, corresponds to a coupling between a vacancy (absence of a carbon atom in the lattice) and a nitrogen atom, adjacent to the vacancy. It can be easily created by irradiation of NDs crystallites with a helium (He^+) or proton (H^+)[6] ion beam. The processed NDs (equivalent to n-doped NDs) emit stable red fluorescence when excited by a laser. The intensity of the fluorescence depends on the concentration of NVs, which itself is higher the higher the energy of the He^+ ion beam used for their manufacture (the concentration of NVs can vary between 10 and 30 ppm in relation to the number of carbon atoms; when unprocessed their concentration is less than 1 ppm)[7].

Their biocompatibility, high stability, absence of toxicity as well as their high fluorescence intensity in the red make them diagnostic tools of choice in biology rather than inorganic quantum dots (QDs), which are most often made up of toxic elements Cd, Se, Pb, etc. (Chambers et al. 2008).

6 The sub-micron diamond powders used to make FNDs are obtained at high temperature and high pressure and contain approximately 100 ppm of nitrogen atom impurities dispersed throughout the diamond structure. Under the effect of ion bombardment, vacancies (V) are produced, their migration and their proximity to nitrogen atoms (N) leads to the formation of NVs centers.

7 NVs centers can exist in two forms NV° and NV^-, with NV^- dominating. NV^- determines the magneto-optical properties due to its different spin numbers $m_s = 0$ and ± 1. NVs are characterized by a strong absorption band centered at 550 nm and an intense fluorescence band around 685 nm, with a high quantum yield (70%) (Schirhagl et al. 2014).

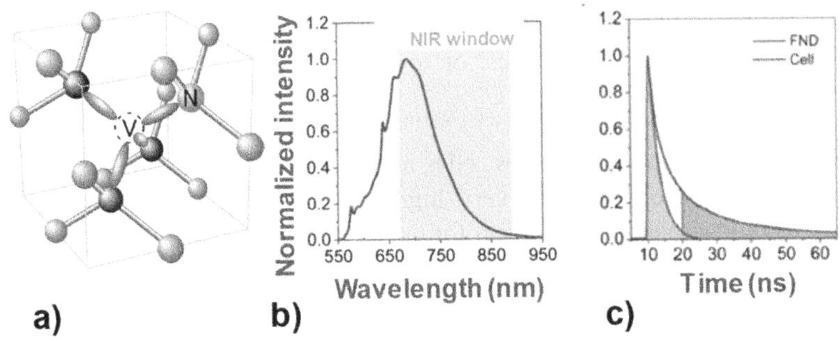

Figure 1.3. *Fluorescence of ND nanocrystals with NV defects. (a) Structure of an NV defect. (b) Fluorescence spectrum of FNDs (size 35 nm) obtained by laser excitation at 532 nm. (c) Differences between the fluorescence lifetimes of an FND (pink) and that of biological tissue (cell) due to the presence of endogenous fluorophores (green). According to Hsiao et al. (2016). For a color version of this figure, see www.iste.co.uk/lacaze/nanotechnologies.zip*

COMMENT ON FIGURE 1.3.– *An NV defect results from the proximity between a nitrogen atom (impurity due to the preparation of NDs) and a vacancy corresponding to the absence of a carbon atom. The ovals shown in purple (a) correspond to the dangling bonds of the carbon atoms neighboring the vacancy. The rectangle colored in pink (b) represents the spectral operating window for NDs, for which the autofluorescence of a biological tissue (green curve in c) is negligible. The fluorescence of the NDs can thus be detected by imposing a delay of 10 ns between the excitation and the detection of the emission (c).*

1.1.3. *Carbon quantum dots*

Less than 10 nm in size, they comprise a graphitic sp^2 carbon core, corresponding to fragments of graphene or graphene oxide (GO), linked together by sp^3 carbons. They contain significant amounts of oxygen on their surface, mainly as COOH, carbonyls and hydroxyl groups (Lim et al. 2015) (Figure 1.4).

Discovered by chance during the preparation and purification of CNTs (Xu et al. 2004), they quickly generated great interest because of their fluorescence properties, which make them potential biomedical imaging reagents. As with NDs, their non-toxicity is also an additional advantage, favoring their use in the medical field over inorganic QD semiconductors, which are often made with heavy metals.

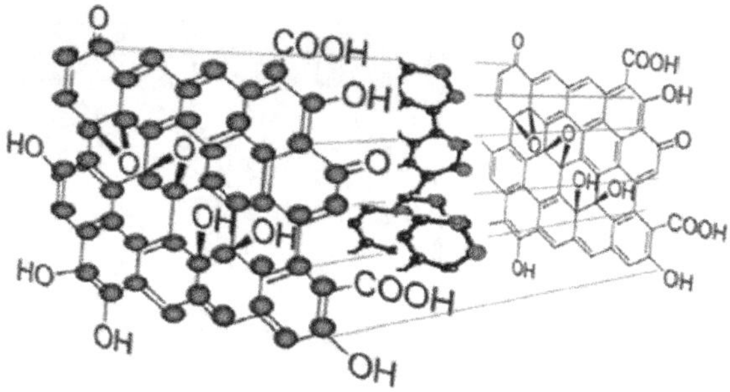

Figure 1.4. *Structure of carbon quantum dots (CQDs) showing the association between graphene oxide and graphene fragments. According to Lim et al. (2015) and Demchenko and Dekaliuk (2013)*

Unlike inorganic QDs, for which the absorption is determined by the band gap, CQDs are characterized by a very wide absorption band resulting from plasmonic absorption of the π[8] orbitals. This absorption band covers a large part of the UV-visible spectrum and extends into the near infrared spectrum, thus intercepting a large part of the solar spectrum (Fernando et al. 2015).

Two main emission mechanisms, not always easy to identify, are responsible for the fluorescence of CQDs. The first results from transitions between energy levels of sp^2 carbon bonded domains. The second is less obvious and it involves the surface defects present on the CQDs. A third type of mechanism, corresponding to up-conversion, has also been identified, where an emission is observed in the visible due to an excitation in the red (Figure 1.5).

The photoexcitation of CQDs is said to produce a separation of charges in the carbon cores. Their recombination with surface defects produces fluorescence, which also explains the great diversity of emissions, different from that observed with inorganic semiconductor QDs, characteristic of their band gaps.

One of the most striking characteristics of CQD fluorescence emission is the dependence of the emission spectrum on the excitation source. This has been attributed to different factors, such as the quantum size effect, the defects and surface states present in CQDs, the presence of fluorophores with different conjugation lengths, the radiative recombination of electron-holes due to sp^2 carbon clusters within a sp^3 carbon matrix, each of which are decisive and can contribute to fluorescence (Wang and Hu 2014).

---

8 Plasmonic absorption corresponds to a collective excitation of π electrons delocalized over several benzene cycles.

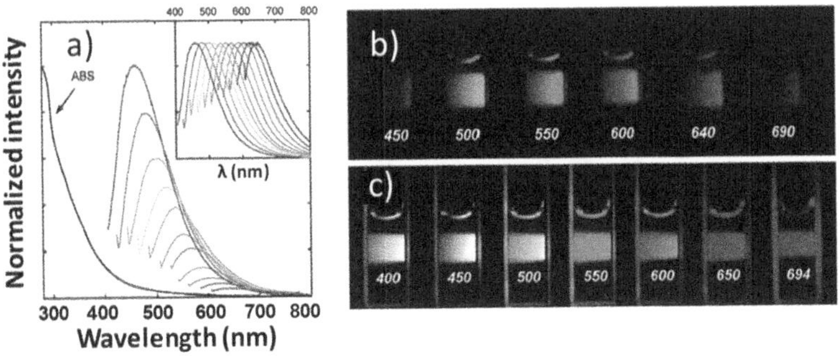

Figure 1.5. *Luminescence of an aqueous solution of CQDs produced by laser ablation of a graphite powder. (a) Absorption (ABS) and luminescence spectra obtained by excitation at different wavelengths (from 400 to 600 nm in 20 nm increments – from left to right). (b) Photographs of the luminescence of solutions of CQDs functionalized with polyethylene glycol (PEG) with a molar mass of 1,500 and comprising two amino groups at its chain ends* $–CH_2NH_2$ *(PEG_{1500N}). (c) The same solutions of CQDs are directly photographed after excitation at 400, 450 and 694 nm. According to Sun et al. (2006). For a color version of this figure, see www.iste.co.uk/lacaze/nanotechnologies.zip*

COMMENT ON FIGURE 1.5.– *CQDs (a) are functionalized by poly(propionylethyleneimine-co-ethyleneimine) (PPEI-EI), and the intensities are normalized to the quantum yields. Normalization is performed relative to peak intensity. The photographs in (b) correspond to an excitation at 400 nm of the CQDs and are obtained after passing the emitted light through color filters of 450, 500...690 nm.*

The fluorescence quantum yields of most CQDs are typically around 10%; in some cases, however, high yields of around 80% have been observed and seem to be linked to the presence of nitrogen resulting from the precursors used during pyrolysis (citric acid and ethylene diamine) (Zhu et al. 2013).

Finally, for certain CQD preparation conditions, multiphoton absorption was also observed, corresponding to the up-conversion of the emitted photons, which is an essential property for biomedical imaging, where excitation can be produced in the near-infrared spectrum (800–900 nm) with very high penetration of biological tissues (several mm). Such a property was discovered by Sun and colleagues (Cao et al. 2007), with CQDs of sizes of around 5 nm, passivated by a nitrogen-containing polymer (polypropionylimine-co-ethylenimine).

Given the ease of their production and the numerous functionalization possibilities of their surface, the production of various multifunctional probes can be considered. By combining, for example, magnetism and luminescence properties, MRI and fluorescence imaging can be performed simultaneously. Such a result has been demonstrated with the fixation of nuclear resonance contrast agents (gadolinium complexes) on fluorescent CQDs (Bourlinos et al. 2012).

Their solubility in an aqueous medium makes them diagnostic agents and very good vectors for transporting drugs to diseased biological cells. For instance, platinum salt-based anticancer drugs have been bound onto CQDs (Zheng et al. 2014a). Other applications have also been described (Wang and Hu 2014; Hu et al.2019a) and relate to optoelectronics (photovoltaics and electroluminescence), energy and its storage (supercapacitors and electrodes for lithium batteries) and analytical chemistry with the design of various sensors based on optoelectronic phenomena (Molaei 2020; Nazri 2021).

1.1.4. *Carbon nanotubes*

Iijima (1991) was the first to show the formation of multi-wall carbon nanotubes (MWCNTs) through the electrical discharge between carbon electrodes under a reduced argon-pressurized environment. He thus obtained a mixture of MWCNTs with diameters between 4 and 30 nm, approximately 1 μm long, made up of two to seven concentric tubes. Single-wall carbon nanotubes (SWCNTs) were obtained shortly after using cobalt-based catalysts (Bethune et al. 1993). These nanotubes have walls with a structure identical to that of graphene, which is made up of an assembly of sp^2-hybridized carbon atom monolayers, forming regular hexagons between them.

Since the first productions and characterizations of CNTs, numerous manufacturing and separation techniques have been introduced by several teams and are described in several articles (Grobert 2007; Hersam 2008; Komatsu and Wang 2010; Chen et al. 2014a). Essentially, the vaporization of carbon by laser ablation, the electrical discharge between graphite electrodes, catalytic chemical vapor decomposition (CCVD), the HiPco process (decomposition of CO under high pressure)[9] and the use of silica nanoparticles doped with Co (Co-MCM 41)[10] are the

9 The HiPco (high-pressure carbon monoxide) process was developed in the 1990s at Rice University. It makes it possible to obtain high purity SWCNTs by reacting carbon monoxide mixed with iron pentacarbonyl $Fe(CO)_5$ at high temperature (900–1,100°C) under a pressure of 30–50 atmospheres. The amount of CNTs produced using this process is around 450 mg per hour and provides a SWCNT mixture with 10 distinct chiralities (Bronikowski et al. 2001).

main CNT manufacturing methods. The choice of catalyst, as well as the carbon precursor fragmentation technique, is crucial. The CCVD technique is the most widespread as it is more likely to produce SWCNTs in large quantities, which is more favorable for production on a commercial scale (Figure 1.6).

The CVD technique typically involves introducing a gaseous carbon source (ethylene, acetylene, toluene, CO, etc.) mixed with a carrier gas (H_2, argon) into a reactor. The gas mixture, heated to a high temperature of at least 600°C corresponding to the decomposition of the carbon precursor, comes into contact with the catalyst (in the form of metal nanoparticles) on which the CNTs will develop.

The old electrochemical route is also the subject of research into the production of large quantities of carbon nanomaterials, such as graphene, carbon fibers, CDs and CNTs. The reduction of CO_2 in the form of molten carbonate is a solution that was recently developed by the Pint group to produce CNTs and has the merit of converting CO_2 into high added value products (Douglas and Pinta 2017).

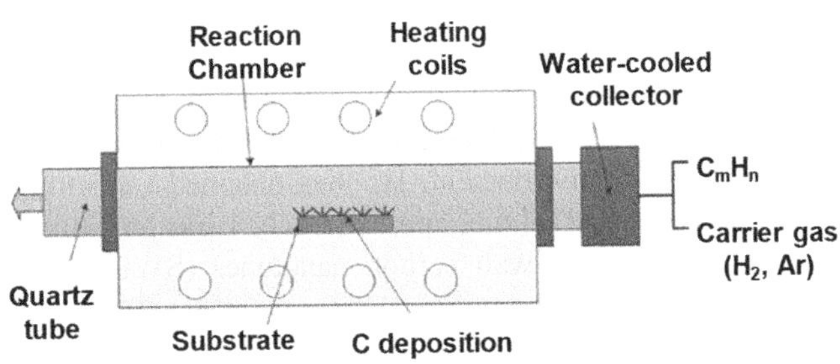

Figure 1.6. *Simplified schematic of the CVD device used to produce CNTs. Adapted from Gore and Sane (2011)*

Most manufacturing techniques produce CNT mixtures of multi-walled (MWCNTs) and single-walled nanotubes (SWCNTs) with different architectures, corresponding to distinct chiralities. The improvement of separation methods and the development of new catalysts, making it possible to increase the selectivity in the production of CNTs with a determined configuration or chirality, are important research topics.

10 MCM-41 is a particular zeolite whose pores can be adjusted between 2 and 10 nm. After depositing a thin layer of cobalt on its internal walls, the zeolite can be used as a nanoreactor to produce SWCNTs (Lim et al. 2003).

1.1.4.1. *Chirality of carbon nanotubes*

The formation of SWCNTs can be thought of as the result of rolling up a graphene fragment in a particular direction, defined by a C_h vector (Figure 1.7). This C_h vector, known as the chirality vector, is defined by the sum of vectors $\mathbf{a_1}$ and $\mathbf{a_2}$, forming a 60° angle between them. C_h is equal to n $\mathbf{a_1}$ + m $\mathbf{a_2}$, and the CNT with a chirality of (n, m) is thus obtained by rolling up the graphene fragment according to the direction of the vector C_h to connect the two ends of the vector. This makes the SWCNT circumference equal to the length of the C_h vector (Bachilo et al. 2002).

This chirality is particularly important because it determines the electrical properties of CNTs and, consequently, their use in fields such as nanoelectronics, photovoltaics, energy storage and sensors.

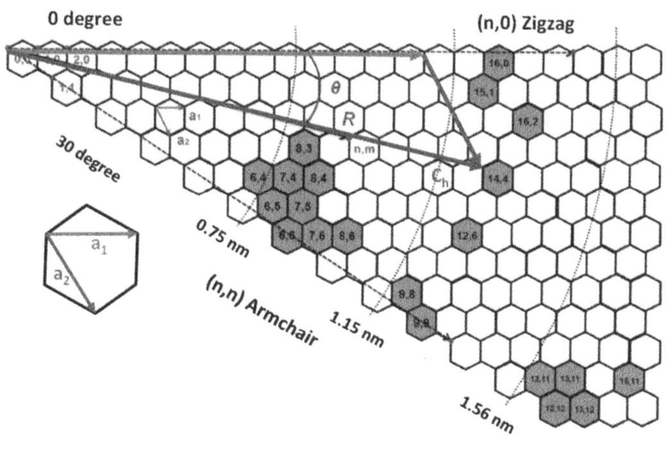

Figure 1.7. *Formation of an SWCNT defined by the rolling up of a graphene fragment along the direction of the chiral vector $C_h = na_1 + ma_2$. Adapted from Wang et al. (2015). For a color version of this figure, see www.iste.co.uk/lacaze/nanotechnologies.zip*

COMMENT ON FIGURE 1.7.– *The C_h vector (red arrow) equal to $14a_1 + 4a_2$ corresponds to the formation of an SWCNT with the chirality (14,4). An angle of Θ with respect to the a_1 vector axis defines an SWCNT (4,14) with R-type chirality (rectus), which has the same diameter as S-SWCNT (14,4) (sinister (S) chirality). The blue hexagons correspond to the most common CNT chiralities obtained synthetically with diameters between 0.7 and 2 nm.*

For SWCNTs with different chiralities (n,m), two limit forms, Armchair (n,n) and Zigzag (n,0) structures, are important and correspond to Θ angles of 0° and 30°, respectively (Figure 1.8). The Armchair (n,n) configuration has a metal-like conductivity; the Zigzag configuration (n,0) and SWCNTs of chiralities (n,m) also have metal-like conductivities but only when the difference between n and m is a multiple of 3 (Maultzsch et al. 2005) and are semiconductors when it is not the case. Band gaps are less than 1 eV and decrease as the CNT diameter increases (Odom et al. 1998).

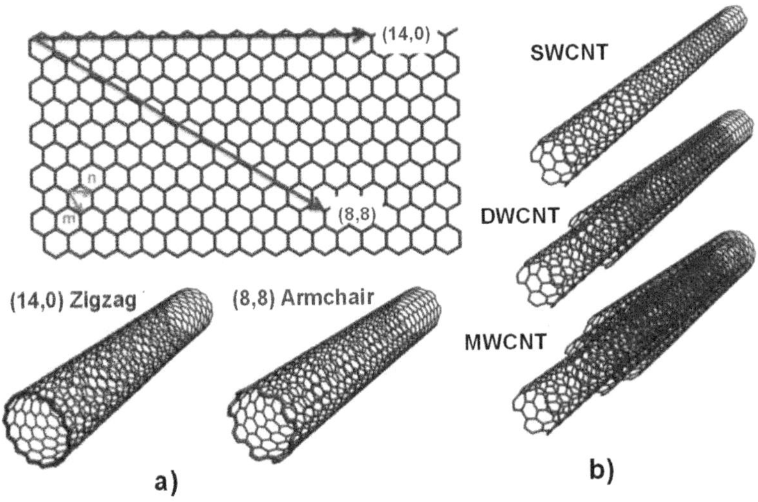

Figure 1.8. *Examples of CNTs with different configurations. (a) Examples of two SWCNTs (14,0) and (8,8) constructed from a graphene sheet, rolled according to the vectors (14,0) (red) and (8,8) (green) and corresponding to the two Zigzag and Armchair configurations. (b) Images of single-walled (SWCNT), double-walled (DWCNT) and multi-walled (MWCNT) CNTs. Adapted from Schnorr and Swager (2011). For a color version of this figure, see www.iste.co.uk/lacaze/nanotechnologies.zip*

Each configuration (n, m) corresponds to a distinct Raman spectrum and a particular CNT diameter, which allows them to be identified. The frequency of the Raman Radial Breathing Mode band (RBM), corresponding to the radial mode of vibration of the nanotube (ω_{RBM}), is linked to the diameter (d) by the relation:

$$\omega_{RBM} = A/d + B$$

A and B are experimentally determined constants (Bachilo et al. 2002; Maultzsch et al. 2005).

1.1.4.2. *Some properties and applications of CNTs*

The list of their properties is particularly long and well summarized in several review articles (De Volder et al. 2013; Yang et al. 2015; Ferrier 2021; Dai 2022).

CNTs are characterized by their high conductivity due to their metallic character. Wei et al. (2001) showed that MWCNTs could withstand current densities greater than 10^9 A/cm^2 over 2-week periods at temperatures between 200 and 250°C without observing any CNT degradation or variation in conductivity. CNT semiconductors are also characterized by extremely high electron and hole mobility properties, far beyond silicon, reaching up to 79,000 cm^2 V^{-1} s^{-1} (Dürkop et al. 2004). The band gap widths are inversely proportional to the CNT diameter and are approximately 1 eV for diameters of 1 nm, a value comparable to that of silicon (Kataura et al. 1999).

The thermal conductivity in the axial direction of a CNT is very high (3,500 Wm^{-1} K^{-1} for a SWCNT 2.6 μm in length and 1.7 nm in diameter, while for copper, a very good thermal conductor, it is only 385 Wm^{-1} K^{-1}; Pop et al. 2006). On the other hand, the thermal conductivity in a radial direction is very low at only 1.52 Wm^{-1} K^{-1}, almost a thermal insulator in this direction. This difference between the two orientations explains that for films made up of CNT assemblies arranged disorderly, the thermal conductivity is only 1,500 Wm^{-1} K^{-1} (Koziol et al. 2017). Note that, under a vacuum, these CNTs resist up to temperatures of 2,800°C and 750°C under an air atmosphere.

Ideal CNT structures lead to exceptional mechanical characteristics, with a tensile strength of 100 GPa (10^{11} Pascal), Young's modulus of 1 TPa (10^{12} Pascal), and an elongation at break of 18%. Such values were obtained with CNTs that were several centimeter long and defect free. The use of CNTs as polymer reinforcement fibers with the production of CNT/polymer composite materials is a key application area that has been the subject of numerous studies (Sun et al. 2013; Chen and Yan 2017a).

Obtaining self-supported films of agglomerated CNTs, leading to a buckypaper structure, coupling very good mechanical and electrical properties, is a commercially available product with numerous potential applications. Note that 6 μm thick buckypaper sheets can be made simply by vacuum filtration of an MWCNT dispersion; the film obtained has an elastic modulus of 3 GPa and a conductivity of about 2×10^4 S/cm, a function of the deformation state, which makes it a possible sensor for detecting movements (Degraff et al. 2017).

CNTs are widely used in nanoelectronics, energy, mechanics, biomedicine and sensors. Their use in the production of biosensors has been the subject of recent research (Ferrier 2021; Dai 2022). Their exceptional properties have made CNTs successful and have encouraged many companies to develop large-scale preparation techniques. Over a few years (between 2004 and 2011), global production of CNTs has increased from 200 kg to 4,500 tons per year, a 22,500-fold increase (De Volder et al. 2013), reaching around 20,000 tons in 2022[11]. The cost of SWCNTs, however, remains high: it depends on the packaging method and varies depending on the companies that manufacture them. In 2020, the lowest prices were between $80 and $100 per kg, and the price of industrial-grade MWCNTs varies between $200 and $400 per kg.

1.1.4.3. *Conclusion*

CNTs are a family of nanomaterials whose strategic importance has continued to grow. Due to their exceptional electrical and mechanical properties, they have given rise to many applications thanks to very broad commercialization. However, obtaining large quantities of highly pure CNTs with a specific chirality is still challenging.

1.1.5. *Graphene*

Geim et al. were the first to isolate a fragment of graphene by mechanically exfoliating a sheet of graphite. The prepared graphene film is extremely stable, and a range of electrical and mechanical measurements can easily be done[12]. Moreover, monolayer of graphene can be easily identified by Raman spectroscopy. A characteristic line (G) appears at 1,584 cm^{-1}, as well as another line (2D or D') close to 2,700 cm^{-1}, whose width increases as the number of graphene layers increases (Ferrari et al. 2006; Graf et al. 2007; Bonaccorso et al. 2010).

The physical, chemical and mechanical properties discovered for graphene were exceptional and it immediately became the "magic" material at beginning of the 21st century. Its unique electrical properties make it a material of choice for electronics and allowed considerable improvements in the performance of various devices used in the fields of electronics, optoelectronics, energy and biomedical analysis. Since the 2010s, graphene-based transistors have been manufactured. However, the fact

11 A 2021 Fortunes Business Insights report forecasts an increase in the global CNTs market from 5.32 to 10.52 billion USD between 2021 and 2028, representing a CAGR of 10.2% for this period (ID: 5685490).

12 The technique may seem simple but, in fact, the graphene sheet deposited on a Si substrate covered with a thin layer of SiO_2 only becomes visible if an optical contrast occurs between the silica and the graphene. This implies that the thickness of the silica layer must be comparable with that of the graphene sheet. Geim and Novoselov (Nobel Prize in Physics in 2010) recognized that it was a fortunate combination of circumstances that allowed them to highlight the graphene monolayer (Geim and Novoselov 2007).

that graphene is a zero-bandgap material does not allow it to be used as a field-effect transistor for digital logic operations (Schwierz 2010). A band gap opening is necessary: One solution is to use graphene ribbons for which 200–300 meV band gaps have been obtained. Moreover, the variations in conductivity caused by the surface adsorption of chemical compounds are significant and can be exploited to produce various sensors (Pumera 2011).

1.1.5.1. *Extreme electrical properties of exfoliated graphene*

Graphene, made up of an assembly of sp^2 carbons organized according to a perfect two-dimensional hexagonal lattice, is a material with a zero bandgap.

The conductivity of graphene and the mobility of electrical charges are extremely high and have been extensively described in numerous articles by Geim, Novoselov, and their colleagues (Novoselov et al. 2004, 2005b; Geim and Novoselov 2007). It can be doped with positive or negative charges by applying an electric field. In contrast, with other two-dimensional semiconductors, an electric field only modulates the electron density (Novoselov et al. 2005a) (Figure 1.9(a)).

In the case of graphene, the charge carriers no longer obey the laws of non-relativistic classical quantum mechanics deduced from the Schrödinger equation, but instead those of Dirac's relativistic physics, where low-energy electrons are assimilated to massless quasi-particles with 1/2 spin. Conical-shaped valence and conduction energy bands intersect at a single point with zero energy and zero charge in the k vector wave space associated with these particles (Dirac point or neutrality point). One of the paradoxes of the theory is the existence of a non-zero conductivity at this neutrality point, even though the corresponding charge density is zero.

Figure 1.9(b) shows the differences in conductivity of graphene and two-dimensional semiconductor materials such as $NbSe_2$ and MoS_2, each incorporated as a field-effect transistor channel. In the case of graphene, the conductivity variation curve is perfectly symmetrical with respect to the axis $V_g = 0$ V. On the contrary, for $NbSe_2$ and MoS_2, a monotonic variation in conductivity versus Vg is observed.

The very high mobility values of over 150,000 cm^2/Vs cited in the literature can only be obtained for extremely pure graphene and correspond to a ballistic type of charge transport mechanism, for which the mean free path is approximately a micron. With impure graphene, the mobilities are much lower (around 15,000 cm^2/Vs) and have a transport mechanism partially controlled by diffusion; the mean free path of the charges is also smaller and is only around 150 nm (Bolotin et al. 2008a)[13].

---

13 The mean free path of electrons, a function of the applied electric field, is for most conductive elements between a few nm and a few tens of nm. In the case of copper, it is around 40 nm (Gall 2016).

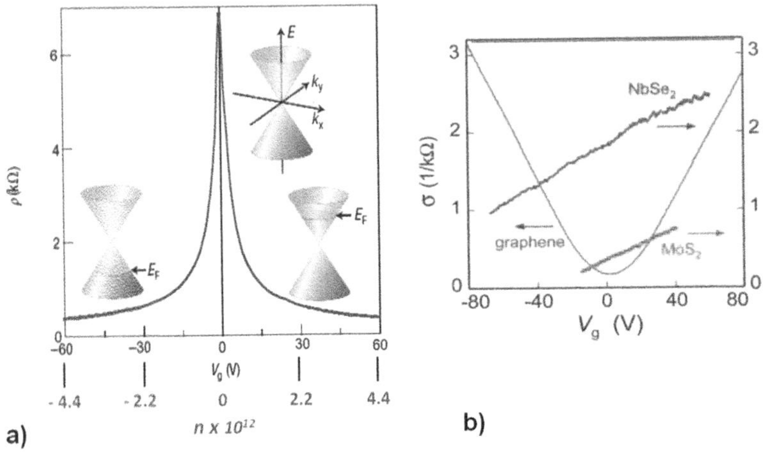

Figure 1.9. *Resistivity variations of graphene and 2D materials as a function of charge densities. (a) Resistivity ρ of a graphene sheet inserted as a channel in a field effect transistor subjected to a gate potential Vg. Adapted from Geim and Novoselov (2007). (b) Comparative conductivity curves of graphene and the 2D semiconductors MoS_2 and $NbSe_2$. According to Novoselov et al. (2005a, 2005b). For a color version of this figure, see www.iste.co.uk/lacaze/nanotechnologies.zip*

COMMENT ON FIGURE 1.9.– *The perfect symmetry of the curve $\rho_{xx} = f(V_g)$ relative to the axis $V_g = 0$ V corresponds to graphene completely free of impurities and is obtained at very low temperatures. Note that for a zero charge density, the resistivity of graphene maintains a finite value. The different conductivity regions are accompanied by diagrams of the energy bands and the Fermi level (E_F) corresponding to each polarization; negative and positive values of n correspond to the hole and electron densities. According to the device used, the electron or hole charge carrier densities vary linearly with the voltage according to the relationship $n = \alpha V_g$ with $\alpha = 7.2 \times 10^{10}$ $cm^{-2}V^{-1}$. Adapted from Geim and Novoselov (2007).*

1.1.5.2. *Graphene preparation techniques*

Several techniques for producing graphene for commercial purposes have been developed. This mainly involves chemical vapor deposition (CVD) and chemical and electrochemical exfoliation, with the latter having the advantage of producing graphene at a lower cost but of lower quality.

Of the many proposed techniques, growing graphene epitaxially on a SiC crystal under atmospheric argon pressure at high temperatures (1,650°C) avoids graphene transfer operations. This makes it possible to obtain a graphene film of the same size as the crystal directly deposited on an insulating support, a necessary condition for producing field-effect transistors (Emtsev et al. 2009).

1.1.5.2.1. CVD on a solid catalyst

The first graphene deposits obtained by CVD were made on mono- or poly-crystalline nickel and copper substrates. The process uses a methane and hydrogen gas stream, heated to approximately 900–1,000°C, decomposing on contact with nickel or copper and releasing carbon atoms. These two processes have undergone numerous modifications. CVD formation conditions are like CNT formation conditions and only differ by the nature of the catalysts and higher temperatures. However, the deposits obtained on a copper substrate are of better quality than those obtained on nickel (Figure 1.10).

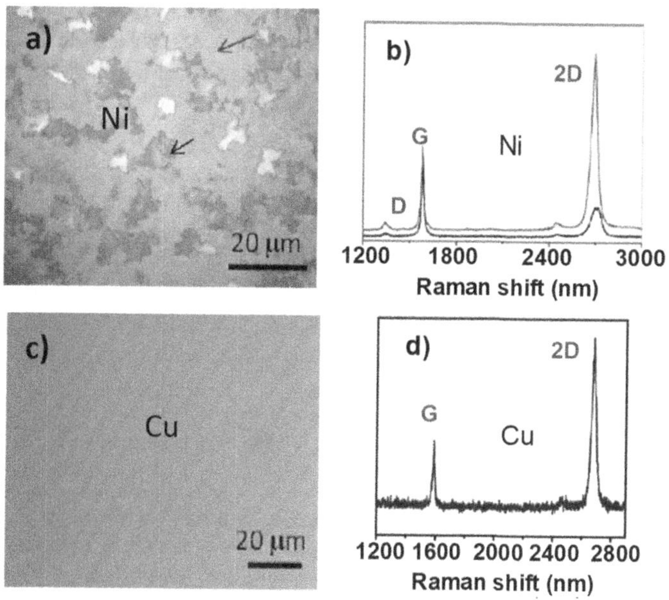

Figure 1.10 *Graphene deposits on Ni and Cu and transferred onto a Si/SiO2 support. (a and c) Optical images of graphene fragments deposited on Ni and Cu, respectively. (b and d) Raman spectra of graphene for the graphene deposits on Ni and Cu. Adapted from Zhang et al. (2013a). For a color version of this figure, see www.iste.co.uk/lacaze/nanotechnologies.zip*

COMMENT ON FIGURE 1.10.– *Unlike the deposit synthesized on Ni, the graphene film obtained on Cu has great homogeneity. In the case of nickel (image a), the heterogeneity of the deposits (black and red arrows) is revealed by two distinct Raman spectra (b), one corresponding (in red) to a monolayer of graphene, with* *some defects (D band around 1,300 cm^{-1} non-zero), the other (in black) a multilayer deposit (characterized by a very low intensity of the 2D band). In the case of copper, the perfect homogeneity of the graphene film (image c) is confirmed by the unique presence of the two Raman bands 2D and G, with an I_{2D}/I_G intensity ratio equal to 2, characteristic of a monolayer of graphene (d).*

The difference comes from the fact that in the case of copper, and although the experimental conditions (CH_4, H_2 and temperature) are comparable to those used with Ni (Li et al. 2009), the carbon atoms resulting from the dissociation of methane are poorly soluble in copper and adsorb directly to the surface of the metal to give graphene (Zhang et al 2013a). On the contrary, in the case of Ni, the carbon atoms coming from the dissociation of methane solubilize inside the nickel and, upon cooling, re-diffuse to the surface, then recombine to form a graphene film, which in the case of polycrystalline nickel, with many grain boundaries, gives poorer film homogeneity than with copper. This difference appears clearly in Figure 1.10, showing images of graphene films and their Raman spectra[14] for each metal.

1.1.5.2.2. CVD on a liquid catalyst

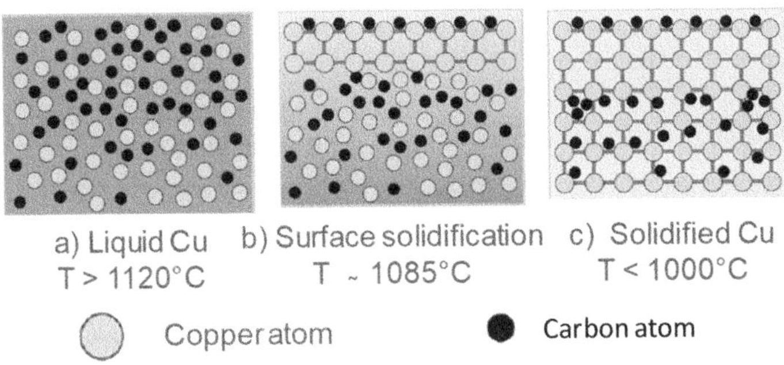

Figure 1.11. *Different stages of the catalytic process of forming a graphene monolayer. (a) Copper is in a liquid state with carbon atoms in solution; (b) initial cooling to 1,085°C produces surface crystallization of copper. (c) Further cooling blocks the graphene monolayer on the copper surface. Adapted from Zeng et al. (2014). For a color version of this figure, see www.iste.co.uk/lacaze/nanotechnologies.zip*

---

14 The different Raman bands of graphene films (G, 2D and D) make it possible to assess their thickness and the presence or absence of defects. A monolayer of graphene is characterized by a band intensity ratio I_{2D}/I_G close to 2, just as the thickness (evaluated by the number n of layers) can be measured from the position of the band G, which verifies the wave number relationship $\omega_G = 1581.6 + 11/(1 + n^{1.6})$ cm^{-1}. D Band only appears when graphene has defects and is located around 1300 cm^{-1}; 2D band, sometimes considered as a harmonic of the D band is very intense for a graphene monolayer and is located at 2750 cm^{-1}. It decreases in intensity and widens as the number of graphene layers increases (Wall 2011).

COMMENT ON FIGURE 1.11.– *During the initial cooling (b), the crystallization of Cu on the surface prevents the C atoms within the liquid from diffusing toward the surface. Thus, it is possible to maintain an organized monolayer of carbon on the copper surface to form graphene.*

The use of liquid metal catalysts in CVD is a new trend in the manufacturing of 2D materials. Applied to graphene, this method might improve its quality. This technique, described in detail by Zeng et al. (2014), is comparable to CVD methods, but differs in the use of the catalyst in the liquid state. Subjected to slow cooling, leading to the phase change from a liquid state to a solid surface state, the carbon atoms coming from the dissociation of the carbon precursor (CH_4), initially dissolved in the molten metal core, stop diffusing toward the surface as soon as it solidifies. This surface solidification limits the concentration of surface carbon atoms, thus promoting the formation of graphene monolayers (Figure 1.11).

1.1.5.2.3. The CVD process adapted to large-scale production

Given the strategic importance of graphene, numerous improvements to the CVD preparation process have been proposed in recent years to produce graphene on a large scale and at the lowest cost. The CVD process is in fact the only one suitable for the mass production of graphene by allowing two successive operations to be carried out continuously and automatically: the synthesis of graphene and the transfer of graphene to a flexible support. Adapting the roll-to-roll (R2R) printing process to the production of graphene, in the form of 30 cm wide several meter long films, is one of the most impressive examples. Many industrial groups have invested in this applied research topic, and it has been the focus of numerous university-industry consortia[15]. Several variants of the CVD process applicable to R2R technology have been proposed in recent years and have been well described by Xin et al. (2018). Large graphene films, adhering to flexible transparent plastic supports, were produced by this technique.

Bae et al. (2010) were among the first, in collaboration with Samsung, to produce rolled hybrid graphene glued to plastic using the R2R technique, up to 1 m long. Later, Kobayashi et al. (2013) from Sony produced graphene films using the R2R technique with qualities comparable to ITO plates and lengths over 100 m. They used a copper ribbon heated by the Joule effect to around 1,000°C that passed slowly into a chamber where a gas mixture of methane and hydrogen circulated at reduced pressure. Accurately adjusting the gas flow and the rate of the copper ribbon is decisive for obtaining a good quality single-layer graphene film. After forming the graphene film, it is automatically transferred onto plastic support, and the copper

15 Europe has funded a consortium between universities and companies in the industry for the development of large-scale graphene production techniques.

automatically dissolved, producing graphene coils 100 m long and 20 cm wide, characterized by a square resistance between 150 and 250 Ω/□[16] and an optical transmission of around 85% (Kobayashi et al. 2013).

1.1.5.2.4. Chemical methods to produce reduced graphene oxide (rGO) from GO

The starting product is graphite oxide, a multi-layer precursor of GO with very low conductivity. A wide variety of graphite oxidation methods have been described, the most frequently used is the oxidation of graphite powder using potassium permanganate[17]. Exfoliation is then done using strong ultrasound agitation in a liquid medium (Eigler et al. 2013). This chemical reaction leads to significant degradation of the crystal lattice, following the breaking of C-C bonds, resulting from oxidation with CO_2 release (Eigler et al. 2013).

Once GO is obtained, in the form of a suspension in a water-alcohol mixture (0.1 mg/mL), the GO is dispersed onto a solid support, then reduced with a mixture of hydriodic acid (HI) and trifluoroacetic acid (CF_3CO_2H) as a vapor, which gives reduced graphene rGO, deposited in the form of flakes sized between 500 nm and 1 µm[18]. A statistical Raman analysis of these rGO flakes shows that after reduction, according to the full width at half maximum (FWHM) of the 2D Raman band at 2,700 cm^{-1}, they can be classed into three distinct fractions: rGO with few.

FWHM <50 cm^{-1}) and a high proportion of rGO with many structural defects (FWHM > 50 cm^{-1}).

Graphene obtained by the reduction of GO does not have the same high purity as graphene obtained by mechanical exfoliation or CVD. However, although its conduction properties are less good than those of previous graphene, it can be produced at a lower cost without the need for large investments in equipment. Its mechanical and conduction characteristics are sufficient for many applications.

---

16 This notation, specific to electronic engineers, corresponds to a "square resistance" (R□ = ρ/e, ρ being the resistivity, and e the thickness). For example, a material of width W = 2 mm and length L = 3 mm, with a "square resistance" of 500 Ω/□, (500 Ω per square) will have the resistance R = (L/W) × R□ = 750 Ω.

17 Hummers' method involves gradually adding potassium permanganate to graphite powder suspended in a solution of silver nitrate and sulfuric acid. The reaction is exothermic and can become explosive if the temperature is not controlled (Hummers and Offeman 1958). Starting from 1 g of graphite powder, Eigler et al. obtained 500 mg of GO and 200 mg of graphene after chemical reduction.

18 Using a mixture of hydriodic acid ($I^- + H_3O^+$) and trifluoroacetic acid as a GO reducing agent is a non-toxic alternative to hydrazine.

1.1.5.3. *Some applications of graphene and its derivatives*

The properties of single-layer graphene are exceptional. In addition to its low surface mass of 0.77 mg/m^2, it is in fact characterized at the mechanical level by a tensile strength of 130 GPa and by an elasticity modulus (E) of 1,100 GPa, whereas for stainless steel, E is around 200 GPa and its tensile strength is 0.5 GPa; its electrical conductivity (10^8 S m^{-1}) is greater than that of copper (6×10^7 S m^{-1}) and it can withstand current densities between 10^{10} and 10^{11} A m^{-2} without degrading. Its thermal conductivity is also very high with a record value of 5,300 Wm^{-1} K^{-1} while that of copper is only 400 Wm^{-1} K^{-1}. These values are obviously only obtained for graphene free of defects and impurities. However, for large-scale applications, which require graphene to be produced in large quantities, the record characteristics indicated previously are not achieved, but these are not always necessary. Reevaluating the required graphene production conditions in terms of quality, uniformity and reliability is necessary for more competitive products and devices, in terms of cost and performance[19].

For many applications, such as electrode materials, solar energy conversion devices, or energy storage with different battery systems, multilayer graphene sheets of medium quality may be sufficient. This is also the case for producing fibers used as electrical cables.

1.1.5.3.1. Graphene oxide

GO, obtained by the oxidation of graphite, can be seen as a stack of graphene sheets, functionalized in their plane (phenolic and epoxy) and at their ends (carboxylic) by oxygenated groups. Ultrasonic agitation in an aqueous medium exfoliates the sheets and produces a colloidal suspension in water of GO nanosheets with a thickness of approximately 1 nm and lateral dimensions varying by a few nanometers to several hundred micrometers. The solution can be filtered very slowly through a porous disk, producing a thick compact film. The thickness of these films can vary from 1 to 30 µm. The resulting paper-like product has very good mechanical properties due to the strong interactions between the hydroxyl groups. For films a few micrometer thick, the elasticity modulus (E) is around 30 GPa (Dikin et al. 2007), while for buckypaper, resulting from the agglomeration of CNTs, it is only around 7 GPa.

---

19 Consulting the Manchester Graphene Engineering and Innovation Centre (GEIC) website, as well as that of the Graphene Council, created in 2013, devoted to the development and exploitation of graphene, gives an idea of the vast number of current and potential applications for graphene prepared in different ways. Refer to the interview of T. Barkan (Graphene council), "The future of graphene", February 2019 (AZOnano): https://www.azonano.com/article.aspx?ArticleID=5135.

1.1.5.3.2. Graphene paper and fibers obtained from rGO sheets

By reducing a suspension of GO with hydrazine followed by vacuum filtration, Li and colleagues obtained films with a metallic appearance and controllable thickness (from a few tens of nanometers to several micrometers) (Chen et al. 2008) (Figure 1.12). These films, which look like sheets of paper, are like the type of graphene, rGO. Very mechanically resistant, rGO is also a material with good electrical conduction properties, differentiating it from insulating GO.

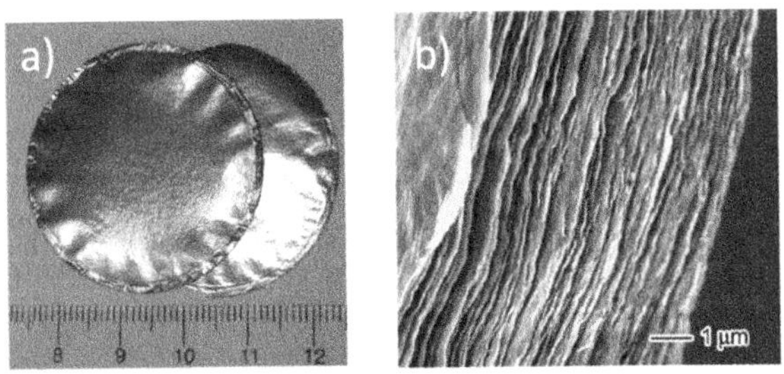

Figure 1.12. *Graphene paper (rGO) obtained after reducing graphene oxide (GO) sheets. (a) Photograph of two metallic-looking graphene sheets (front and back views). (b) Scanning electron microscopy (SEM) image of graphene paper showing the stacking of graphene sheets. According to Chen et al. (2008)*

Graphene paper's mechanical and electrical properties depend on the strength of the cohesion between the different layers. This cohesion is weakened by residual gases and liquids resulting from previous chemical processes. Located between the graphene sheets, these impurities are difficult to remove due to the impermeability of the sheets. An American team proposed solving this problem by intentionally creating holes on the surface of the graphene sheets (making holey graphene and h-graphene) to release these impurities and reconstruct the graphene network's hexagonal structure after suitable heat treatment (Chen et al. 2019).

Along with its mechanical and electrical properties, we must also consider its biocompatibility, which makes rGO usable in the biomedical field as biosensors.

Another interesting application of rGO is manufacturing fibers. Qu et al. were among the first to propose a manufacturing method using a suspension of GO in water (Dong et al. 2012). Their very simple technique involved heating a suspension of GO (8 mg GO/mL) contained in a glass tube (inner diameter 0.4 mm), closed at both ends, at 230°C for 2 h. This produced fibers with diameters of 150 µm when wet and only 35 µm when dry.

From a mechanical point of view, the stress–strain curves considerably improve after heat treatment at 800°C for 2 h under argon, increasing the tensile strength from 120 to 420 MPa. The electrical conductivity of these fibers is 10 S/cm, like fibers made from CNTs. It is easy to knot or twist the fibers, illustrating their great flexibility (Figure 1.13).

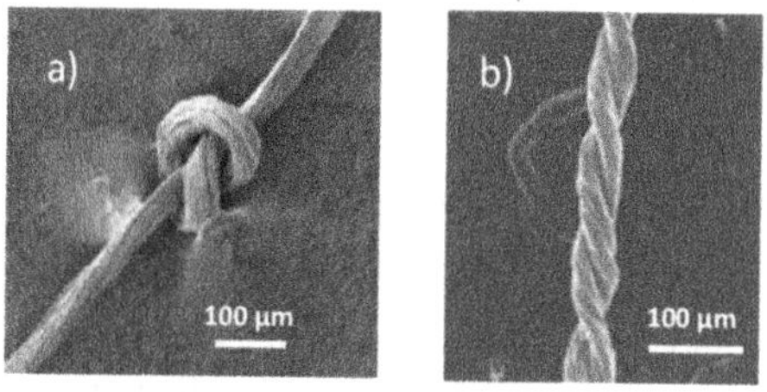

Figure 1.13. *SEM images of graphene fibers. (a) Knot formation; (b) two-strand twisted graphene fiber. Adapted from Dong et al. (2012)*

Given the technological importance of these fibers, many teams have become involved in this field of research to produce graphene fibers with properties as close as possible to those of default-free monolayer graphene films.

Gao's group recently continuously produced very high-quality graphene fibers on a large scale (Xu et al. 2016). The high quality was due to the optimal orientation of the GO micro-sheets. Initially obtained as a dispersion in water (from 1 to 10 mg of GO/mL in solution), the GO dispersion was then transferred into a solution of N,N dimethyl formamide (DMF), which has the effect of creating a liquid crystal structure with the GO micro-sheets in a parallel arrangement. The fibers were extruded through a small diameter die (60 µm) and then wound onto a spool using rotating cylinders. Heat treatment under nitrogen at 3,000°C transforms the GO fibers (GOFs) into graphene fibers (GFs). These GFs have very good mechanical properties as well as very good electrical and thermal conductivities. Their Young's elasticity modulus is between 385 and 400 GPa (measured with a 5 mm long wire), and their ultimate tensile strength is around 1.8–2.2 GPa. Conductivity at room temperature reaches up to 0.8×10^6 S/m and is independent of the length of wire tested; the ampacity is extremely high and corresponds to current densities of 2.3×10^{10} A/m^2. These values are comparable and sometimes better than those measured with wires formed from CNTs.

This large-scale production of extremely flexible and mechanically resistant GFs, whose electrical properties are comparable to those of metals, such as copper, but with very low mass densities, makes them suitable for numerous applications. The authors thus demonstrate how graphene fiber winding can replace copper wires in the manufacture of an electric motor, with a significant mass advantage, considering the differences in density between GFs (0.23 g/cm^3) and copper wires (8.96 g/cm^3) (Dong et al. 2012).

In parallel with this work, the same group showed that the conductivity of GFs could be greatly increased when doped with certain elements. Liu et al. (2016) obtained GFs with conductivities of 0.77 × 10^7 S/m (GF-$FeCl_3$), 1.5 × 10^7 S/m (GF-Br_2) and 2.24 × 10^7 S/m (GF-K), values close to those of copper (~6 × 10^7 S/m to 20°C).

1.1.5.3.3. Graphene-polymer nanocomposites

The idea of using a small amount of graphene to significantly improve the mechanical, electrical and thermal properties of composite polymers is attractive. Ruoff and his research group were the first to investigate this concept, starting with GO nanoparticles (GO-NPs) dispersed in water to develop large-scale polymer composite materials (Stankovich et al. 2006).

The main obstacle to overcome is that the graphene nanoparticles (G-NPs) inserted into the polymer matrix are generally obtained by reducing GO-NPs, also dispersed in water. After reduction, the G-NPs immediately aggregate together, given the hydrophobic nature of the nanoparticle core. This makes it difficult to individually insert G-NPs into a polymer solution because most polymers are insoluble in water. The solution proposed by Ruoff et al. involves making the GO-NPs less hydrophilic by transforming the OH and COOH groups of the GO sheets into amide and carbamate functional groups, which are weakly hydrophilic. Exfoliation is no longer possible in an aqueous medium but readily occurs in an aprotic solvent like DMF. The authors obtained a dispersion of functionalized graphene oxide nanoparticles (fGO-NPs) in an organic medium, where numerous polymers can be solubilized. Astonishingly, the reduction of fGO-NPs by hydrazine in the presence of a polymer such as polystyrene only leads to an aggregation of the G-NPs and the polymer and not to an aggregation of NPs, ultimately forming a graphene–polymer composite.

Since this initial research, many teams have invested in this area of research and made numerous improvements and modifications to the graphene–polymer composite manufacturing process. Several articles have been published on this topic (Potts et al. 2011; Das and Prusty 2013; Hu et al. 2014a; Bera and Maji 2017; Malucelli 2017), which provide details of the composites obtained for many polymers of technological interest and the variations in manufacturing methods, including the functionalization of G-NPs, which have led to significant improvements on a mechanical and electronic level. Numerous applications exploited graphene–polymer composites in the design of biosensors, as described in the review article by Yusaf et al. (2022).

1.1.5.3.4. Conclusion

Graphene is one of the few materials that exhibit exceptional electrical and thermal conductivity, resist extremely high temperatures and high current densities greater than those of the most conductive metals such as copper. They also have mechanical properties such as elasticity and tensile strength far exceeding those of steel and a low mass density of 0.23 g/cm^3. Such properties have only been obtained for mechanically exfoliated graphene sheets, which are defect free and can only be obtained in very small quantities, making it unrealistic to imagine large-scale production.

CVD methods have evolved considerably and can now continuously produce good-quality graphene ribbons. However, this technique requires large equipment that can only be used in an industrial setting. On the other hand, the chemical transformation methods of GO into G prove to be easier to access and have progressed considerably too. GO, obtained in large quantities by the simple oxidation of graphite, is the precursor to graphene, which can be obtained in large quantities and various forms. Graphene can be produced as very long, highly homogeneous fibers a few micrometers in diameter and can even be used to manufacture composite materials. The quality of the graphene produced is considerably improved by heat treatment at very high temperatures in an inert atmosphere, which restores its graphitic structure.

Given these properties, graphene's potential areas of application are vast (Figure 1.14), falling under the following: energy (storage and conversion), biomedicine (medical imaging and vectorization of drugs), electronic devices (sensors, transistors, electrodes, electromagnetic shielding) and mechanics (artificial muscles).

The cost of graphene, however, remains high and its price, which varies greatly, fluctuates between 100 and 300 USD per gram. Furthermore, it is also available in the form of sheets adsorbed onto a Si/SiO_2 substrate, making it possible to directly manufacture various sensors. Fortunes business estimated the graphene market at 430 million USD in 2023, reaching an expected 2.96 billion USD in 2030. This represents a compound annual growth rate (CAGR) of 31.6% for this period[20].

20 Researchers at Rice University proposed significantly lowering the cost of graphene thanks to a new process called "flash graphene" (FG), which involves irradiating ordinary carbon sources at 2,700°C for 10 ms. They believe it is possible to obtain 1 g of the FG product with high purity while consuming little electrical energy, equivalent to 7.2 kJ (Luong et al. 2020).

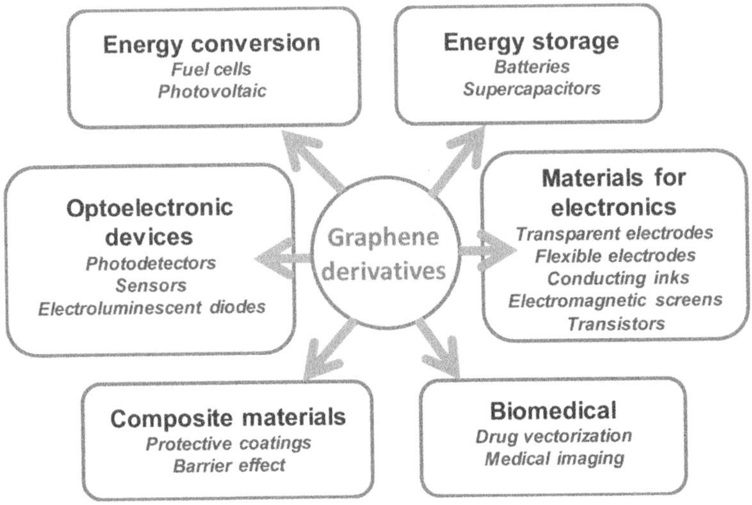

Figure 1.14. *Implications of graphene derivatives in different fields. For a color version of this figure, see www.iste.co.uk/lacaze/nanotechnologies.zip*

1.1.6. *Graphene quantum dots*

Graphene quantum dots (GQDs) are nanometric-sized graphene fragments like point nanoparticles (0 D) with characteristics similar to graphene and CQDs. As for CQDs, one should expect quantum confinement effects, edge effects and some structural properties characteristic of graphene. Less than 10 nm in size and less than 10 layers thick, GQDs are fluorescent materials due to the confinement effect. Unlike graphene, GQDs have a non-zero band gap, making them semiconductors and suitable for applications in optoelectronics.

1.1.6.1. *GQD manufacture methods*

Novoselov et al. first manufactured GQDs by CNT fragmentation in 2008 (Ponomarenko 2008), and many techniques have since been proposed. These techniques have been classified into two main categories: top-down and bottom-up. Top-down techniques involve cutting fullerenes, CNTs, carbon fibers or graphene into fragments, while bottom-up techniques involve synthesizing GQDs by chemically transforming carbon molecules (precursors). Citric acid, glucose or other compounds containing benzene rings have been used as precursors for manufacturing GQDs (Valappil et al. 2017; Tian et al. 2018).

The fragmentation of a carbon-based material, preferably with aromatic nuclei, is frequently used in top-down methods. Graphite is preferred, but other materials, such as cotton fibers, have also been transformed into GQDs (Zuo et al. 2017). For example, chemically transforming graphite in an acidic environment together with microwaves produces large quantities of GQDs and does not require large logistical investments. This was the method used by Luo et al. (2016) to obtain emissive GQD films used in white light-emitting diodes.

Unlike previous methods, bottom-up methods can produce GQDs from growth reactions of carbon compounds and form condensed aromatic structures by thermolysis in an aqueous medium. The carbon precursors can be polycyclic aromatic compounds (pyrene, 1,5-dinitro naphthalene) or non-aromatic compounds.

Bottom-up methods have the undeniable advantage of simplicity when obtaining GQDs doped with elements other than carbon. Nitrogen-doped GQDs (N-GQDs) can be produced in a single step by directly treating a mixture containing a nitrogen derivative and the carbon precursor. Nitrogen-doped nanomaterials have significantly better photoluminescence (PL) properties than undoped GQDs and are particularly interesting for medical imaging due to their luminescence properties, high stability in an aqueous medium and their non-toxicity. They have also found other applications in the energy and sensors fields (Kaur et al. 2018).

Through the thermolysis of citric acid in ethylene diamine, Yang and his colleagues obtained N-GQDs sized between 5 and 10 nm with thicknesses between 0.8 and 2.5 nm, the equivalent of two to five superimposed layers. Curiously, the high internal quantum yield of the blue PL obtained with these N-GQDs, around 75%, far exceeds the quantum yields usually observed with this type of materials (Gu et al. 2016).

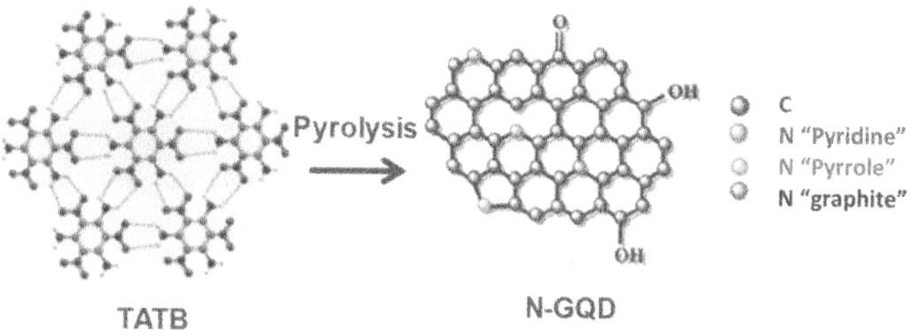

Figure 1.15. *Obtaining N-GQDs by pyrolysis of TATB. The TATB precursor is made up of a supramolecular planar network of TATB molecules linked by hydrogen bonds between the H of the NH_2 and the O of the NO_2 (the red and blue spheres represent the oxygen and nitrogen atoms, respectively, and the gray spheres, the carbons). In the N-GQD, the nitrogen atoms are incorporated into the graphitic network and occupy different positions corresponding to different functionalities (pyridine, pyrrole or graphite). Adapted from Li et al. (2016a). For a color version of this figure, see www.iste.co.uk/lacaze/nanotechnologies.zip*

Another way to obtain N-GQDs is to start from a single compound like 1,3,5-triamino-2,4,6-trinitrobenzene (TATB), which has carbon and nitrogen necessary for forming N-GQDs (Li et al. 2016a). A pyrolysis reaction at a high temperature (750°C for approximately 20 min) in a nitrogen atmosphere synthesizes N-GQDs with good yields. The doping rate atomic value of nitrogen is about 10% compared to carbon atoms (Figure 1.15).

1.1.6.2. *Properties and applications of GQDs*

GQDs, unlike graphene, have a non-zero forbidden energy band, the width of which depends on the size, shape and edge effects of these nanoparticles. Given the great diversity of GQDs and their lack of homogeneity in size and shape, their optoelectronic and photoemission properties closely depend on the abovementioned parameters. Photoemission is one of the main properties of GQDs, exploited in various detection systems, particularly in medical imaging.

The PL of GQDs is controlled by the nanomaterial core (which may be assimilated to a graphene fragment), the surface states corresponding to functional groups positioned at the ends of the nanoparticles and the heteroatoms incorporated after doping. Like CQDs, the fluorescence emission of GQDs strongly depends on the excitation wavelength (Figure 1.16).

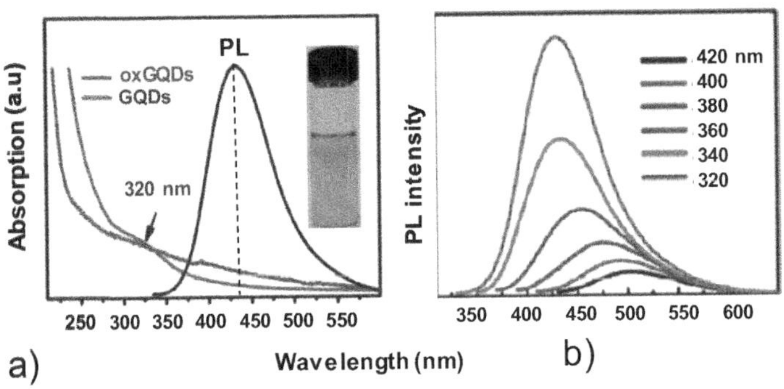

Figure 1.16. *Photoluminescence (PL) intensity curves of GQDs as a function of excitation wavelength. (a) Absorption curves of GQDs and oxidized GQDs (oxGQDs) and PL curves resulting from the excitation of GQDs at 320 nm. Visible light photographs of an aqueous solution of GQDs. (b) PL intensity curves emitted for excitation wavelengths between 320 and 420 nm. Adapted from Pan et al. (2010). For a color version of this figure, see www.iste.co.uk/lacaze/nanotechnologies.zip*

Zhu et al. (2014), in a very detailed spectroscopic study (static and dynamic), showed that the photoemission properties of GQDs are determined by the relative position of the GQD intrinsic energy (IE) levels and the energy levels due to edge effects (EE). IE energy levels correspond to the energy of the orbitals π* in graphene fragments and the EE levels to the functional groups grafted to the edge of the GQDs.

Using GQDs of increasing size ($C_{42}H_{18}$–1.2 nm, $C_{96}H_{30}$–1.9 nm, $C_{132}H_{34}$–2.5 nm), the authors show that the energy of the intrinsic transition $\pi \rightarrow \pi^*$ decreases with the GQD size along with the energy difference ΔE between IE (π*) and EE levels. The energy difference ΔE determines the nature and intensity of the luminescence, which originates from the deactivation of the EE state. The emission is fluorescence for a non-zero ΔE and can be combined with phosphorescence when ΔE is zero. The fluorescence intensity can be very weak or absent when the IE and EE energy levels are similar, which is observed in GQDs with formulas $C_{132}H_{34}$ and $C_{222}H_{42}$ (Zhu et al. 2014) (Figure 1.17).

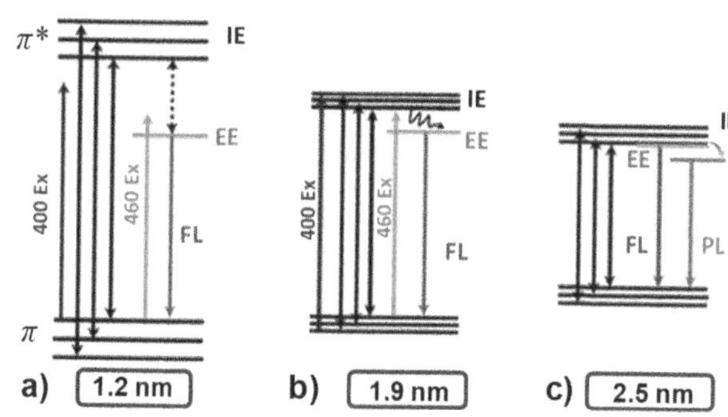

Figure 1.17. *Interpretation of the photoluminescence of GQDs according to their size and luminescence excitation energy. (a) GQD ($C_{42}H_{18}$, size 1.2 nm): excitation at 460 nm only gives fluorescence emission. (b) GQD ($C_{96}H_{30}$, size 1.9 nm): the two excitations at 400 and 460 nm, after energy transfer from IE (intrinsic energy state π*) toward EE (energy level due to functional groups located on the edges of the GQD) emit high-intensity fluorescence FL. (c) GQD ($C_{132}H_{34}$, size 2.5 nm): EE merges with IE; possible intersystem crossing (ISC) of IE toward its triplet state simultaneously emits fluorescence and phosphorescence PL. Adapted from Zhu et al. (2014a). For a color version of this figure, see www.iste.co.uk/lacaze/nanotechnologies.zip*

Recent years have seen the development of many research studies proposing the use of GQDs for energy and biomedical applications. In the field of energy, GQDs can significantly improve the operation of battery electrodes and the performance of supercapacitors. Their PL properties have also been used to improve the operation of OLEDs and some photovoltaic devices since they can intervene favorably in photocatalytic processes. GQDs have found the most potential applications in the biomedical field, including biomedical imaging, vectorized drug delivery, various forms of phototherapy, antibacterial products and biosensors (Kumar et al. 2020).

1.2. Inorganic nanomaterials

The metals Au, Ag and their alloys, one of the largest families of inorganic nanomaterials currently studied, have by far the most studies devoted to them, followed by oxides (TiO_2, MnO_2, Fe_3O_4) and chalcogenides (CdS, CdSe and Bi_2Te_3). The latter are involved in manufacturing QDs due to their semiconductor properties (Figure 1.18) (Portehault et al. 2018).

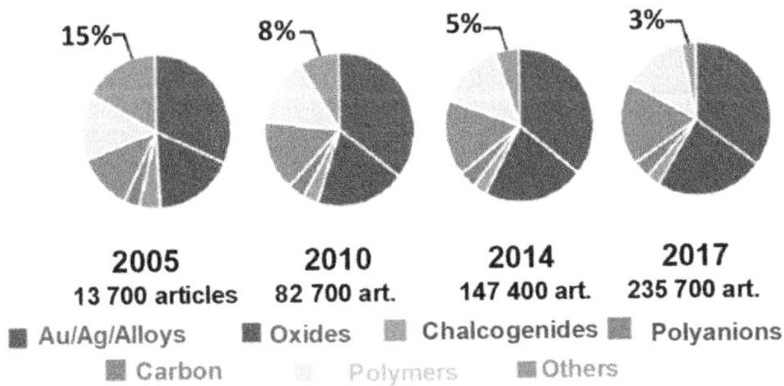

Figure 1.18. *Cumulative progression of the number of articles published on the different types of NPs between 2005 and 2017. Adapted from Portehault et al. (2018). For a color version of this figure, see www.iste.co.uk/lacaze/nanotechnologies.zip*

COMMENT ON FIGURE 1.18.– *The number of publications devoted to noble metals and oxides increased continuously between 2005 (around half of the total) to two-thirds of the number of publications in 2017. Publications on materials other than carbon derivatives, polymers and inorganics declined significantly from 15% to 3% during the same period; those concerning chalcogenides and polyanions remained unchanged and formed only a small part of all publications.*

All these nanomaterials can be easily functionalized with organic molecules or polymers, making them soluble in aqueous environments in the form of colloids and sensitive to their environment. It also gives them suitable transport and vectorization properties for biomedical applications. They can also be organized into lattices, giving them new properties such as plasmon resonance and exaltation phenomena, mainly observed in the UV-visible spectral range for gold and silver.

1.2.1. *Metal nanoparticles*

Many studies have been devoted to the NPs of noble metals, particularly gold, due to the vast potential applications in analytical chemistry, catalysis, in developing optoelectronic devices, and, due to their safety, in the biomedical field.

The properties of metallic NPs as well as their preparation methods have been the subject of numerous studies, reviewed in the mid-2000s by the Astruc (Daniel and Astruc 2004) and El Sayed (Eustis and El-Sayed 2006) groups. More recently, the Jin and Gilroy groups also discussed metal nanoclusters (NCs) and bimetallic nanocrystals and other specific domains of NPs (Gilroy et al. 2016; Jin et al. 2016). Producing monodisperse NPs with a controlled shape and accurately defined physicochemical criteria for good reproducibility are among the main challenges of current research. There are currently two contending techniques for preparing these nanomaterials. The most widespread and accessible is the chemical preparation of colloidal NP solutions by reducing metal salts in an aqueous medium with surfactants. Other methods, more extensive in terms of equipment and well known to physicists, are based on laser ablation techniques for materials in the gaseous state or solution (Kortshagen et al. 2016) and, more recently, on sputtering techniques for metallic materials (Ishida et al. 2017), well suited for the manufacture of metallic NCs.

1.2.1.1. *Gold nanoparticles*

This is an area of research that has given rise to the greatest number of publications, most of which have been devoted to studying the growth mechanisms of gold NPs and the techniques for producing them. The main challenge has been selecting the physicochemical parameters, which will allow control of the NP size and morphology.

In the 1950s, Turkevich et al. (1951) were the first to propose a precise protocol for obtaining "citrated" spherical gold NPs with controlled sizes by reducing Au^{3+} ions in an aqueous solution using trisodium citrate[21]. The Au^{3+} ions are solubilized in water in the form of the tetrachloroauric acid complex ($HAuCl_4$) and the citrate is gradually added to the Au^{3+} ion solution to produce a double effect: on the one hand, it reduces the Au^{3+} ions to the Au° metal state, and on the other hand, it stabilizes the formed NPs by adsorbing to their surface, resulting in the formation of a stable colloidal solution of gold NPs dispersed in water. The concentrations of the reagents

21 Gold in a divided form has been used as a color pigment since ancient times. Only in the form of nanospheres was it used and mixed with different materials such as glasses and polymers, and it was Michael Faraday (1850) who was the first to recognize that their colors were due to their extreme divisions.

are decisive for obtaining spherical NPs of a determined size. The reaction mechanism involves several stages of reduction and complexation, later the subject of various interpretations (Kumar et al. 2006).

Different techniques have been developed to increase the monodispersity of metallic NPs. The seed-mediated growth technique, where NPs are grown from nucleation seeds, was developed by El Sayed and Murphy and their teams, and adopted by numerous groups. Developing various shapes (rods, cubes, discs, rectangular or triangular platelets, etc.) has become possible thanks to these nucleation seeds (Figure 1.19).

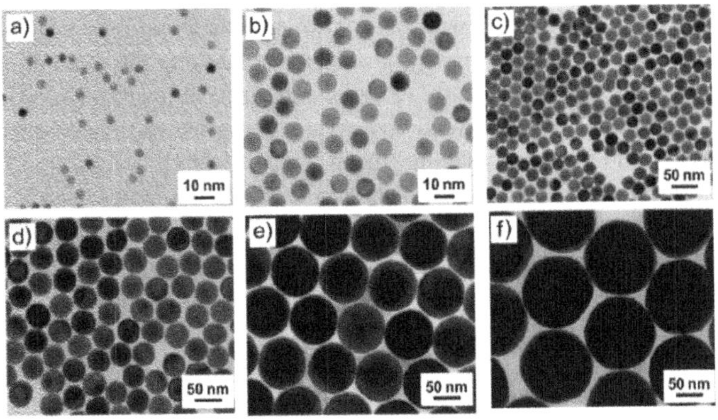

Figure 1.19. *Transmission electron microscopy (TEM) images of gold NPs of different sizes. (a) 5 nm, (b) 10 nm, (c) 23 nm, (d) 46 nm, (e) 100 nm and (f) 150 nm. According to Zheng et al. (2014b)*

Spherical gold NPs with controlled sizes between 5, 10, 30, 50, 100 and 150 nm were also synthesized using this method by injecting increasingly larger gold seeds into tetrachlorauric ($HAuCl_4$) acid solutions, together with ascorbic acid as a reducing agent (instead of citrates) and a surfactant such as cetyltrimethylammonium chloride (CTAC). Gold NPs of approximately 30 nm in diameter are made using seeds of 10 nm gold NPs; the 30 nm gold NPs can then be used as seeds to obtain larger NPs, and so on until NPs of 150 nm are obtained. Producing NPs in the form of nanorods (NRs) began with El Sayed (Huang and Wan 2009) and Murphy (Sau and Murphy (2004) in the 2000s and has also been the subject of numerous studies. The anisotropic growth technique using nucleation seeds makes it relatively easy to produce gold NPs, but the experimental conditions differ from those defined for obtaining nanospheres. Such a mode of growth has been described by Song and his collaborators using gold decahedrons as seeds, synthesized in the presence of high concentrations of poly(vinylpyrrolidone) compared to that of $HAuCl_4$ (Seo et al. 2008, 2009) (Figure 1.20).

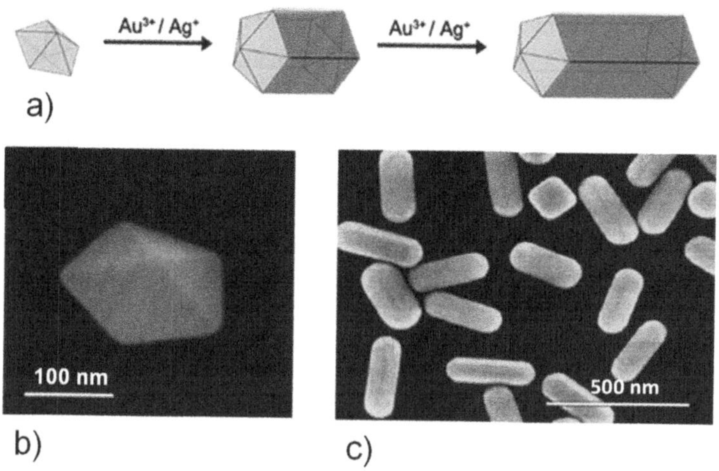

Figure 1.20. *SEM images of NRs obtained from decahedral seeds. (a) Schematic diagram of the NR growth from decahedral seeds in the presence of silver salts. Ag^+ ions are known to block the crystal faces of gold nanopolyhedra and promote the anisotropic growth of NRs. (b) SEM image of a large gold decahedron (80 nm). (c) SEM images of NRs obtained from gold decahedra whose edges have average lengths of 60 nm. (a and c) Adapted from Seo et al. (2009); (b) according to Seo et al. 2008)*

This method of growing gold NPs from metal seeds has made it possible to produce varied forms of NPs and is applicable to many metals besides gold. Numerous variations in the method have also been proposed and varying forms of gold NPs have been obtained. These are described in the review article by Xia et al. (2017) together with the other metals produced via this method.

The optical and physical properties of NRs depend on their dimensions and particularly on their aspect ratio, defined as the length/width ratio, due to their elongated shape. The higher the aspect ratio, the more the light absorption is shifted toward the red, which corresponds to a shift in the plasmon resonance of these NPs toward lower frequencies (Eustis and El-Sayed 2006). Gold NPs are widely used in sensor applications because their optical extinction spectrum closely depends on the environment surrounding them. The sensor applications of gold NPs, based on their plasmonic properties, will be discussed in Chapter 4.

1.2.2. *Metal nanoclusters (MNCs)*

These are very small NPs, aggregates of atoms (between 10 and 300 atoms), whose diameters are between a fraction of a nm and a few nm (maximum 2.2 nm)[22]. Compared to larger NPs (beyond 10 nm), the optical properties of MNCs vary even for very small fluctuations in size (1 Å), whereas for NPs bigger than 10 nm, their optical properties are little affected when size variations of 1 nm occur. This is because, as soon as we reach dimensions less than 2–3 nm, the quantum effect due to confinement causes the band structure characteristic of metal in the bulk state to be replaced by discrete energy levels analogous to those of an atom or molecule. In the case of large NPs, the absorption of light corresponds to a plasmon resonance of the electrons located in the conduction band (see Chapter 4), while for very small NPs the absorption of light corresponds to electronic transitions between occupied (highest occupied molecular orbital [HOMO]) and vacant energy levels (lowest unoccupied molecular orbital [LUMO]).

The field of MNCs is vast, but it is gold (AuNCs) and silver (AgNCs) NCs that have found the greatest applications, mainly due to their fluorescence properties (Russier-Antoine et al. 2014). Since the 2000s, the luminescence of gold NPs has been demonstrated with AuNCs sized between 1 and 2 nm, with very low quantum yields (between 10^{-5} and 10^{-3}) (Zheng et al. 2012). These low quantum yields appear to be the general rule for most AuNCs stabilized by thiolated ligands.

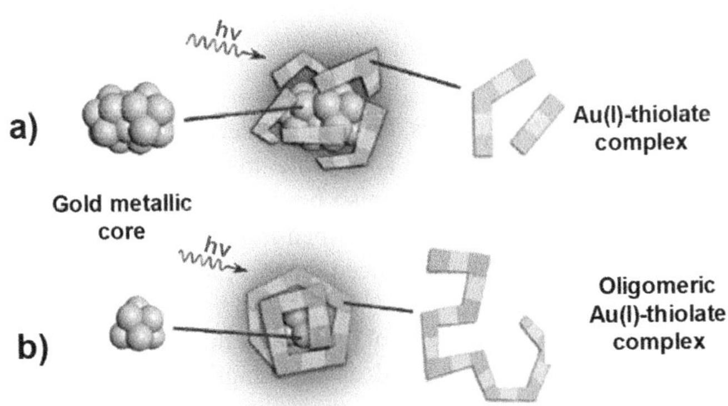

Figure 1.21. *Comparative luminescence of gold NCs chelated by simple Au(I)-Thiolate complexes or by oligomers of the Au(I)-thiolate complex. (a) UV excitation (360 nm) of the Au(I)-thiolate complex does not induce any luminescence. (b) Excitation under the same conditions of the Au(0)@Au(I)-thiolate NC complex induces strong luminescence around 600 nm (orange). Adapted from Luo et al. (2012). For a color version of this figure, see www.iste.co.uk/lacaze/nanotechnologies.zip*

22 The word nanocluster describes very small atomic aggregates.

Thanks to the development of new preparation methods, luminescence quantum yields of over 10% were then achieved by Xie and his team, by producing core-shell structures in the case of Au_{25}-NCs complexed with glutathione (GS) ligands (Luo et al. 2012). According to the authors, this increase in quantum yield corresponds to a new phenomenon of luminescence known as aggregation-induced emission (AIE)[23] resulting from the aggregation of emissive units transformed into oligomers around the metal core (Figure 1.21). This increase in luminescence compared to conventional AuNCs is interpreted by Luo et al. as the consequence of a stiffening of the NC, weakening all non-radiative deactivation processes.

These results apply to a whole series of ligands with one atom (sulfur, nitrogen, etc.) capable of firmly attaching to a metal core of atoms. The quantum yields observed with different gold NCs, listed in several review articles, vary between 1% and several tens of percentage. Very high quantum yields (from 10 to 70%) were obtained by complexing AuNCs with poly(amidoamine) dendrimers. Dickson and his team observed luminescence emissions ranging from UV-visible for the smallest sizes to near-IR for the largest (Zheng et al. 2004).

AuNCs are currently considered luminescent materials particularly suited to the analysis and detection of metal ions, molecules of biological interest, proteins, and nucleic acids, as well as the fluorescent labeling of biological molecules used in imaging. Their ease of preparation and functionalization, the emission of a stable and intense luminescence and color that can be easily modulated according to the size of the MNCs, their good biocompatibility and their absence of toxicity constitute many favorable properties, which have found many applications in the detection and analysis of metal ions and biological molecules, as reported in the review article by Halawa et al. (2018a).

1.2.3. *Semiconductor QDs*

The 2023 Nobel Prize in Chemistry was awarded to Moungi Bawendi, Louis Brus and Aleksey Yekimov for developing semiconductor QDs. Their discovery and

---

23 This AIE effect was described in the 2000s by Tang and team by comparing the luminescence of 1-methyl-1,2,3,4,5-pentaphenylsilole in solution and in the solid state (Luo 2001). In solution in alcohol, the luminescence quantum yield is very low (0.6×10^{-3}), whereas in the form of a solid film, the luminescence quantum yield is 0.2, an increase of more than 300 times. This increase in luminescence is due to a greater flatness resulting from the formation of aggregates and stronger electron conjugation. Note that this enhanced luminescence phenomenon by the aggregation of emissive units is unique and opposes the accepted general law, which states that the condensed state leads instead to the extinction of fluorescence.

the ability to synthesize such materials with simple chemical methods was an important stage in the evolution of nanoscience and nanotechnology[24].

Considered to be semiconductor NCs, resulting from binary, ternary and multinary combinations of different elements, they are a family of inorganic nanomaterials less than 10 nm in size. Over the last two decades, they have been the subject of many studies, focusing on their preparation techniques and physicochemical properties. Essentially, their fluorescence is used in biomedical imaging, electroluminescence applied to LEDs (light emissive diodes), television display screens, photovoltaics, etc. (Kovalenko et al. 2015).

Figure 1.22. *Photoluminescence of some QDs showing that their color emission domains cover the entire visible range. For a color version of this figure, see www.iste.co.uk/lacaze/nanotechnologies.zip*

Semiconductor QDs are characterized by a band gap that varies depending on their size and composition, which gives color emissions ranging from UV-Visible to IR and far-IR. The most typical QDs are binary combinations of elements belonging to columns II and VI of the periodic table, corresponding to metal chalcogenides (CdS, CdSe, ZnS, ZnSe, PbS, TeS, etc.). However, an evolution has occurred toward ternary and quaternary combinations of metals that are much less toxic and correspond to combinations of type I-III-VI_2, ($CuInS_2$, CIS; $CuIn_{1-x}Ga_xSe_2$, CIGS corresponding to chalcopyrites) and I_2-II-IV-VI_4 (Cu_2ZnSnS_4, CZTS corresponding to kesterites), whose band gap, well adapted to the solar spectrum, are well suited for photovoltaic applications.

QDs are typically characterized by broad excitation bands, narrow fluorescence emission peaks and high intensities. Their quantum yields are comparable to those of organic fluorophores but their extinction coefficients are 10 to 50 times larger.

24 Available at: https://www.nobelprize.org/prizes/chemistry/2023/press-release/.

This strong propensity to absorb light makes them ideal candidates for use in energy conversion systems. Unlike organic fluorophores, they are more chemically stable and resist photobleaching (loss of fluorescence during irradiation), 100 times better than that of organic molecules (Yao et al. 2014).

The first colloidal solutions of QDS such as CdS and CdTe were synthesized in the 1980s, where the QDs were stabilized by the adsorption of organic molecules, but high fluorescence yields were not obtained. In the 1990s, significant progress toward controlling the dimensions of CdS, CdSe and CdTe NCs was made (Murray et al. 1993). Subsequently, synthesis methods producing large quantities of QDs were proposed in the 2000s. Cao and his group (Yang et al. 2005) thus developed a method for synthesizing CdSe based on the reaction of a mixture at high temperature of two precursors (Myristate of Cd and Se) in solution[25].

Their stability with respect to the external environment can be considerably improved using core-shell structures, such as ZnS@CdSe, where the fluorescent QD (CdSe) is covered by a SC layer, such as ZnS, with a band gap much greater than CdSe (Kortan et al. 1990). Unfortunately, this strategy makes the fluorescence of QDs insensitive to their environment and hence unsuitable for sensor applications.

Currently, the techniques for preparing QDs via the colloidal route have made considerable progress. It is now possible to produce large quantities of QDs at costs compatible with their commercialization, covering a whole range of wavelengths, from visible to near IR (NIR; Bang et al. 2017; Owen and Brus 2017; Pu et al. 2018b). Lead-based QDs that fluoresce in the NIR are particularly important for applications in biomedical imaging.

In conclusion, colloidal QDs are currently considered nanomaterials with exceptional fluorescence properties, including emission wavelengths that vary depending on their size, high quantum yields and pure color emissions. They are also chemically stable and resistant to photobleaching, unlike organic fluorophores. They can be prepared on a large scale and at low cost. These properties make QDs particularly suitable for display and lighting applications requiring electroluminescence. Their luminescence properties make them attractive in sensor applications, for which different emission and quenching mechanisms by resonance energy transfer can be considered for analytical purposes (see Chapter 4).

---

25 Cd Myristate (Cd salt of tetradecanoic acid – $CdC_{28}H_{54}O_4$) decomposes at 226°C; Selenium powder has a melting point of 221°C and becomes slightly soluble in a solvent such as 1-octadecene (ODC-$C_{18}H_{36}$) beyond 190°C. By choosing a reaction temperature of 240°C, QDs can be obtained with very good monodispersity (size differences less than 5%) and very good fluorescence properties (quantum yield around of 30–40%) (Yang et al. 2005).

1.2.4. *Two-dimensional inorganic lamellar nanosheets*

Graphene has extraordinary conductivity properties due to its two-dimensional lamellar structure. Since its discovery, there has been renewed interest in other 2D inorganic materials (2DMs) with similar lamellar structures, such as transition metal dichalcogenides (TMDs) and some oxides and hydroxides, since it was shown that they could be obtained in the form of two-dimensional nanosheets with one or more atomic layers. Strano et al. nicely described the optoelectronic properties and applications of the main TMDs formed of MoS_2, $MoSe_2$, WS_2 and WSe_2 nanosheets (Wang et al. 2012).

Alongside the vast family of bidimensional TMDs, other lamellar materials such as tin sulfide (SnS_2) (Qu et al. 2014), and phosphorene, a semiconductor made of an assembly of black phosphorus (BP) atoms can also be exfoliated into nanosheets. Also, hexagonal boron nitride (hBN), a two-dimensional material isomorphic to graphene but insulating, is another 2D material structured as a hexagonal network of boron and nitrogen atoms (Zhang et al. 2017a).

Finally, a new family of 2DMs, called MXenes, was discovered in 2011 by Gogotski et al. (Naguib et al. 2011). They have the generic formula $M_{n+1}X_nT_x$, where M is a transition metal, X is a carbon or nitrogen atom, and T is an OH, O, or F group. Also considered as transition metal carbides or nitrides, they have very good conductivity and are often compared to graphene. Their properties depend on the nature of the surface Tx terminations.

1.2.4.1. *Dichalcogenides*

Unlike graphene, which is formed only of carbon atoms, the great chemical diversity of TMD atoms gives a much wider range of properties, including materials that can be semiconductors (MoS_2 and WS_2), semi-metallic (WTe_2 and $TiSe_2$), metallic (NbS_2 and VSe_2) and superconductors ($NbSe_2$ and TaS_2) (Lv et al. 2015). Furthermore, depending on the number of layers and composition, semiconductor TMDs can have a direct or indirect energy band gap and n or p-type charge carriers (Chhowalla et al. 2013, 2015). These properties have opened the field to catalysis, energy storage and electronics applications. New field effect transistors (FETs) have been made from TMDs and used for analytical purposes to detect chemical compounds and create optoelectronic devices and new logic circuits, as described in several review papers (Tan and Zhang 2015; Anichini et al. 2018; Xiong et al. 2019).

Metallic dichalcogenides such as molybdenum sulfide (MoS_2), tungsten sulfide (WS_2) and titanium sulfide (TiS_2) are the most widely studied, with MoS_2 being considered the most accessible (Anichini et al. 2018). Unlike graphite, formed by the stacking of identical graphene layers, TMDs comprise an assembly of hybrid layers in which a layer of a transition metal M is arranged like a “sandwich” and linked covalently to two outer layers of sulfur atoms. The external planes of these hybrid layers are held together by Van der Waals forces, which explains (as for graphite) their easy separation in exfoliation techniques (Figure 1.23).

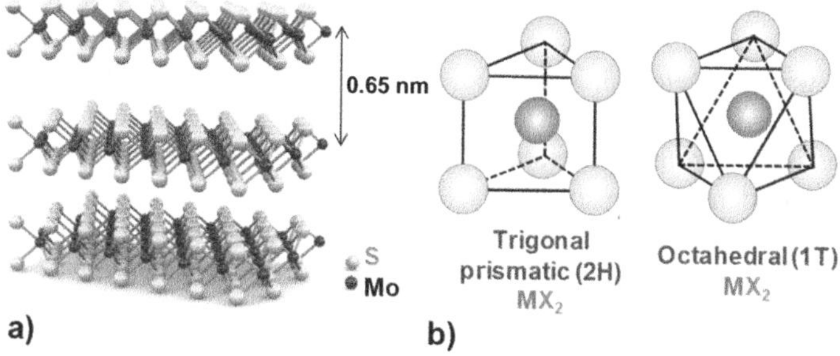

Figure 1.23. *Representation of the crystal structures of TMDs (MX2 structure). (a) Three-dimensional representation of the different planes of a MoS_2 crystal (distance between two 0.65 nm layers). According to Radisavljevic et al. (2011). (b) Representation of the two crystal forms 2H (trigonal) and 1T (octahedral, the most stable). According to Lv et al. (2015). For a color version of this figure, see www.iste.co.uk/lacaze/nanotechnologies.zip*

Two main methods are used to produce these nanosheets: exfoliation of the material from the bulk state or CVD. The first is an electrochemical reaction, lithium-ion intercalation-based exfoliation, where Li+ ions are intercalated into the bulk of TMD using a similar mechanism to that found in a lithium battery. After the intercalation of lithium, the product obtained is subjected to sonication, which releases dihydrogen, causing TMD to disintegrate and nanosheets to form, collected as a suspension in water with a yield greater than 90% (Huang et al. 2013). The same direct sonication process can be used to cause the disaggregation of TMDs with a solvent such as N-methylpyrrolidone, which exhibits good wettability with the TMD. MoS_2, WS_2, $MoSe_2$, $NbSe_2$, $TaSe_2$, $NiTe_2$, $MoTe_2$, h-BN and Bi_2Te_3 nanosheets with one or more atomic layers can be prepared according to this method.

The second method for preparing TMDs and CVD produces large lateral crystals with controlled thicknesses of one or more atomic layers. MoS_2 nanosheets are obtained by heating sulfur and MoO_{3-x} precursors in powder form. Sulfur in the vapor state reacts with the sub-oxide MoO_{3-x}, forming MoS_2 condensed as a film on a Si/SiO_2 substrate. Numerous variations in the choice of precursor have been proposed, making it possible to obtain large nanosheets.

TMDs, particularly MoS_2, have interesting properties. Easy surface functionalization by metals and modification of their composition by doping or forming alloys; hybrid nanostructures can also be synthesized by combining TMDs nanosheets with carbon-based or inorganic nanomaterials (Tan and Zhang 2015). TMD nanocomposites doped on the surface by precious metals (Pt, Pd, Au, Ag) or alkalis, TMD alloys of the $MoS_{2(1-x)}Se_{2x}$ and $Mo_{1-x}WxS_2$ type, or assemblies between TMD and graphene, CNTs or carbon fiber nanosheets have been fabricated.

Doping with a metal chemisorbed to the surface of a TMD modulates the electronic properties of TMDs. MoS_2 and WSe_2 nanosheets doped with potassium can be transformed into degenerate n^+ semiconductors, characterized by very high surface electron densities (respectively, ~$10^{13}/cm^2$ and $2.5 \times 10^{12}/cm^2$ for MoS_2 and WSe_2) (Fang et al. 2013). FETs with very good electrical characteristics can be formed with a channel made of MoS_2 or WSe_2 nanosheets, whose extremities in contact with the electrodes (source and sink) are n^+ doped. One advantage of MoS_2 or $WeSe_2$ nanosheets over silicon is that their surfaces are extremely homogeneous, and the thicknesses of the nanosheets are perfectly uniform over the entire area.

In summary, TMD nanosheets are a new family of 2D nanostructures that are easy to produce and functionalize. Their electrical and photonic properties make them potential competitors of all other 2D nanomaterials, requiring more difficult production conditions. Their semiconductor properties make them more operational than graphene for the design of transistors. The few application examples described in this section show that their application prospects are immense and have already brought significant achievements in plastic electronics, energy and analytical chemistry with the design of innovative sensors using FETs (see Chapters 4–6).

1.2.4.2. *MXenes*

MXenes have the general formula $M_{n+1}X_nT_x$, where M is a top-row transition metal (Ti, Sc, V, etc.), X carbon or nitrogen and T is the terminating surface groups corresponding to groups OH, F, Cl, and =O. MXenes are obtained from lamellar 3D materials with the general MAX structure, comprising alternating MX and A sheets strongly bonded to each other (A is generally aluminum [Al]). Unlike other TMD materials, such as 3D chalcogenides, for which van der Waals forces weakly bond the sheets, obtaining MXenes sheets from MAX requires strong chemicals to break the bonds to obtain thin layers.

Ti_3C_2 is one of the most used MXenes. It was synthesized for the first time by Barsoum and Gogotsi group (Naguib et al. 2011) and obtained by reaction of hydrofluoric acid HF on Ti_3AlC_2. This reaction breaks the bond that connects Al to two MX sheets and replaces the Al atoms with hydroxyl groups. With only hydrogen bonds holding the sheets together, exfoliation can be done simply by sonication, producing Ti_3C_2Tx sheets, with T representing surface OH groups (Figure 1.24).

Subsequently, a whole series of MXenes were synthesized and are described in several review articles (Naguib et al. 2012, 2014; VahidMohammadi et al. 2021; Alwarappan et al. 2022; Chen et al. 2023). The properties of these nanomaterials (high conductivity, high chemical stability, ease of implementation, biocompatibility and ability to produce large quantities) make them particularly attractive for many applications (catalysis, batteries, supercapacitors, electromagnetic shielding, medical treatments and detection devices). Although often compared to graphene because of their high conductivity, with properties that depend on their surface Tx terminations, their use for detecting various analytes is only envisaged in certain composites. This generates heterojunctions, which increase chemical interactions with the analyte and induce strong variations in conductivity.

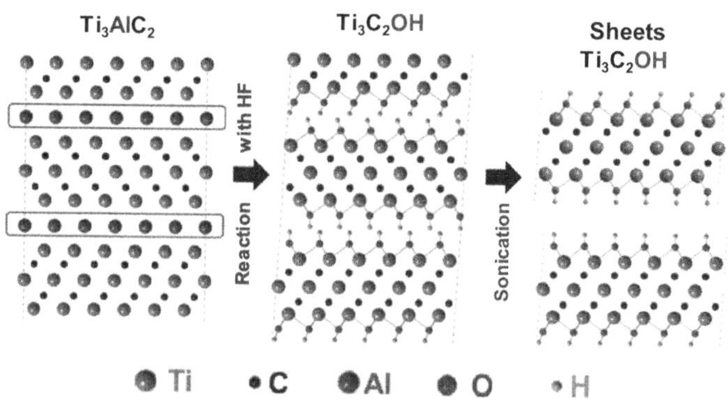

Figure 1.24. *Principle of the method of obtaining Ti_3C_2OH sheets from Ti_3AlC_2. Reacting it with HF replaces the Al atoms, which bond the Ti_3C_2 sheets with weakly binding OH groups, facilitating the exfoliation process by sonication. Adapted from Naguib et al. (2011). For a color version of this figure, see www.iste.co.uk/lacaze/nanotechnologies.zip*

1.2.4.3. *Phosphorene*

Phosphorene is a p-type semiconductor (Cho et al. 2016) with one or more BP monolayers. It was first isolated in 2014 by the mechanical exfoliation of BP, using the scotch tape method developed by Geim and Novoselov to extract graphene from graphite. Subsequently, chemical exfoliation methods of BP in the liquid phase, liquid phase exfoliation (LPE), were proposed using various organic solvents, mainly N-methyl pyrrolidone (NMP), dimethylformamide (DMF), as well as aqueous media with added surfactants. These were each subjected to vigorous sonication. A good description of these preparation techniques can be found in recent papers (Gusmao et al. 2017; Yang et al. 2018).

One of the unusual characteristics of phosphorene is that its band gap varies between 0.3 and 2.2 eV when it passes from a compact solid state (BP) to the final single layer state (Figure 1.25).

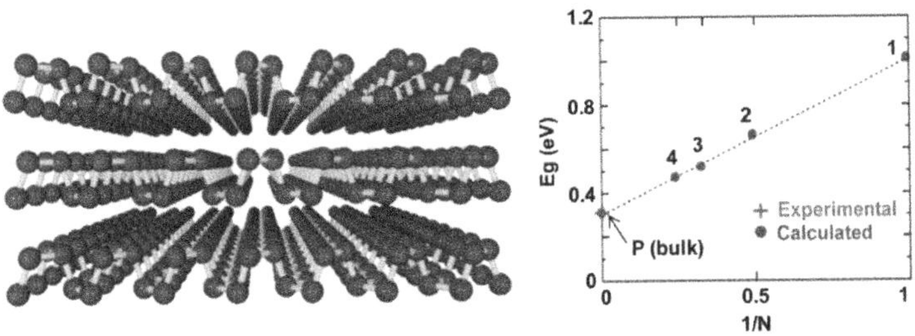

a) b)

Figure 1.25. *Structural and energetic characteristics of phosphorene. (a) Perspective view of phosphorene sheets. (b) Variation in band gap as a function of the number N of layers (4, 3, 2 and 1). Adapted from Liu et al. (2014a). For a color version of this figure, see www.iste.co.uk/lacaze/nanotechnologies.zip*

Phosphorene is characterized by a high hole mobility, between 600 and 1,000 $cm^2 V^{-1} s^{-1}$, and, when incorporated in an FET structure, the "On/Off" ratio of the drain current is also high, around 10^3–10^5 (Liu et al. 2014a). These properties make phosphorene an intermediate material between graphene and TMDs, well suited to produce chemical sensors. However, its sensitivity to air humidity requires extra precautions during use and protection of the sensitive layer. Dispersing gold or platinum NPs onto the surface of phosphorene modifies its electronic properties, making it comparable to an n-type SC with very good stability over several months (Cho et al. 2017). Recently, the pulsed laser of a fine BP powder suspended in NMP produced phosphorene with greater stability when the BP precursor is obtained by the high energy mechanical milling (HEMM) of red phosphorus (the most stable form of phosphorus) (Zhang et al 2019a). Other methods including exfoliation by anodic electrolysis of BP in a saline environment (0.5 M Na_2SO_4) have also been proposed and give stable phosphorene nanosheets.

1.2.5. *Metal organic frameworks (MOFs)*

MOFs are formed directly by self-assembly reactions between metal ions or metal clusters and organic compounds, with reactive functions in determined directions (multitopic ligands). Like porous crystals, they have the advantage over other porous inorganic materials (zeolites, silicas, etc.) in that it is possible to modulate porosity and "host–guest" interactions by the choice of reagent used in the synthesis process. The interest in MOFs has continued to grow since their discovery, giving rise to many applications which range from the enhanced recovery of hydrocarbons to the capture of gases or their separation, as well as their use in the design of sensors and specific capillary electrophoresis devices (see Chapter 2). Compounds similar to MOFs, but entirely organic, have also been developed and differ from MOFs in that their bonds are covalent, hence their name covalent organic frameworks (COFs) (Geng et al. 2020).

1.2.6. *Methods for manufacturing MOFs*

MOFs are obtained according to two main methods. The first involves a crosslinking reaction between predefined inorganic blocks and organic ligands, which leads to hybrid crystal structures with controlled porosities. The second uses an existing MOF and transforms it through ligand substitution to obtain new functionalities.

Directly obtaining MOFs using this crosslinking method was developed on a large scale by Yaghi (USA) and Férey (France) and their teams. The synthesis of hybrid, crystalline and robust materials with controlled and predefined porosities is described in several review articles (Yaghi et al. 2003; Férey 2008). This method generates an MOF result from the self-assembly between preconstructed inorganic blocks, known as secondary building units (SBU), and rigid and directional organic ligands.

Remarkably, SBU blocks retain their entire geometric structure during the crosslinking reaction. The wide range of rigid organic compounds and inorganic SBUs available makes it possible to obtain an infinite number of materials whose porosity can be determined in advance.

These synthesis principles are described by Yaghi et al. (Li et al. 1999) for one of the very first examples of MOF production (MOF-5) where zincates arranged in the form of a super-tetrahedron-like cluster (SBU) react with rigid organic ligands, such as 1,4-benzene dicarboxylate (BDC) (Figure 1.26).

Crosslinking takes place in a hot organic medium in a closed reactor. An MOF cubic network, where the four zincates are linked together by a common oxygen (O) and six carboxylate ligands (-CO_2) coming from the six BDCs, gives the stoichiometry $Zn_4(O)(BDC)_3$.

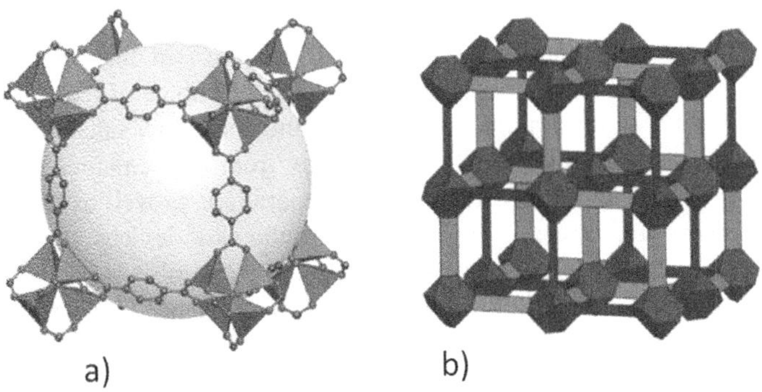

Figure 1.26. *Structure and topology of MOF-5. (a) Representation of the unit cell obtained by reaction between the zincates (blue tetrahedra) and the carboxylate groups of the BDCs (the oxygen atoms are in red, the carbons in black). The yellow sphere represents the maximum occupancy volume of the internal cavity defined by eight groups of four zincates and twelve BDCs (diameter of the occupancy sphere ~1.2 nm). (b) Representation of the cubic network of MOF-5. Assemblies formed by four zincates (SBU) are represented in red by truncated tetrahedra; the links between zincates correspond to the BDCs and are represented in blue. Adapted from Yaghi et al. (2003). For a color version of this figure, see www.iste.co.uk/lacaze/nanotechnologies.zip*

These porosity properties can be further enhanced by choosing larger organic ligands, while retaining the topology of the crystal (isocrosslinking principle). A few years later, Yaghi et al. synthesized a series of MOFs, producing porous systems whose cavities have pores that can be predetermined by the choice of the ligands binding the inorganic SBUs. Thus, by using the same inorganic cluster formed by the four ZnO_4 derivatives (identical to that of MOF-5) and by choosing linear, rigid organic ligands of increasingly longer lengths, ranging from fumarate to TPDC (terphenyl- 4,4'-dicarboxylate), the pore size is then considerably enlarged. Its volume is 7.8 times greater than for MOFs obtained with fumarate (Figure 1.27)[26]. Intermediate sizes are obtained with ligands like BDC (phenyl-1,4-dicarboxylate and MOF-5) and BPDC (biphenyl-4,4′-dicarboxylate and IRMOF-10).

---

26 Compared to the various usual microporous materials such as zeolites (alumino silicates AlO_4^{5-} and SiO_4^{4-}), silicas (SiO_2) or active carbons for which the gas adsorptions measured by the Brunauer–Emmett–Teller (BET) method are between 900 and 1200 m^2/g, and their porosity volumes are, respectively, 0.3 g/cm^3, 1.15 g/cm^3 and 0.60 g/cm^3, MOFs have significantly higher gas absorption possibilities ranging from more than 2,000 m^2/g with MOF-5 to values over 7,000 m^2/g with NU-110, with very low mass densities (Furukawa et al. 2013). The BET method is used to determine the equivalent surface area of a porous material from the quantity of nitrogen adsorbed in the pores of the material at 77 K.

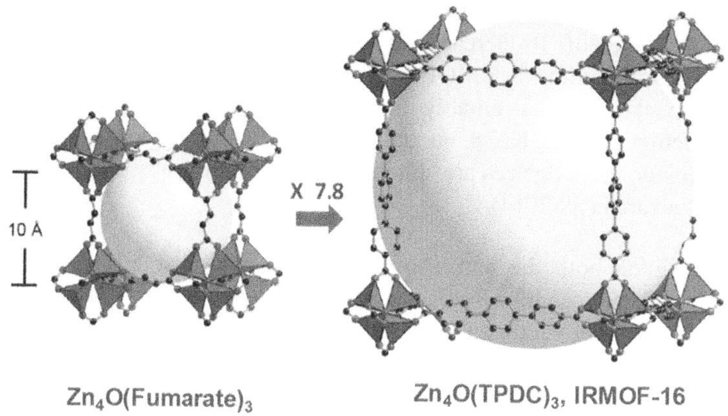

Figure 1.27. *Principle of synthesis by isoreticulation. The use of fumarate then terphenyl-4,4''-dicarboxylate (TPDC) increases the bond lengths between the SBUs (ZnO_4) while retaining the initial topology and increasing the size of the pore in the MOF cubic network. As with the previous figure, the yellow sphere represents the maximum size of a chemical species that can be inserted into the cavity. Adapted from Furukawa et al. (2013). For a color version of this figure, see www.iste.co.uk/lacaze/nanotechnologies.zip*

More than 20,000 MOFs have been synthesized over the last 10 years by changing the nature of both the ligands and the inorganic SBUs, whose varying structures give very varied application possibilities. Records have been set for BET adsorption measurements reaching more than 7,000 m^2/g, with predictable theoretical values of more than 10,000 m^2/g, (Farha et al. 2012). Finally, it is important to point out that these MOFs can be used to synthesize nano-MOFs (n-MOFs) in a micellar constrained medium, which introduces the possibility of luminescent n-MOFs, and again, potential application in the field of biomedical imaging (Carne et al. 2011).

Another layer-by-layer synthesis technique was also used to manufacture MOF films adsorbed onto a solid support (Ohnsorg et al. 2015), by functionalizing the substrate with long hydrocarbon chain molecules, capable of surface self-assembly. Their reactive end (such as –COOH) can bind with the ligand and induce the growth of an MOF film and thus obtain a surMOF attached to the substrate. For example, a surMOF film with the global formula $Cu_3(TMA)_2$, known as HKUST-1[27], is often used in the design of optoelectronic devices in thin layers. It is obtained after a series

27 Most MOFs are identified by acronyms that describe the original laboratory or university where they were discovered and produced. HKUST is the acronym for Hong Kong University of Science and Technology.

of immersion cycles, rinsing and drying a gold surface functionalized by a self-organized layer of 16-mercaptohexadecanoic acid[28]. This technique is very interesting because it allows an MOF film of controlled thickness and dimensions to be locally attached onto a suitably functionalized surface. We can thus design various detection devices based on optical or electronic responses. Adding their modular porosity, these devices are suitable for the selective absorption of gases and molecules (Falcaro et al. 2014).

By playing on both the nature of the ligands and that of the SBUs, many structures have been manufactured, giving them varied applications in fields such as medical imaging, drug vectorization (Horcajada et al. 2010), catalysis, the creation of optoelectronic devices, or the improvement of proton conduction in H_2/O_2 fuel cells. The storage of gases and their separation, the identification of molecules and the functionalization of MOFs to create multiple luminescence centers are other application areas. Together an adjustable porosity and luminescence, obtained according to the choice of ligands and metal clusters, make extremely well-suited MOFs for the development of photoluminescent devices used in the detection of chemical species. They are also involved in the formation of electrophoresis columns where porosity plays a determining role (see Chapter 2).

The luminescence of an MOF can have very different origins: it can be due to the ligand, the metal, the insertion of fluorescent species into the MOF or even antenna effects between the metal and the ligand corresponding to energy transfers. These luminescence properties can be used in a variety of ways. When the ligand and the metal are fluorescent, we can use the MOF porosity to insert a particular chemical species. If this species quenches the fluorescence of the ligand or the metal (quenching effect), then the MOF can be used to detect the species in question. This property can be used to detect explosives (Allendorf et al. 2009; Lan et al. 2009; Hu et al. 2014b).

In summary, synthesized by assembling metal ions with organic ligands, MOFs are a new class of crystalline and hybrid three-dimensional porous materials, distinguished from traditional porous materials (zeolites, activated carbons, porous silicas, etc.) by their modular porosity and by the possible introduction of various functionalities into their pores. This makes them suitable for a wide range of applications. Their relatively simple manufacture, which does not require very high temperatures (condensation generally takes place in an organic medium and for

28 The first cycle involves first immersing it in a copper acetate solution, then, after rinsing and drying, immersing it in a 1,3,5-benzene tricarboxylic acid solution (trimesic acid [TMA]), which leads to the first surMOF layer of approximately 2 nm thick when the reactions are carried out at 25°C. This production cycle can be repeated several times and the thickness of the film increases linearly with the number of cycles (Ohnsorg et al. 2015).

temperatures between 100 and 200°C), makes them more accessible than many comparable materials. Among their numerous applications, three applications are now proving extremely important: the separation and storage of gases, proton conduction with a potential application to hydrogen fuel cells and applications linked to the detection of chemical species due to optical properties of fluorescence and controlled porosity.

1.3. Conclusions

The last 20 years have seen the emergence of new families of carbon nanomaterials that have established themselves through their properties and the new solutions they provide in energy, electronics, biology and sensors.

Fullerenes, CNTs, graphene and their derivatives have become essential materials despite having a simple structure based solely on sp^2 hybridized carbon chains. Their properties differ depending on their dimensional characteristics, 0D for fullerenes, 1D for CNTs and 2D for graphene. CNTs, graphene and their functionalized derivatives have been particularly successful because of their exceptional electrical and mechanical properties, which have given rise to new plastic hybrid materials with improved mechanical and electrical properties and to the production of composite electrodes used in electrocatalysis, energy storage and conversion, and analysis.

Graphene and its derivatives (GO) are undoubtedly the compounds with the greatest number of applications and the greatest development with the manufacture of plasticized ribbons, foams, papers, or even fibers, bringing together both excellent electrical and mechanical properties. Nanometric graphene fragments are a new variety comparable to QDs (GQDs), characterized by their fluorescence properties and by non-zero forbidden energy bands, which make them suitable for the construction of various optoelectronic devices.

NDs, carbon nanoparticles in the form of CQDs, are a new family of materials; they are distinguished by their small size, less than 10 nm, and by different types of carbon–carbon bonds involving, in variable proportions, hybridized sp^2 and sp^3 carbons. For NDs, these are mainly bonds between sp^3 carbons, while for CQDs, the particle nucleus is essentially composed of graphitic sp^2 carbons. These are also commercialized materials, much easier to obtain than CNTs, fullerenes and graphene, and therefore accessible at lower costs. For certain electrochemical applications (energy storage, production of electrolytic capacities), CQDs compete with activated carbon. Their use is also linked to their fluorescence properties, differentiating them from intrinsically non-luminescent fullerenes, CNTs and graphene. Furthermore, the presence of hydrophilic functional groups on the surface makes them soluble in an aqueous medium, and due to their non-toxicity, they are considered excellent candidates for medical imaging.

In the case of inorganic nanomaterials, three main families can be distinguished: metallic nanoparticles (MNPs), semiconductor nanoparticles or QDs, and, more recently, two-dimensional nanomaterials formed from transition metal chalcogenides (TMDs). This last category also includes Mxenes and phosphorene, which are promising in various applications ranging from drug delivery to energy storage. The most common applications relate to electronic devices (especially transistors), optoelectronic devices and gas sensors.

Metallic gold NPs (AuNPs) are the most widely studied and have given rise to many analytical applications. Their optical properties, however, differ considerably depending on their size. For sizes less than 3 nm, their light absorption is like that of a molecular compound, while for larger sizes, the metallic character becomes predominant and their absorption is plasmonic. We observe luminescence properties for very small NPs (NCs), whose intensities depend on the NP functionalization mode. As with carbon nanomaterials, and due to their lack of toxicity, these gold NCs are used essentially for their fluorescence properties in biomedical imaging applications and the detection of compounds in very low concentrations. On the contrary, larger gold NPs are mainly used for their quenching properties and ability to enhance numerous spectroscopies.

Semiconductor QDs, also wet synthesized, are another family of nanomaterials, particularly sought after for their exceptional fluorescence properties, far superior to organic fluorophores. In the core/shell form (CdSe/ZnS type), the photofluorescence quantum yields are close to one, their photostability is very good and without intermittency, and very pure fluorescence colors are obtained depending on the composition and the size of the QDs. These PL properties are preserved in electroluminescent devices and mean that QDs are now considered among the most efficient nanomaterials to produce color screens and for the more general design of optoelectronic devices, also applicable in analytical chemistry.

The abundant and easily synthesized two-dimensional nanosheets of TMDs, including MoS2 and WS2, are currently 2D nanomaterials of growing interest. Their enthusiasm is comparable to that experienced by graphene when it was discovered. Their semi-conduction and intercalation properties make them materials of choice for applications ranging from plastic electronics to energy storage. They also lend themselves easily to the production of various optoelectronic systems. They are therefore used to produce various sensors for detecting gases and pollutants in trace amounts.

Of the hybrid and composite materials, three-dimensional organometallic networks (MOFs) are also an essential family of nanoporous compounds, which can be synthesized with a moderate energy cost compared to zeolites. The architecture of these compounds can be modulated infinitely, which makes it possible to define the dimensions of the pores and the crystallographic characteristics in advance and, thus, to best optimize the structure for a specific function in gas separation or analytical chemistry.

2

Separation and Amplification Techniques

Identifying an analyte in trace amounts or mixed with many compounds sometimes requires techniques that allow either an increase in concentration or the separation of the mixture's constituents. Such situations are frequently encountered with biological compounds. Increasing the concentration is necessary with DNA samples, which are obtained in trace amounts, for which a specific concentration amplification method known as polymerase chain reaction (PCR) is required. Likewise, identifying the different oligonucleotides within DNA and establishing the order in which they are assembled is also achieved using sequencing techniques. Physical separation techniques are also necessary to isolate the different analytes from a gas or liquid mixture. Typically, these may be chromatographic techniques for the gaseous or liquid state or electrophoretic techniques for electrically charged compounds.

2.1. Principle of the PCR technique applied to the concentration amplification of DNA traces

The PCR technique, which revolutionized biological sciences, was developed by K. Müllis in the 1980s, earning him the Nobel Prize in 1993. Based on a cycle of transformations of a DNA molecule, after n repetitions, it gives 2^n identical DNA molecules, which, in a few hours, that is, after some 20 cycles, produces over 1 million molecules.

A DNA molecule, whose structure was revealed in 1953 by Watson and Crick, is formed by assembling two complementary strands coupled with hydrogen bonds, which are easily broken when the temperature is raised to around 100°C (Figure 2.1).

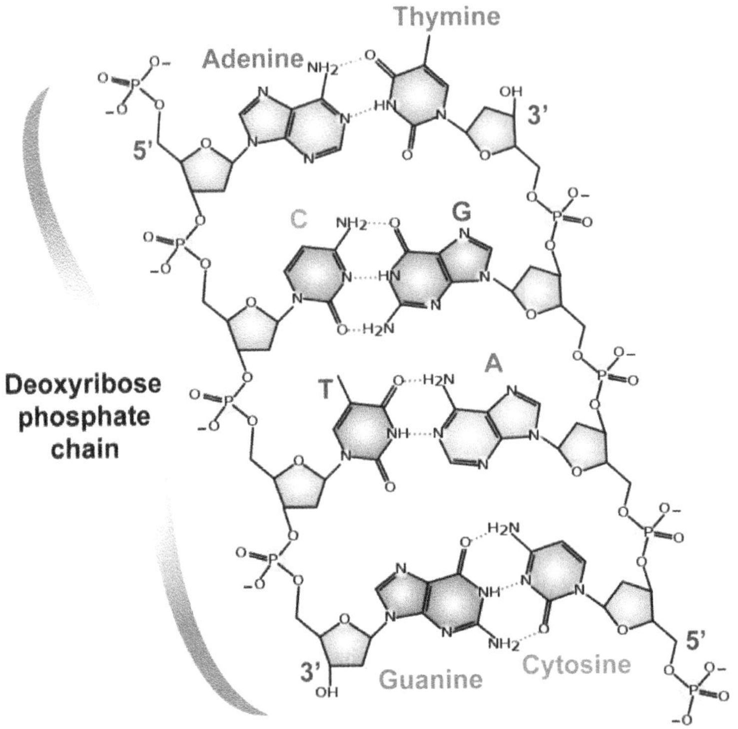

Figure 2.1. *Structure of a DNA sequence, containing two strands linked together by the bases adenine (a) and thymine (T), and guanine (G) and cytosine (C). Adapted from Wikipedia. For a color version of this figure, see www.iste.co.uk/lacaze/nanotechnologies.zip*

COMMENT ON FIGURE 2.1.– *The reactive parts of DNA are the OH groups of the C3' carbons, which explains why when the two strands are separated, and new nucleotides react in solution to lengthen the chains, the numbering (5'→3') progresses from top to bottom for the left strand, and from bottom to top for the right strand. Note that the specific pairing between the bases C and G has three H bonds, while A and T have only two. Furthermore, DNA strands are linked by two deoxyriboses, involving condensation between an OH on a C3' carbon of one of the deoxyriboses and the triphosphate on the C5' carbon of the other. This explains why the two strands of DNA are numbered 3'-5' and 5'-3', respectively.*

Each strand of DNA (single strand DNA [ss-DNA]) is formed by the random sequence of four nucleotides (dNTP) corresponding to deoxynucleotide triphosphates of the four bases adenine (A), thymine (T), cytosine (C) and guanine (G). It differs from an RNA strand (ribonucleic acid) in that it contains thymine instead of uracil. Unlike DNA, which is made up of deoxyriboses (an H in C2'), RNA contains only oxyriboses, characterized by the presence of a hydroxyl group on the C2' carbon, which also explains its lower stability compared to that of DNA (Figure 2.2).

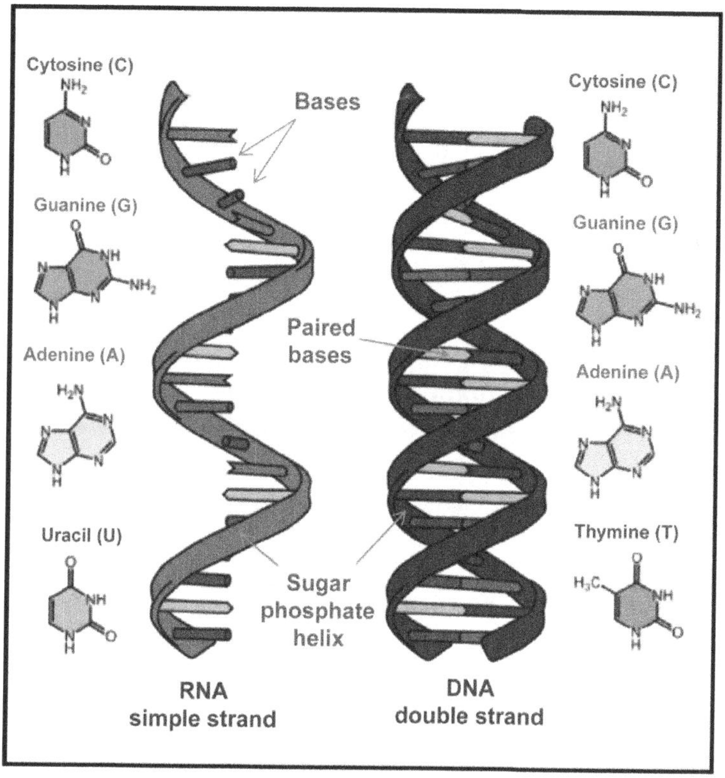

Figure 2.2. *Representation of single-strand RNA and double-strand DNA helices. The different nitrogenous bases are represented by horizontal-colored segments identifiable with bases C, G, A and U for RNA and C, G, A and T for DNA. Adapted from documents on the internet. For a color version of this figure, see www.iste.co.uk/lacaze/nanotechnologies.zip*

Each PCR cycle comprises three steps corresponding to different temperature cycles (Figure 2.3).

First, maintaining the temperature at 92°C for 30 s, the initial double-stranded DNA molecule (ds-DNA) is denatured into two complementary single strands (ss-DNA1 and ss-DNA2). By lowering the temperature to between 50 and 60°C, each strand is hybridized with a primer (short-length oligonucleotide, complementary to a sequence of the initial DNA), specifically binding to a sequence of nucleotides of each DNA strand (approximately 30 s).

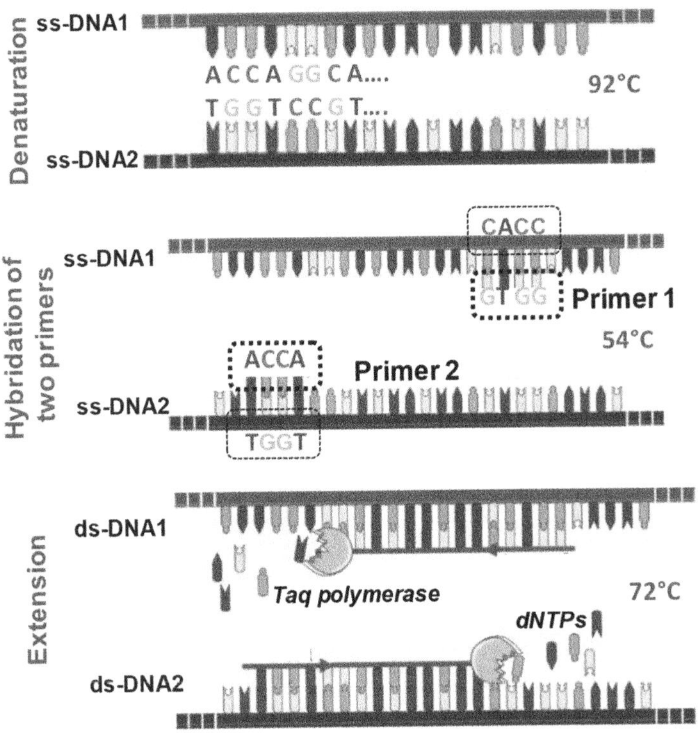

Figure 2.3. *The three steps of the first PCR cycle leading to the doubling of a DNA molecule. Adapted from a document from Université de Strasbourg. For a color version of this figure, see www.iste.co.uk/lacaze/nanotechnologies.zip*

COMMENT ON FIGURE 2.3.– *The two primers 1 (GTGG) and 2 (ACCA) complementary to the single-strand sequences ss-DNA2 (TGGT) and ss-DNA1 (CACC) are chosen in such a way that they elongate the ss-DNA1 and ss-DNA2 strands, respectively, from right to left and vice versa from left to right, each time hybridizing a nucleotide complementary to the last nucleotide on each ss-DNA1 and ss-DNA2 strands. This requires that the DNA sequencing has first been determined to synthesize primers capable of hybridizing with the ends of the two strands. Taq polymerase is a thermostable enzyme that resists high temperatures of around 100°C. It makes it possible to link the free OH group carried by the C3' carbon of a nucleotide (dNTP) in solution to the first, not yet hybridized, nucleotide of the ss-DNA1 or ss-DNA2 strand.*

Finally, by again increasing the temperature of the medium to an intermediate value of 72°C, in the presence of the four deoxyribonucleotides[1] adenine (dATP), thymine (dTTP), cytosine (dCTP), guanine (dGTP) and DNA polymerase (*Taq* or *Thermus aquaticus*), the strands are extended until double strands (ds-DNA) are obtained.

Currently, this type of transformation is carried out automatically in commercial devices called thermal cyclers; many variations of the method have been developed, particularly real-time PCR, also known as quantitative PCR (q-PCR), which makes it possible to measure the quantities of DNA produced in each amplification cycle using fluorescent markers. The details of these different PCR variants and their applications are nicely described in a recent review article by Zhu et al. (2020).

2.2. Sequencing techniques

Well before the discovery of PCR, the sequencing of biomolecules and that of DNA was of major interest and, in the 1960s, the first results describing the order of a small sequence of bases in DNA strands were obtained but using fragmentation methods that had extremely long experimentation times. It was only in the 1970s that Gilbert and Maxam (1977) and Sanger et al. (1977) simultaneously proposed faster methods, making it possible in a few hours to specify a DNA sequence of around 100 bases based on degradation and inhibition reactions.[2]

The uniqueness of Sanger sequencing is its use of fluorescent markers instead of radioactive markers (Gilbert's method) as well as particular nucleotides, inhibitors of chain propagation, characterized by the absence of hydroxyl groups on the C2' and C3' carbons, that is, dideoxyribonucleotide triphosphates (ddNTP) (Figure 2.4).

These ddNTPs are marked by four different fluorophores (ddNTP*), emitting color fluorescence specific to each base under UV irradiation. According to Sanger, sequencing a DNA strand involves hybridizing a primer comprising a small number of bases on the DNA strand, leading to the hybridized compound in Figure 2.5(a).

1 Deoxyribonucleotide triphosphates (dNTPs) are the building blocks of a DNA strand and correspond to the formula of a deoxyribose bearing on carbon C1' one of the four bases C, G, A and T and on carbon C5' a triphosphate group. The N of dNTP represents one of the previous bases.

2 Gilbert and Sanger received the Nobel Prize in Chemistry in 1980 for these inventions.

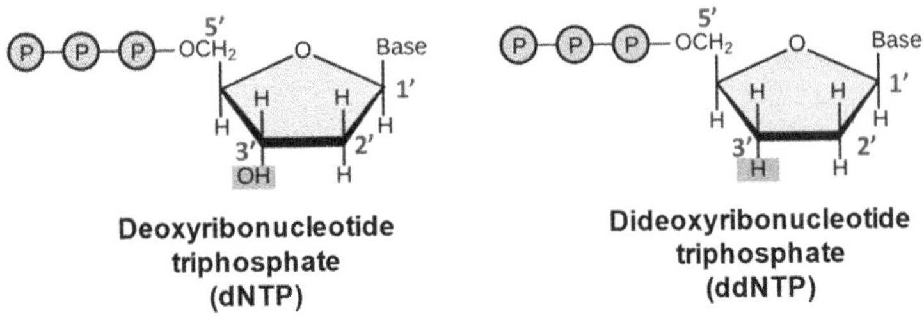

Figure 2.4. *Structures of nucleotides dNTP and ddNTP. For a color version of this figure, see www.iste.co.uk/lacaze/nanotechnologies.zip*

COMMENT ON FIGURE 2.4.– *The two nucleotides dNTP and ddNTP differ in the presence or absence of an OH hydroxyl group on 3' carbons. The absence of an OH group at 3' (ddNTP) blocks the propagation of the chain between different dNTPs leading to the formation of a DNA strand. This reaction results from the condensation between the OH on C3' of a nucleotide and the phosphate (P) on the 5' carbon of another dNTP.*

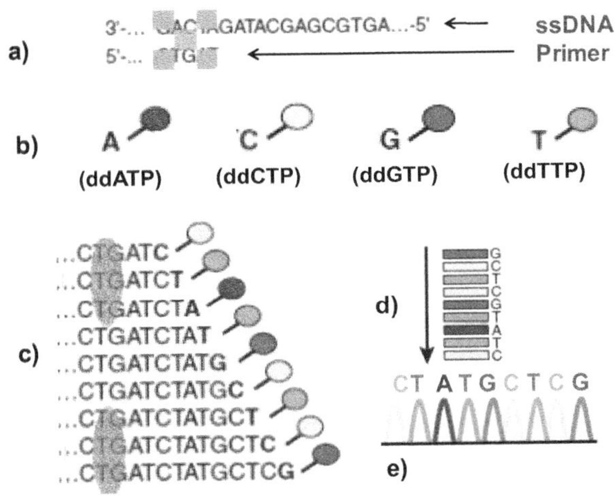

Figure 2.5. *Diagram of the sequencing of a DNA strand using the Sanger method. (a) Hybridization of the CTGAT primer complementary to a GACTA sequence of a DNA strand. (b) Marking by different fluorophores (red, yellow, green and blue) of the four ddNTP). (c) The different fragments obtained reflect the order of nucleotides in the DNA strand. (d) Electrophoretic separation of molecular masses in decreasing order. (e) Fluorometric visualization of the nucleotide sequence. Adapted from Shendure and Ji (2008). For a color version of this figure, see www.iste.co.uk/lacaze/nanotechnologies.zip*

In the presence of DNA polymerase and a mixture of the nucleotides dNTP and ddNTP* – the latter in a small proportion – hybridization lengthens the DNA strand with the dNTPs. Still, it stops each time a ddNTP* reacts with the DNA strand, ultimately leading to the different fragments shown in Figure 2.3(b). Detached from the DNA strand by denaturation, the different oligonucleotide fragments are separated by capillary electrophoresis (CE) and identified through their fluorescence color. The order in which the different fragments emerge makes it possible to establish DNA sequencing, as per the diagram shown in Figures 2.3(c)–(e).

This technique has undergone many successive improvements, increasing the sequencing automation speed and considerably reducing its cost, and, after several years of effort, revealing the sequence of the human genome comprising several billion base pairs. The production of micro-devices adapted to DNA sequencing by CE and equipped with a fluorescence detection system has also developed and, very early on, gave rise to numerous commercial devices (Liu et al. 1999; Schmalzing et al. 1999; Krylov and Dovichi 2000).

However, marking oligonucleotides with fluorophores introduces elements of complexity, which contribute to the increased cost and duration of sequencing. Rothberg et al. (2011) had the idea of replacing the method of identifying fluorophore-labeled nucleotides by directly measuring the pH resulting from their insertion into a DNA strand.

This pH variation originates from the natural physiological process of DNA replication. Incorporating a dNTP monomer into a growing DNA strand involves the formation of a covalent bond and the release of pyrophosphate and an H^+ ion, with the dNTP being incorporated only if it is complementary to a nucleotide in the matrix. The method, marketed by Ion Torrent, is a DNA sequencing method based on detecting H^+ ions produced during replication. It is particularly advantageous in terms of costs, complexity and time. The details of the technique are well described in the book Clinical Genomics, edited by Kulkarni and Pfeifer (2015).

Ion Torrent machines are today marketed as fast benchtop sequencers (Figure 2.6). Sequencing 200 base pairs takes approximately 2 h 30 min with such a device.

Other techniques have also been developed to simplify sequencing procedures by proceeding directly to DNA or RNA fragment analysis without requiring pre-amplification and labeling oligonucleotides for their identification. One increasingly efficient technique is based on measuring disturbances of a transmembrane ionic current resulting from the passage of ss-DNA or ss-RNA through a polarized nanopore membrane (~200 mV) (Figure 2.7). Known as Nanopore sequencing, it was developed by a group of researchers from Oxford, who founded Oxford Nanopore Technologies (ONT). They developed a commercial sequencing product called MinION, the history of which is described in the review article by Deamer et al. (2016). This device has benefited from numerous improvements, resulting in the development of increasingly precise portable devices. Its operating principle is described in Figure 2.7.

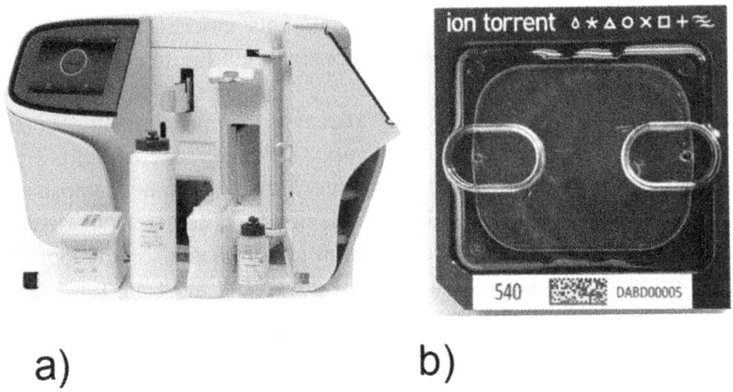

Figure 2.6. *IonTorrent sequencer and chip. (a) IonTorrent sequencer HID Ion GeneStudio S5, whose sale price is between 100 and 150 k€. It is a fully automated benchtop machine that requires dedicated reagents. (b) The IonTorrent Ion540 chip is used in the IonTorrent sequencer. The price of the chip is between 1,500 and 2,000 € (2023 values). For a color version of this figure, see www.iste.co.uk/lacaze/nanotechnologies.zip*

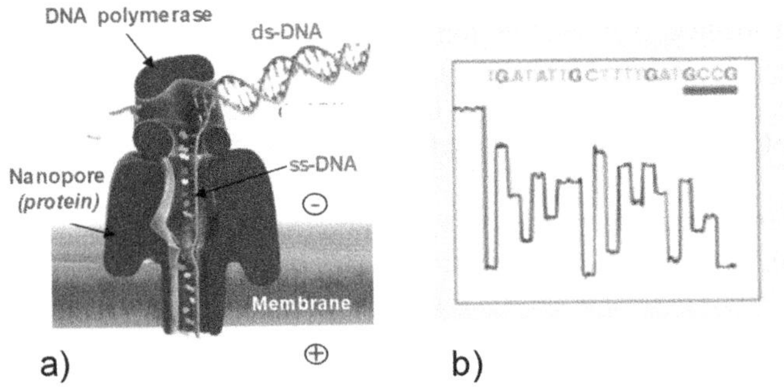

Figure 2.7. *Principle of nanopore sequencing. (a) Diagram of the device comprising a nanopore inserted into a positively polarized membrane on its internal face. (b) Ionic current measured as a function of time during the passage of the ss-DNA strand through the nanopore. Adapted from ONT diagrams. For a color version of this figure, see www.iste.co.uk/lacaze/nanotechnologies.zip*

COMMENT ON FIGURE 2.7.– *The DNA polymerase coupled to the nanopore splits the ds-DNA into ss-DNA introduced into the nanopore under the effect of polarization (between 100 and 200 mV). The internal dimensions of the nanopore cavity are around 1 nm, and it includes a restriction element placed in the cavity, which slows down the speed of the ss-DNA (around 30 bases per second) and induces variations in the intensity of the ionic current measured through the wall of the membrane. These variations in intensity are processed by computer software. They are thus correlated with the sequential passage of the different complementary nucleotides of the ss-DNA strand-resulting in DNA sequencing.*

In conclusion, the sequencing technique combined with PCR-type preamplification is still the reference technique. Although it has evolved toward increasingly efficient technology, it nevertheless remains complex. Although the Ion Torrent technique simplifies the process by eliminating markers, the technique proposed by ONT is even more advantageous due to the direct sequencing of DNA strands at very low concentrations and the elimination of the pre-amplification step.

2.3. Separation techniques for product mixtures

The previous two sections deal with DNA-specific techniques, the main objective of which is the decoding and concentration amplification of oligonucleotide sequences. Separating the constituents of a mixture of analytes is a more general objective and forms an essential field of analytical chemistry, with chromatography and electrophoresis as the main techniques. Chromatographic separation, applicable to gas mixtures and analytes in solution, is based on a partition equilibrium specific to each analyte. In contrast, separation by electrophoresis is based on migration times of charged species subjected to an electric field. The two techniques can also be coupled with mass spectrometry (MS), which, in addition to separation, allows the identification of each analyte. Analyte sizes are highly variable and are measured in Daltons and kDa[3].

2.3.1. *Chromatography techniques*

The principle of liquid chromatography (LC) or gas chromatography (GC) consists of forcing the flow of a mixture of products (in the liquid or gas state) through a chemically inert porous material (stationary phase) inside a column. Initially, the moving of the solution (containing the mixture of solutes) through the

---

3 Dalton (Da) is the name used by biologists to designate what chemists call "u", a unified atomic mass unit (i.e. 1/12th of the mass of a carbon atom ^{12}C). John Dalton, 1766–1844, was a British physicist and chemist.

column is carried out under pressure or by gravity. Still, it can also be carried out more quickly using electro-osmotic flow, as shown for the first time by Pretorius et al. (1974). Each product is characterized by a different retention time, corresponding to their elution through the column and resulting from a partition equilibrium between the analyte (dissolved in the mobile phase) and the stationary phase. Several types of column packing are used, each specific to a particular type of chromatography.

The simplest filling is that of a porous stationary phase that produces a sieve effect with respect to the analytes. Small analytes penetrate the pores and are, therefore, slowed down, while larger analytes, which cannot penetrate the pores, are carried away by the eluent solution without slowing down. The result of this discrimination is that the elution times are inversely correlated with the size of the analytes. This type of chromatography, referred to as size exclusion chromatography (SEC), has been widely used to separate polymers and proteins of very different molecular masses since the 1960s. A more detailed study of how it works can be found in the article by Sun et al. (2004) devoted to the relationships linking elution times to polymer mass (linear and branched).

Stationary phases, adapted for other modes of interaction, are finding use in new types of chromatography. The stationary phase can be either a hydrophilic polar material (microparticles of silica or alumina), or a non-polar hydrophobic material (generally microparticles of silica functionalized with a hydrophobic compound), or even an ion exchange resin in the case of the analysis of anionic or cationic compounds.

In the first case, the polar molecules in the mobile phase are strongly adsorbed by the stationary phase and are characterized by very strong retentions, leading to very long elution times, unlike the hydrophobic molecules that interact weakly with the stationary phase and move through the column quicker, which corresponds to shorter elution times. This type of chromatography is often referred to as normal phase chromatography (NPC).

When the stationary phase is hydrophobic, the mobile phase containing the analytes is then generally a polar aqueous solution, and the retention times between polar molecules and the hydrophobic stationary phase are, in this case, reversed compared to NPC. This type of chromatography is known as reversed phase chromatography (RPC) (Sander and Wise 1987).

Electrically charged analytes (cations or anions) can be separated using ion exchange resins as stationary phases. This is also an old method, still preferred for preparative applications, and as such can compete with the much more popular electrophoresis techniques. Known as ion exchange chromatography (IEX), it is frequently used for the separation of proteins and peptides, transformed in a cationic or anionic form, after controlled pH acidification[4]. Complexed to varying degrees by the resin, they are then eluted from the column by a concentration-gradient electrolyte solution. Analytes weakly complexed by the resin are eluted first, while highly complexed analytes are eluted last, at high electrolyte concentrations of the mobile phase. Recently, it has been used to analyze degradation products (charge variants) of monoclonal antibodies[5] (Liu et al. 2019a).

All these forms of chromatography are based on strong interactions between analytes and the stationary phase. They give differentiated analyte retentions; elution times are longer as retention increases (Dorsey and Dill 1989). Figure 2.8 illustrates a typical liquid phase separation in the case of a column filled with a stationary phase, on top of which a solution containing a mixture of analytes is poured.

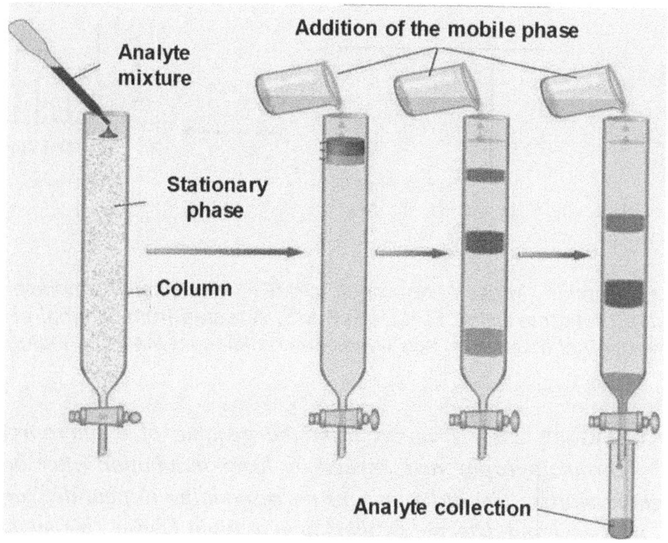

Figure 2.8. *Basic principle of liquid chromatographic separation. Drawing adapted from Wikipedia. For a color version of this figure, see www.iste.co.uk/lacaze/nanotechnologies.zip*

4 A good description of the IEX technique can be found in "Chemistry Libre Texts", chapter "Ion Exchange Chromatography" by Kevin Ahern and Indira Rajagopal (Oregon State University).

5 Monoclonal antibodies (mAbs) are an important class of proteins for therapeutic use. They are susceptible to severe chemical degradation during handling or prolonged storage, which can affect their biological properties. It is important to control their purity and separate the main product from the degradation products, which can be achieved by various separation techniques, including IEX chromatography.

Alongside liquid and gas chromatography, chromatography carried out on thin layers of gel (thin layer chromatography [TLC]) is also longstanding; it is still widely used due to its simplicity and the very wide variety of commercially available gel plates, allowing both qualitative and quantitative characterization of families of samples. Their use is widespread in biochemistry, clinical analysis, pharmacy, toxicology and cosmetics (Issaq and Barr 1977).

Finally, for all these chromatography techniques, in the same way as for gases, the products resulting from LC or TLC can also be coupled to MS analysis by adding a nebulizer and ionization system to the chromatographic device. This equipment ionizes the eluted samples, transforms them into ionized gas particles and detects them (Figure 2.9).

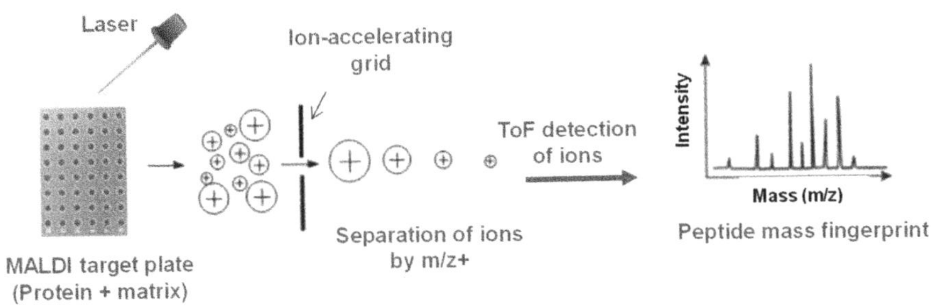

Figure 2.9. *Peptide mass fingerprint (PMF) identification principle of the microorganism proteome using MALDI-ToF-MS. Adapted from Singhal et al. (2015). For a color version of this figure, see www.iste.co.uk/lacaze/nanotechnologies.zip*

COMMENT ON FIGURE 2.9.– *Proteins from the genome of a microorganism are separated by chromatography and ionized by laser irradiation after being mixed with an organic matrix. The different ions corresponding to peptide fragments are accelerated by a grid and detected by their time of flight (ToF). This leads to a mass spectrum corresponding to a peptide fingerprint resulting from the protein fragmentation. The matrix used is a mixture of solvents and organic compounds chosen to form a solid deposit with the proteins.*

In the case of gel chromatography, the most common procedure involves taking and solubilizing the different chromatogram "plots" into a drop of liquid, then producing a kind of electrically charged aerosol obtained by applying a high voltage to the drop of liquid that contains the analytes. The technique, complementary to chromatography, is known as electrospray ionization (ESI), very widely used for the

MS analysis of macromolecules (proteins, lipids and saccharides) (Rejšek et al. 2016).

Other, gentler ionization methods have also been developed involving desorption-ionization of the analytes by a LASER followed by MS coupling, which corresponds to the technique known as matrix-assisted LASER desorption ionization-mass spectrometry (MALDI-MS). MS detection is often based on measuring the time of flight of ionized species, which corresponds to MALDI-ToF-MS analysis. It is a means of identification that has proven effective for analyzing the proteome of microorganisms by comparing the experimental mass spectra of peptide derivatives with those contained in databases (Singhal et al. 2015).

2.3.2. *Electrophoresis techniques*

Electrophoresis is also a very old separation method based on the movement of electrically charged macromolecules under an electric field. It has undergone numerous improvements since its appearance in the 1950s, until the widespread adoption of polyacrylamide gels in chromatography, which earned it the name Polyacrylamide Gel Electrophoresis (PAGE). Many variations in this technique have since been made. It was a widely used technique for analyzing protein mixtures prior to two-dimensional electrophoresis, originally proposed by O'Farrell 50 years earlier (O'Farrell 1975). CE, which appeared in the 1980s, shortly after the commercial production of silica capillary tubes, can be considered a variant of the previous technique. Thanks to their use, very high electric fields can be applied, enabling faster separations and reduced product use.

2.3.2.1. *2D-electrophoresis*

Two-dimensional electrophoresis (2D-electrophoresis) is a paradigm shift compared to the initial PAGE technique, based on the unique application of a strong polarization voltage which induces a separation of proteins according to their masses and electrical charges. It consists of performing two successive electrophoresis steps in two mutually orthogonal directions and for different chemical conditions[6].

First, electrophoretic separation of the proteins takes place according to their isoelectric point (pI), the pH at which the electric charge of the amphoteric molecule

---

6 Two-dimensional liquid chromatography (2D-LC) is also done in the liquid state. Compared to the SDS-PAGE technique on a gel, it differs by the use of two columns characterized by distinct stationary phases, the eluents from the first column being directly injected into the second, which refines the initial chromatographic separation. A detailed description can be found in the review article by Stoll and Carr (2017).

is zero). This corresponds to isoelectric focusing (IEF) based on the isoelectric point of proteins[7] corresponding to their zero charge state (Figure 2.10).

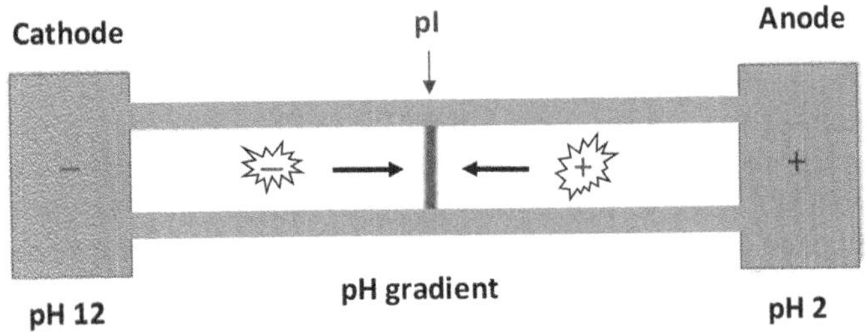

Figure 2.10. *Schematic representation of the immobilization mechanism of an ampholyte deposited on a gel or any other chromatographic support. For a color version of this figure, see www.iste.co.uk/lacaze/nanotechnologies.zip*

COMMENT ON FIGURE 2.10.– *The two ends of the support are connected to two reservoirs, one containing a base (NaOH) with a high pH (12), the other an acid* (H_3PO_4) *with a low pH (2), establishing a pH gradient along the entire length of the gel. Depending on the pH region in which the ampholyte is located, its electrical charge is that of the anion* $H_2N\text{-}R\text{-}COO^-$ *when the pH is greater than its pI or otherwise that of the cation* $H_3N^+\text{-}R\text{-}COOH$ *(pH < pI). The application of the electric field causes the ampholyte to move and immobilize toward the pH value equal to its pI. Since these ampholytes have a strong buffering capacity, it's easy to imagine that the pH gradient in the gel will become more gradual as the number of ampholytes increases. Adapted from Weinberger (2000).*

This involves pre-establishing a pH gradient on the plate, obtained by impregnating the chromatographic support with a mixture of amphoteric products (ampholytes or zwitterions) chosen so that their isoelectric points cover a determined pH interval. The chromatographic support is in electrolytic contact with two reservoirs, one containing the catholyte (NaOH) and the other the anolyte (H_3PO_4). Under the effect of the electric field, an ampholyte moves and approaches the pH zone equal to its pI, with the consequence being its neutralization corresponding to the zwitterion state and its immobilization (Liu et al. 1996).

---

7 It is a set of 50–100 polyaminocarboxylic acids with molecular masses between 300 and 1000, characterized by two pKa values corresponding to the two equilibria $H_3N^+\text{-}R\text{-}COOH \leftrightarrow H_3N^+\text{-}R\text{-}COO^- + H^+$ (Zwitterion, $pKa_{,1}$) and $H_3N^+\text{-}R\text{-}COO^- \leftrightarrow H_2N\text{-}R\text{-}COO^- + H^+$ ($pKa_{,2}$), with $pKa_{,2}$ greater than $pKa_{,1}$. The isoelectric point (pI) of the zwitterion occurs at a pH equal to $(pKa_{,1} + pKa_{,2})/2$. The first and second equilibrium take place at an acidic and basic pH, respectively.

Adding the mixture of proteins (also amphoteric) to the pH gradient plate and applying a high electric field moves the proteins to the pH zone that corresponds to their isoelectric point. A second electrophoresis is then carried out in a direction perpendicular to the previous one but after treatment of the proteins with SDS (sodium dodecyl sulfate). The addition of the surfactant SDS induces, on the one hand, the denaturation of the proteins (obtaining a linear structure) and, on the other hand, the attribution of the same electric charge per unit of mass, resulting in a separation according to sizes (Figure 2.11).

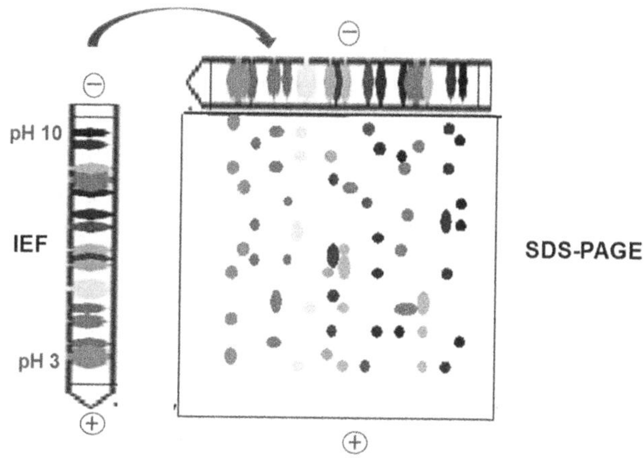

Figure 2.11. *Principle of 2D electrophoresis. Adapted from a diagram Wikipedia. For a color version of this figure, see www.iste.co.uk/lacaze/nanotechnologies.zip*

COMMENT ON FIGURE 2.11.– *First, the proteins, peptides, or DNA in the mixture are separated on a pH gradient gel according to their isoelectric point and under the effect of a strong electric field (Isoelectric focusing – IEF). Each analyte (identified by a fictitious coloring) reaches a pH region corresponding to its isoelectric point. After treatment with SDS, this first separation is followed by a second electrophoresis in a direction perpendicular to the first orientation. SDS treatment negatively charges the different proteins, leading to their migration toward the positive pole of the plate and separation according to their masses.*

Using a suitable developer reagent (generally methylene blue), a series of spots appear distributed over the gel surface, corresponding to the different proteins, characterized according to their size and, therefore, their mass, with the gel acting as a sieve.

It should be noted, however, that separating and identifying analytes using this technique requires large quantities of products and a series of manipulations that are difficult to automate, before mass spectrometric analysis can be performed. CE, proposed in the 1980s by Jorgenson and Lukacs (1981), offers other possibilities; it can be automated and can operate with very small quantities of products at faster speeds.

2.3.2.2. *Capillary electrophoresis*

Used in various forms, CE requires the injection of only small amounts of analyte solutions into the capillary (a few tens of nanoliters, or approximately 1% of the total capacity of the capillary) and provides fluorescence detection limits, which can reach concentrations around fM (10^{-15} M) (Lu et al. 2018); migration times are short and redesigning the device can allow analysis automation. All these characteristics mean that CE has quickly established itself as the leading technique for identifying and separating a large number of families of compounds; it has given rise to numerous automated commercial devices applicable to the detection and separation of various biomolecules (oligonucleotides, DNA, fatty acids, lipids, peptides and proteins, enzymes, carbohydrates) and metal ions, described in several review papers (Warner 1994; Hu and Dovichi 2002; Quigley and Dovichi 2004; Breadmore et al. 2011; Lu et al. 2018; Voeten et al. 2018; Kristoff 2020). It has continually improved over the last decade, also promoting the design of miniaturized microfluidic devices (Kuban and Kuban 2023).

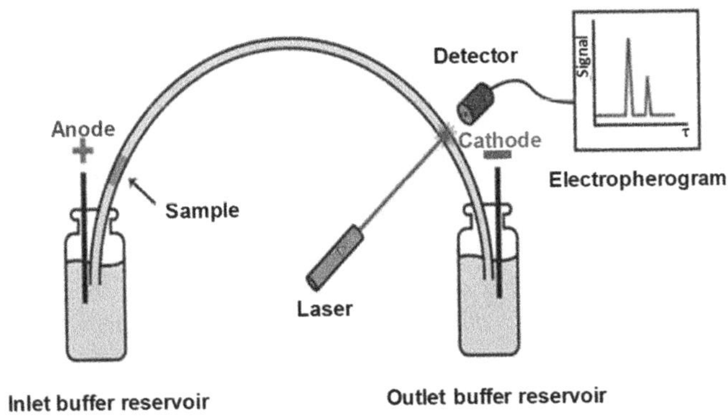

Figure 2.12. *Basic setup of a capillary electrophoresis device. For a color version of this figure, see www.iste.co.uk/lacaze/nanotechnologies.zip*

COMMENT ON FIGURE 2.12.– *The capillary contains a buffer electrolyte into which the sample to be analyzed is injected as a plug approximately 1 cm long. The capillary plunges through its two ends into inlet and outlet reservoirs, where several kV is applied. In this case, the device is equipped with an optical detection system near the capillary outlet, with the analyte signal being detected by UV absorption or laser-induced fluorescence (LIF). The corresponding electropherogram represents the signals (S) detected as a function of migration time (τ) of different analytes. Adapted from an online document from the Raghavendra Institute of Pharmaceutical Education and Research (RIPER –Anantapur).*

In its simplest configuration, a CE device is composed of a capillary tube of fused silica (approximate length between 60 and 80 cm) with a small internal diameter (between 20 and 100 μm), plunging at both ends into two reservoirs, one containing the solution to be analyzed in a buffer solution, and the other the outlet reservoir, also containing an electrolyte buffer solution (Figure 2.12).

By applying a voltage of around 20–30 kV to the CE device, the induced electric field (between 300 V/cm and 500 V/cm) causes the migration of the ions with speeds that depend on their charge and mass, which leads to an electropherogram, where each analyte is characterized by a specific migration time.

The general method is sometimes referred to as capillary zone electrophoresis (CZE), justified by separating the analytes into zones. However, it is typically referred to as CE (Knox 1994). Besides CE, we must mention capillary electrochromatography (CEC), which has similarities with CE and HPLC in that the application of an electric field propels the mobile phase containing the analyte. It moves through a stationary phase composed most often of solid particles. The difference with capillary gel electrophoresis (CGE) – another form of CE – lies in the fact that analyte separation in the case of CEC is due to a partition equilibrium of the sample between stationary and mobile phases, whereas, in the case of CGE, analyte separation is based primarily on differences in migration speeds.

2.3.2.2.1. Some characteristics of CE

Analogous to retention times, which characterize the various forms of chromatography, the ionic species subjected to an electric field are characterized by a migration time t_m, corresponding to the time elapsed between the start of the injection and the appearance of the ion in the capillary detection space. This migration time is used to define the electrophoretic mobility μ_{ep} by the following relation:

$$\mu_{ep} = v_{ep}/E = (L_d/t_m)/(V/L_t)$$

where $\boldsymbol{v_{ep}}$ represents the electrophoretic rate (cm/s), L_d is the distance between the starting point of the sample in the capillary and the location of the detector, L_t is the total length of the capillary and E is the electric field corresponding to the voltage V applied between the two ends of the capillary, that is, $E = V/L_t$. It can also be expressed from friction forces, that is, $\mu_{ep} = q / 6 \pi \eta r$, where q is the ion charge (in coulomb), η is the viscosity of the buffer (kg m^{-1} s^{-1}) and r is the Stokes radius (m)[8].

Ideally, the movement of the ionized species is determined by their electrophoretic mobility and the electric field applied to the capillary tube, which moves the positive ions toward the cathode and the negative ions toward the anode. A second phenomenon, electro-osmosis, characterized by mobility μ_{eo} is superimposed on that of electrophoresis and modifies the overall migration.

This is a ripple effect of the buffer solution resulting from forming a mobile layer of positive ions near the capillary walls. Typically formed of fused silica, the walls contain silanol –Si-OH groups, which, depending on the surrounding pH, can be polar anionic -Si-O$^-$ (pH > 3) or neutral hydrophilic –Si-OH (pH < 3) forms.

When the pH of the buffer solution is high, the silanol SiOH groups of the silica dissociate into SiO$^-$, resulting in the formation of an electrolytic double layer (DL) (see Chapter 4, Figure 4.14); the first layer is a compact anion layer called the Stern–Helmholtz layer, which is fixed to the tube walls, and the second cation layer is called the diffuse Debye layer, which is mobile. The Debye layer moves toward the cathode under the influence of the electric field and carries the solvent molecules, generating an electro-osmotic flow (Figure 2.13).

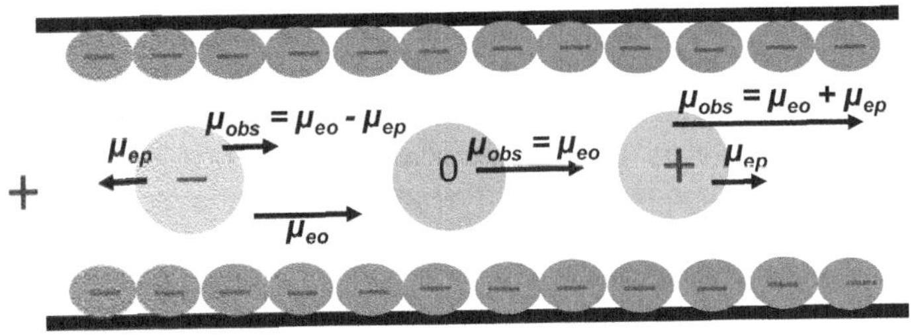

Figure 2.13. *Schematic representation of the mobilities of neutral (0), cationic (+), and anionic (-) molecules in a basic medium, subjected to electrophoretic and electro-osmotic forces, resulting from the application of an electric field in a capillary tube. For a color version of this figure, see www.iste.co.uk/lacaze/nanotechnologies.zip*

COMMENT ON FIGURE 2.13.– *In a basic medium, the internal walls of the silica capillary tube are covered by negative charges corresponding to the ionized silanol*

8 The Stokes radius, also referred to as the hydrodynamic radius, corresponds to the radius of the ion surrounded by its solvation layer.

groups SiO⁻(blue spheres). Under the effect of the electric field, the mobile positive charges of the Debye layer (not shown) formed near the compact layer move towards the cathode, taking with them solvent molecules with electroosmotic mobility μ_{eo}. This electroosmotic flow opposes the electrophoretic movement of anionic molecules (light blue spheres -), adding to the electrophoretic movement of cationic molecules (pink spheres +). The mobility observed for anionic molecules μ_{obs} is μ_{eo} - μ_{ep}, which is weaker than that of cationic molecules ($\mu_{obs} = \mu_{eo} + \mu_{ep}$) and neutral molecules (gray spheres), whose mobility is μ_{eo}.

Overall, as defined by Alizadeh et al. (2021), the phenomenon can be compared to electro-osmotic pumping of the electrolytic solution. The electro-osmotic mobility μ_{eo} and the corresponding rate v_{eo} are defined by the two equations: $v_{eo} = \mu_{eo} E$ and $\mu_{eo} = k(\varepsilon_r \zeta)/\eta$, where ε_r is the relative permittivity of the medium, ζ is the zeta potential measured at the shear plane of the particle, located near the compact Stern–Helmholtz layer, and η is the viscosity of the buffer solution. k is a numerical coefficient, a function of the particle size ($k = 1$ for particles larger than the DL width and $k = 2/3$ for particles smaller than the DL width (Bhattacharjee 2016).

This electro-osmotic effect can be very strong when using strongly basic buffers. This leads to near-total ionization of the silanol groups, with the result that the cationic and anionic forms always move toward the cathode.

However, for slightly acidic pHs, the electroosmotic effect can be almost eliminated by preventing the ionization of the silica walls, which makes it possible to normalize electrophoresis conditions. Overall, the rate of ion movement is equal to $v = (\mu_{eo} \pm \mu_{ep})$ (V/L), with the signs + and – corresponding to the cationic and anionic species, respectively.

In terms of resolution, the separation between the different detection signals resulting from electrophoresis can also be affected by diffusion, which contributes to a scattering of the different eluted peaks. Throughout the duration t of migration of an ionic species, its displacement σ due to molecular diffusion is defined by the Einstein relation $\sigma^2 = 2D_m t = 2D_m L^2/(\mu_{ep} V)$. The separation efficiency N is given by the number of theoretical plates, analogous to chromatography, by the relation L^2/σ^2, which, in this case, is equivalent to $N = \mu V/2D_m$. Remarkably, separation efficiency does not depend on column length; separation efficiency improves as molecular diffusion reduces and the stronger the applied voltage. In the presence of an electro-osmotic flow, the previous equation becomes $N = (\mu_{el} + \mu_{eo})V/(2D_m)$ (Jorgenson and Lukacs 1981)[9].

9 For example, for an applied voltage of 30 kV, electrophoretic mobility of $2. \times 10^{-8}$ cm²/Vs and a molecular diffusion coefficient of 3×10^{-9} cm²/s, the number of theoretical plates is 100,000.

In addition to this separation of the various components of the mixture into zones or peaks, the electropherogram also contains "system zones" (SZ), that is, additional peaks or troughs which cannot be identified with the analytes, and which are inherent to the composition and nature of the electrolyte (Beckers et al. 2001).

In conclusion, the electrophoretic movement of variously charged molecules depends, in part, on the electro-osmotic movement of the mobile phase, which can be modulated by controlling the pH of the buffer solution transporting the analytes. In a very basic medium, the transformation of SiOH groups into SiO^- forms a double electrolytic layer, responsible for the electroosmotic effect, whose migration rate is greater than that of anionic and cationic molecules, the latter being larger than those of the buffer ions. However, the low reagent concentration at the detection window is an important feature of this method; it makes identifying weakly concentrated analytes difficult because it requires specific experimental protocols in place for the pre-amplification to increase their concentration.

2.3.2.2.2. Main pre-amplification techniques applied to CE

Several techniques for the pre-amplification of analyte concentrations have been proposed to increase the sensitivity of CE. The most widespread of these involve increasing analyte concentration by means of field amplified sample stacking (FASS), isotachophoresis (ITP) and pH dynamic junction techniques described in several review articles by Osbourn et al. (2000), Beckers et al. (2000), Breadmore et al. (2009, 2011) and Ramachandran et al. (2022).

a) Localized stacking of the analyte by FASS

This is the most popular and simplest method. Described and explained by Chien and Burgi (1992), it involves injecting a small amount of analyte (a weakly concentrated electrolyte solution of the same nature as the BGE) into the capillary, initially filled with a very concentrated background electrolyte (BGE) concentration. In the form of a liquid "plug" around 1 cm in length, this injection creates two distinct conductivity zones in the capillary, resulting in a strong field zone contiguous to a weak field zone (Figure 2.14).

This difference in electric field intensity in the two zones results from the fact that the migration current density (i) of the ions is constant throughout the capillary. This implies that the electric field is much stronger in the low conductivity region than in the high conductivity region, which follows from the relationship $i = \sigma E$, (σ: conductivity, E electric field). Considering the differences in concentration between the analyte and the BGE, analyte ion velocity is much faster than that of ions in the high concentration BGE buffer, resulting in stacking the analyte at the separation surface between the BGE and the analyte plug. This process is brief and can be carried out by applying a strong bias over a period <100 s; it leads to amplification factors of up to several thousand units (Chien and Burgi 1991).

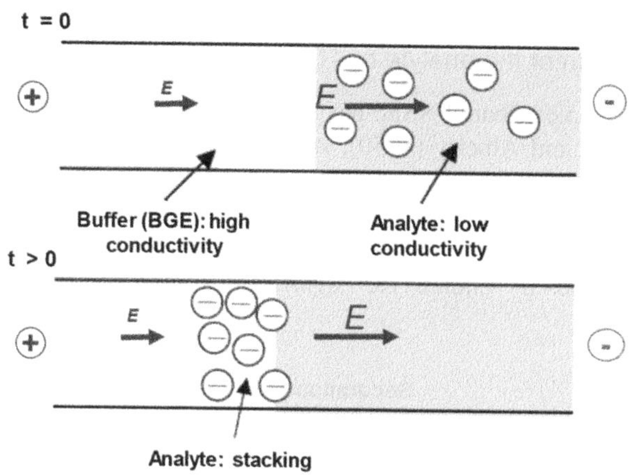

Figure 2.14. *Stacking mechanism of an anionic analyte at the BGE/analyte interface by FASS in the absence of electro-osmotic flow. Adapted from Bharadwaj and Santiago (2005). For a color version of this figure, see www.iste.co.uk/lacaze/nanotechnologies.zip*

COMMENT ON FIGURE 2.14.– *At t = 0, the analyte (in the low conductivity buffer electrolyte) and the high conductivity buffer solution (BGE) are in distinct capillary zones, with a weak electric field in the BGE zone and a strong electric field in the low conductivity zone containing the analyte. As the velocity of analyte ions is higher than that of BGE buffer ions, as soon as they reach the BGE zone, they are slowed down by the weaker electric field and therefore accumulate at the [BGE buffer/analyte] interface, where their velocity becomes equal to that of BGE ions. Suppression of electro-osmotic flow is achieved by covering the capillary walls with a neutral polymer. In the presence of a strong electro-osmotic flow, for which the migration direction of the anions is toward the cathode, the arrangement of the analyte and the BGE electrolyte zones is the reverse of those shown in the figure.*

After the segregation and stacking of the analyte, electrophoretic separation is carried out by distributing the BGE electrolyte in the capillary on either side of the analyte concentration zone and by applying a voltage sufficient for separation times of around 10 min, depending on the experimental conditions. UV absorption detection sensitivity is significantly improved, even more so when the FASS is associated with a laser-induced fluorescence (LIF) detection system, making achieving LODs of 0.1 fM possible. Recently, this method was used successfully for the forensic analysis of morphine – anionized by tetraborate – contained in urine at concentrations ranging from 2 to 2,000 ng/mL (Zarad et al. 2021).

b) Pre-amplification of the analyte by ITP

ITP, more complex than FASS, has its origin in the work of Dole (1945), Longsworth (1945) and Alberty (1950), who, between 1945 and 1950, established the laws that define the movement of the separation planes between different phases of electrolytes in solution under the effect of an electric field. It was not until the 1970s and 1990s that the ITP method of analyte pre-concentration applied to EC became the method of choice for significantly increasing analyte detection sensitivity.

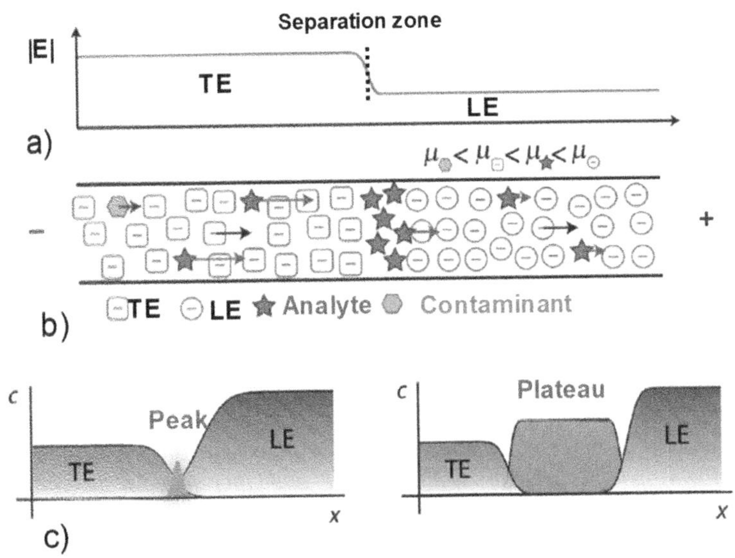

Figure 2.15. *Principle of isotachophoretic analyte concentration. (a) Variation in the electric field between the two electrolyte (TE and LE) zones. The TE electrolyte has a lower conductivity than LE (which induces a stronger electric field in the TE region). Still, the electrophoretic mobility of the TE anion is lower than that of the LE anion. (b) Diagram representing analyte stacking in the TE/LE separation zone (ITP interface) under the effect of the electric field. (c) Segregation of a mixture of analytes in the case of low concentration (peak) or high concentration (plateau). Adapted from Eid and Santiago (2018). For a color version of this figure, see www.iste.co.uk/lacaze/nanotechnologies.zip*

The technique involves inserting the analyte between two electrolyte zones with different electrophoretic mobilities. The capillary tube thus has two zones of electrolytes, one with lower mobility than that of the analyte, referred to as the trailing or terminating electrolyte (TE), and the other, greater mobility, referred to as the leading electrolyte (LE). The LE is also chosen for its better conductivity than the TE, which creates a stronger electric field in the TE zone than in the LE zone. This set of conditions causes the two adjacent TE and LE zones and their separation surface to move at the same speed. The analyte, mixed with the TE, moves more quickly than the TE and, therefore, tends to penetrate the LE zone. With its mobility being lower than that of the LE, it cannot exceed the LE ions; it, therefore, remains localized at the TE/LE separation surface and accumulates there. Ultimately, the different anions migrate at the same speed, a characteristic property of the technique, giving rise to the prefix isotacho (meaning "with equal speeds") (Figures 2.15(a) and (b)).

This type of mechanism applies to anions or cations. In the case of anionic analyte separation, TE and LE are located at the cathode and anode, respectively, opposite in the case of cationic analytes. Furthermore, when the analyte is in very low concentration, stacking occurs in the form of a concentration peak. In contrast, for higher concentrations, the segregation of the analyte at the TE/LE interface produces an intercalation zone between the TE and LE, corresponding to a "plateau" (Figure 2.15(c)).

The pre-concentration can be very high, and amplification factors of around a million have been achieved in some cases (Jung et al. 2006). This implies detection limits at concentrations around 100 attoM (10^{-16} M). Another advantage of the technique is that it is selective because anionic species and contaminants with mobilities higher than the LE or lower than the TE do not concentrate at the ITP interface.

Once this preconcentration has been achieved, separation by electrophoresis can be considered by replacing the initial reservoir, containing the TE and the analyte, with a reservoir containing only the LE electrolyte and replacing the TE electrolyte with the LE in the capillary. A 20–30 kV voltage is then applied for electrophoretic separation.

As with other electrophoresis techniques, numerous experimental variants of ITP have been proposed, both for the injection modes of different electrolytes and their relative arrangements in the capillary tube and the many insertion modes. More recently, the technique has also been implemented in small microfluidic devices and has given rise to numerous devices with varying geometries, very well described in the review articles by Smejkal et al., (2013) and Ramachandra and Santiago (2022).

In conclusion, this "online" analyte pre-concentration technique has many advantages, meaning that the ITP technique is often considered a preliminary step necessary for improving the CE performance. Carefully selecting the mobilities of the LE and TE anions makes it possible to exclude any impurity from the pre-concentration zone, increasing the detection selectivity and sensitivity of the CE. The ITP technique generally obtains several thousand concentration amplifications, greater than those obtained by FASS (Bagha and Santiago 2013).

c) Pre-amplification by dynamic pH Junction

Previous pre-concentration techniques by field amplification often used ionic samples in low concentrations, which is not always possible with biological samples solubilized in a concentrated saline medium. In this case, preliminary desalination of the sample becomes necessary. The dynamic pH junction pre-concentration method, invented by Chen and colleagues, avoids this preliminary method (Britz-McKibbin and Chen 2000).

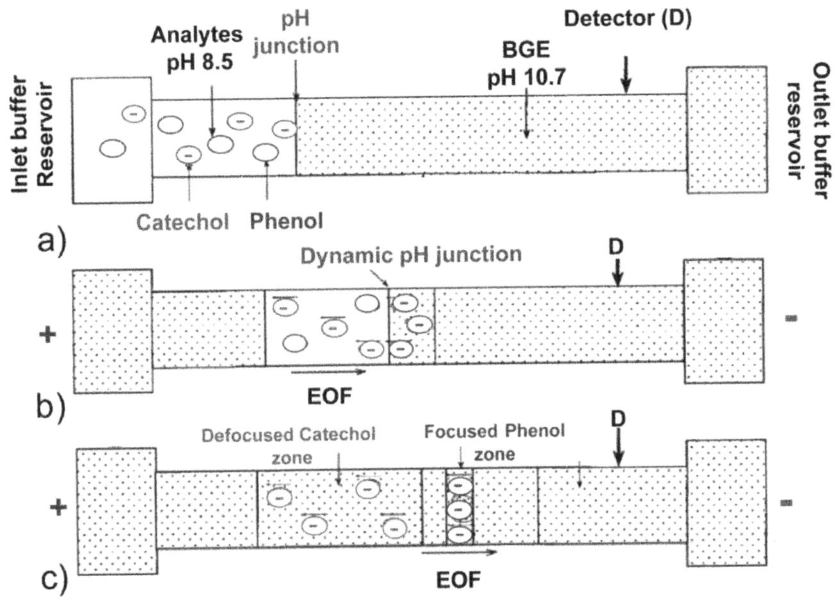

Figure 2.16. *Principle of analyte pre-concentration by dynamic pH junction. (a) Pressurized injection of a plug of solution from the left-hand reservoir containing phenol and catechol at pH 8.5 into the capillary initially filled with the BGE buffer electrolyte (pH 10.7). (b) Applying a voltage causes the analytes to migrate and the phenol to enter the pH 10.7 zone, where it is ionized, slowed down and begins to concentrate. (c) Continued electrophoresis with focused and defocused separations of phenol and catechol, respectively. Adapted from Britz-McKibbin and Chen (2000). For a color version of this figure, see www.iste.co.uk/lacaze/nanotechnologies.zip*

It was established and observed with low acidic analytes, phenol and catechol, characterized by their respective pKa of 10 and 9.14, and solubilized in a slightly basic electrolyte (pH 8.5), chosen to maintain only one of the analytes (phenol) in the neutral state, unlike the second (catechol), which is mainly ionized in its anionic form (Figure 2.16(a)).

In the first stage, a small amount of both analytes is injected into the column containing the buffer electrolyte (pH 10.7). This injection, carried out under pressure, introduces a "solution plug" of pH 8.5 of controlled length, containing the catechol in its dissociated form and the phenol in its virtually neutral form. The capillary then contains two distinct pH zones, one corresponding to the solution containing the two analytes (pH 8.5), the other to the buffer electrolyte (BGE) with a higher pH (10.7).

Applying a voltage of around 20 kV between the two ends of the capillary moves the phenol and catechol toward the cathode, with the neutral phenol being carried along by the electroosmotic flow (EOF) at a higher speed than the catechol, which is slowed down by its electrophoretic force directed toward the anode. The result is rapid penetration of the phenol into the pH 10.7 zone, which ionizes it and slows it down, leading to the process of focusing the phenol in a narrow zone between the two pH regions. The catechol, whose ionization has not changed, continues to move at the same speed and remains defocused. One of the advantages of the dynamic pH junction technique is that, unlike CE, it enables larger quantities of analyte to be injected (a few μL instead of nL), thus improving detection sensitivity.

More recently, the technique has been used to analyze proteins containing the characteristic proteome of cellular organisms (Zhu et al. 2014). Protein separation is achieved with CE by pre-amplifying proteins using the dynamic pH junction method, with their identification resolved by MS. The setup uses a commercial system that includes the CE with dynamic pH junction, a device for nebulization and ionization of the products from electrophoresis, followed by MS analysis, carried out by two quadrupoles, the whole system corresponding to a "tandem" mass spectrometry (MS/MS) analysis[10].

---

10 The device used by Zhu et al. (CZE-ESI-/MS/MS) is an adaptation of a commercial device (LTQ-Orbitrap-Velos mass spectrometer from Thermo Fisher Scientific) combining electrophoretic capillary separation with an ionization and fragmentation system in series, supplemented by two quadrupoles allowing the identification of different fragments by mass spectrometry (Zhu et al. 2014). Generally speaking, these devices, called CE-ESI-MS, include a nebulizer and an analyte ionization system and are well described in the review article by Patel et al. (2021).

2.3.2.2.3. Capillary electrophoresis of neutral molecules

Previously unresolvable by electrophoresis, neutral molecules are either transformed into ionic species making them suitable for electrophoresis or encapsulated in anionic or cationic micelles. In the first case, the technique was tested with sugars onto which anionic fluorophores were grafted. APTS (8-amino pyrene-1,3,6 trisulfonate), which fluoresces at 512 nm after excitation at around 480 nm, was proposed in the 1990s by Evangelista et al. (1995). It has proven to be well suited for CE of neutral molecules, such as glycans[11], on which it can be easily grafted, and therefore directly detectable by LIF (Lu et al. 2018).

The other, older method, proposed by Terabe et al. (1984) and Quirino et al. (2000), differs from traditional CE and is known as micellar electrokinetic capillary chromatography (MEKC). It applies to hydrophobic compounds, usually separated by reversed-phase high-performance liquid chromatography (RP-HPLC).

It is a particular form of electrophoresis based on the formation of micelles and their movement under the combined effect of electro-osmotic and electrophoretic flows. Micelles are amphiphilic molecule aggregates (surfactants) characterized by a long hydrophobic chain and a hydrophilic polar head. These amphiphilic molecules associate in an aqueous medium in such a way as to minimize hydrophobic–hydrophilic repulsion, forming spherical supramolecular structures (micelles). The lipophilic (hydrophobic) chains associate with each other and are directed toward the center of the micelle, with the polar heads on the outside in contact with the aqueous medium. Micelle formation takes place at a critical surfactant concentration (CMC). The main surfactants used for CE are essentially anionic, such as SDS (sodium dodecyl sulfonate), or cationic, such as CTAB (cetyl trimethyl ammonium bromide), but other variants are also possible (Schramm et al. 2003).

The advantage of micelles in electrophoresis is that they solubilize neutral molecules in their hydrophobic core, the micelle itself being stabilized by the aqueous electrolyte. Anionic surfactants, such as SDS, give the micelle a negative charge, and its migration rate in a basic medium is determined, like any anionic species, by two opposing forces: a driving force caused by the electro-osmotic flow toward the cathode (V_{EOF}) and an electrophoretic force toward the anode (V_{el}) due to the negative polar heads of the micelle. In general, V_{EOF} is much higher than V_{el},

11 Glycans are polysaccharides made up of monosaccharides linked together by condensation between two OH groups (glycosidic bond). Cellulose is an example of a glycan with many D-glucoses coupled by condensation of OH on carbons 1 and 2 of two monosaccharides. N-glycans – another family of polysaccharides – differ from the previous compounds by the presence of a nitrogen group, obtained, for example, by the condensation between a polysaccharide and an amino acid or a protein. They are of great interest in biology for their therapeutic applications as well as for their use as biomarkers.

which involves an overall movement of the micelles toward the cathode (Figure 2.17(a)).

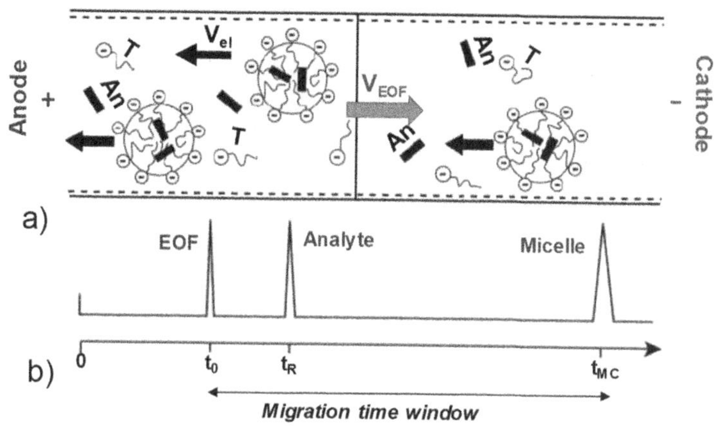

Figure 2.17. *Electropherogram of a neutral analyte in a micellar medium. (a) Schematic representation of the micellar medium in the presence of an analyte (An) and surfactants (T). (V_{el} is the electrophoretic rate of the micelle, V_{EOF} the electroosmotic flow rate greater than V_{el}. (b) Electropherogram corresponding to an analyte partially complexed by micelles, characterized by migration time t_R, intermediate between that of micelles (t_{MC}) and the electroosmotic flow rate t_0. The electroosmotic flow rate t_0 is determined using molecules not complexed by the micelles (methanol), which are considered markers of electroosmotic flow. In the same way, the micelle migration rate t_{MC} is determined by a marker such as the dye Soudan III, entirely complexed by micelles. Adapted from Hancu et al. (2013). For a color version of this figure, see www.iste.co.uk/lacaze/nanotechnologies.zip*

Regarding migration times, two extreme situations are possible. In the case where the micelle affinity of the analyte is zero, it remains in solution and its migration time t_0 corresponds to the EOF. Otherwise, where its affinity for the micelle is maximal, it is entirely encapsulated by the micelle; its migration rate is then identical to that of the micelles, that is, the t_{MC}. For all intermediate cases, characterized by a partition coefficient k (capacity factor)[12] between 0 and infinity, the migration time t_R will be between t_0 and t_{MC}. Qualitatively, highly concentrated analytes in the micelles will be characterized by longer t_R migration times than those weakly complexed by the micelles (Figure 2.17(b)).

12 The partition of an analyte between the micelles and the surrounding aqueous medium is defined by the capacity factor (or partition coefficient), $k = n_{MC}/n_a$, in which n_{MC} and n_a are, respectively, the molar concentrations of analyte in the micelles and in the surrounding aqueous solution.

2.3.2.2.4. Capillary gel electrophoresis

It is a modification of electrophoresis on a gel slab to the capillary technique. Two main methods have thus been developed, one derived from plate electrophoresis (PAGE) becoming CGE, and the other, comparable to two-dimensional electrophoresis (SDS-PAGE), becoming capillary isoelectric focusing (CIEF), which can be carried out on a gel or in solution, the latter being now the most widespread. This is due to recent technical improvements that allow very quick separations.

a) Capillary gel electrophoresis

Developed by Hjerten in the 1980s (Hjerten 1983), the technique takes the basic principles of gel electrophoresis and adapts them to the capillary technique. The many advantages of gel electrophoresis over the traditional method relate to the capillary device, such as easier analysis automation, better separation efficiency and greater execution speed. Problems of gel heating and degradation are eliminated when the gel is in a capillary. Furthermore, depending on the viscosity of the gel, the electroosmotic effect is much weaker than for a CE with a liquid BGE electrolyte. As a result, anionic analytes move toward the anode, rather than the cathode, which also means that for this type of analysis, the detector must be located close to the anode.

Initially designed as an analytical technique, it quickly became a micro-preparative method, competing with HPLC. Primarily used to analyze and separate proteins complexed by surfactants (SDS), numerous improvements have been made since its creation and are described in recent review articles (Sänger-van de Griend 2019; Bhimwal et al. 2022; Hajba et al. 2023).

Compared to the plate-based SDS-PAGE method, which uses a cross-linked polyacrylamide gel, CGE works with non-cross-linked gels consisting of polymers in solution, such as polyvinyl alcohol (PVA), dextran (D-glucose branched polysaccharide) or polyethylene oxide (PEO), transparent in the UV region around 210 nm, for which protein absorption is very strong. Furthermore, as for the SDS-PAGE technique, the association of SDS with proteins means that the migration rate is a function of protein size, which combines a "sieve" effect of the gel with that of analyte migration.

The entanglement of polymer chains achieves this sieve effect and is greater the higher the polymer concentration in the buffer medium. This results in the formation of so-called "dynamic" pores, which can deform and facilitate the passage of analytes. On the other hand, too high a polymer concentration leads to a situation equivalent to that of a cross-linked polymer, resulting in the formation of so-called "static" pores, which cannot be deformed and may be unfavorable to analyte migration.

b) Capillary isoelectric focusing

This is considered an adaptation of the first step of the isoelectric SDS-PAGE technique (see section 2.3.2.1) from plate to capillary and, therefore, applies to amphoteric compounds. First proposed by Hjerten and Zhu (1985), its principle is always the immobilization of an ampholyte at a point in the column corresponding to a particular pH. The stationary phase in which the pH gradient was carried out may be either a gel or an electrolyte in solution.

The process described by Hjerten and Zhu is comparable to that described earlier (see Figure 2.10). It involves injecting the solution containing analytes and ampholytes (pH markers) into the column, one end of the capillary tube in a basic solution (cathode side) and the other in an acidic solution (anode side). Applying a bias voltage establishes a pH gradient. It causes the analytes – amphoteric proteins and peptides – to migrate and immobilize when they reach a pH value corresponding to their IP and zero charge state. Reaching this state is manifested by a drop in the electrophoresis current. The capillary walls are generally treated to suppress the electro-osmotic effect.

After the different analytes are immobilized at different points in the column, detection by UV absorption requires a second step called mobilization, which involves the movement of all analytes toward the detector. Two techniques described by Hjerten and Zhu are possible: one involves uniformly moving all analytes by creating a hydrodynamic flow in the capillary, the other involves moving the analytes by an electrokinetic (electro-osmotic) method. The latter is the simplest and involves replacing the acid solution (H_3PO_4) of the anode compartment with a basic solution (NaOH) and by applying a voltage such that the migration of the ions Na^+ toward the cathode carries with it the solvent and the analytes toward the detector (Hjerten and Zhu 1985). Another more effective way of mobilizing the analytes was proposed by Manabe et al. (1997) to separate proteins in human plasma; it involves replacing the NaOH solution in the cathode compartment with a diluted acetic acid solution and applying an electric field (250 V/cm), causing the migration of acetate ions and ampholytes to be drawn toward the detector.

This mobilization technique, the duration of which is determined by the velocity at which the analytes are carried to the detector, considerably lengthens the analysis time, which is added to that of focusing the analytes. Many alternatives have been proposed to exempt from this mobilization step, which involves sequential spectroscopic analysis of the analytes. Wu and Pawliszyn (1992) thus proposed a new method for identifying analytes after focusing, based on a global spectroscopic analysis of the analytes in the capillary, a method known as capillary imaging IEF (iCIEF). Manufacturers have since adapted electrophoresis devices to this new technique (Mack et al. 2009; Righetti et al. 2013; Wu 2022).

Despite these reservations, the method is still widely used to separate DNA fragments, proteins and therapeutic antibodies (Zhu et al. 2012; Zhang et al. 2023a). Many commercial devices are available to users as well as the most common gels suitable for separating protein mixtures. The production of iCIEF-MS devices coupled to MS is also a new trend, in which microcircuits for the injection of reagents, the capillary containing the gel and the diode network associated with the detection system are brought together in compact form. After isoelectric focusing of the different analytes, their displacement toward the capillary outlet due to electrophoresis is completed by a final spraying and electrostatic ionization step prior to MS analysis (Mack et al. 2019)[13]. Its application to the analysis of therapeutic proteins has proven to be very effective, enabling very short protein identification times of less than 20 min (He et al. 2022).

2.3.2.2.5. Capillary electrochromatography (CEC)

It is generally considered to be a combination of HPLC and CE and was invented by Pretorius et al. (1974), who had the idea of replacing the injection under pressure of the mobile phase of a chromatographic column with an electro-osmotic flow, resulting from the action of an electric field in the capillary tube. The separation of the analytes results from two combined effects: the partition effect of the analyte between a stationary phase and a mobile phase (which is characteristic of chromatography) to which is added the electroosmotic displacement of the mobile phase by an electric field, (specificity of CE). Replacing the conventional HPLC columns filled with microparticles are two types of CEC columns: open tubular (OT-CEC) and monolithic (inorganic, organic, or hybrid). OT-CEC columns have given rise to innovative stationary phases in the form of layers deposited on the capillary walls, made of porous materials, nanoparticles, or even biological compounds. Monolithic columns are, however, made of the same porous material throughout the capillary. A third solution also uses capillaries compactly filled with microparticles, similar to CGEs. Each of these options has been developed by different teams and used in a variety of applications, with the latter solution tending to be less used, due to the greater difficulty of achieving good reproducibility in column manufacture (Mao and Chen 2019).

a) Open-tubular CEC columns

The originality of the technique involves depositing a thin layer of material onto the silica walls of the capillary, acting as a stationary phase but with very different affinities for each analyte.

---

13 This type of iCIEF-MS equipment was developed in the form of a microchip by the company SCIEX (Intabio ZT, (USA, California).

This property relates to a chromatographic retention mechanism, elution thus caused by the electro-osmotic flow and/or the electrophoretic effect in the case of charged analytes. Numerous examples of separation of mixtures containing low molecular weight compounds have thus been carried out in the pharmaceutical sector, the food industry and the environment and are mentioned or described in several review articles, covering large periods of development (Cheong et al. 2013; Tarongoy et al. 2016; Mao and Chen 2019; Hu et al. 2020; Li et al. 2023). Replacing capillaries with microfluidic devices – better suited for the realization of portable devices – does not change the operating principle. Microfluidic devices are characterized by the presence of micro-channels etched in a plastic support, in which electrophoresis takes place, which makes them similar to open-tubular capillaries.

As with open-tubular capillaries, coating the walls of a microfluidic device is crucial for efficient analyte separation. Pumera et al. (2001) show that it is only after covering the microchannel walls with a hybrid coating (polymer/gold NPs) that separating a mixture of three isomers (para, ortho and meta aminophenols) becomes possible[14].

Another example, carried out using a capillary tube, also shows the importance of the choice of coating used to increase the separation specificity. This is the case of a mixture of amino acids whose separation can be considerably increased when the capillary walls are covered by a porous film of proteins (Huang et al. 2006).[15] The innovativeness of the process relates to the coating structure, which has a thin layer of the protein bovine serum albumin (BSA) on its surface, responsible for the selectivity.

---

14 The channel walls are coated with a layer of poly(diallyldimethylammonium chloride) (PDADMAC) onto which gold NPs (10 nm in diameter), stabilized by citrate ions, are grafted. While the bare walls or PDADMAC-covered walls do not allow the three isomers to be separated, it is only after the addition of the gold NPs that the separation of the three isomers becomes possible, their detection being carried out by amperometry. The role of gold NPs is relatively complex. They play a role in establishing the electro-osmotic flow and in enhancing and differentiating the retention processes of the three isomers by the coating.

15 The coating takes several steps. First, the fused silica walls are silanized by ϒ-methacryloxypropyltrimethoxysilane ($C_{10}H_{20}O_5Si$), which has reactive methacryloxy groups on the surface (-O-CO-C(CH_3) = CH_2), on which the polymer is grafted in situ resulting from a co-polymerization between 2-hydroxyethyl methacrylate and 2-vinyl-4,4-dimethylazlactone. This polymerization is carried out in the presence of 1-decanol, used as a porogen, and AIBN, as a polymerization initiator. The last step involves reacting the azlactone groups (very reactive) with a protein such as BSA (molecular mass of 66.5 kDa, spherical, 10 nm in size). It is characterized by a point isoelectric in water of 4.7, which means that in a basic medium, it contains many anionic groups, favorable to the differentiated retention of amino acids.

In a study by Huang et al. (2006), the separation of three amino acids, with similar molecular masses (histidine-155 Da, phenylalanine-165 Da and tryptophan-204 Da) solubilized in a buffer solution (sodium phosphate of pH 8), is significantly improved due to the selective properties of the coating (Figure 2.18).

From this example, it appears that effective analyte separation depends mainly on the selectivity exerted by the coating with respect to the analytes. It is in fact the specific interactions between the amino acids and the protein that are responsible for the selectivity and not the overall thickness of the coating (several microns), which is almost entirely due to the copolymer film. Curve *iii* obtained with the thick copolymer film does not, in fact, allow the three amino acids to be well separated, whereas the addition of a very thin albumin layer (approximately 10 nm) is sufficient to effectively separate the three amino acids.

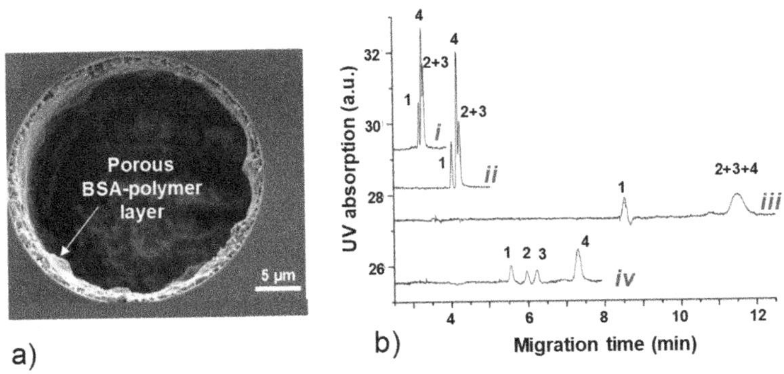

Figure 2.18. *Comparative separation of a mixture of amino acids by CEC according to the capillary tube coating. (a) Transverse SEM image of the capillary tube (internal diameter 20 µm) covered by a porous film (polymer-BSA). (b) Comparison of the migration curves of three amino acids obtained for different internal capillary tube coatings. (i) Bare capillary tube. (ii) Capillary walls silanized by Y-methacryloxypropryltrimethoxysilane. (iii) Silanized walls covered by a polymer film obtained in situ by the copolymerization of 2-hydroxyethyl methacrylate and 2-vinyl-4,4-dimethylazlactone in the presence of 1-decanol (porogenic agent). (iv) Electropherogram obtained with the final coating after the addition of BSA. Adapted from Huang et al. (2006). For a color version of this figure, see www.iste.co.uk/lacaze/nanotechnologies.zip*

COMMENT ON FIGURE 2.18.– *Peaks 1, 2, 3 and 4 correspond to dimethyl sulfoxide (neutral compound, electroosmotic flow marker, 0.2% v/v), histidine (0.5 mg/mL), phenylalanine (1.0 mg/mL) and tryptophan (0.2 mg/mL), respectively. The capillary tube has a length of 40 cm and an internal diameter of 25 µm. The amino acid and dimethyl sulfoxide mixture is injected into the capillary electrokinetically (5 kV for 4 s), and separation is achieved due to the combined effects of electro-osmotic flow and electrophoresis by applying a voltage of 15 kV. UV detection is measured at 214 nm. The mobile phase containing the analytes is a buffered solution (pH 8) of sodium phosphate (20 mM).*

The method developed by Huang et al. may seem complicated, but it shows that the interaction properties between the solute and the film determine selectivity. This explains why research is now very active in developing coatings capable of interacting with specific analytes, the electrophoretic and/or electro-osmotic function, only marginally affecting the migration rate. Using coatings for specific chirality also makes separating enantiomeric molecules possible.

In summary, the OT-CEC technique generally entails finding a stationary phase, with a porosity adapted to the compounds that require separation, that is stable over time and allows good repeatability of the migration processes. Many examples illustrating these concepts have been described in the literature and are well listed in the review article by Hu et al. (2020).

b) Monolithic columns

This is another concept whose greater simplicity and stability of use differentiates it from the previous OT-CEC technique. The first monolithic columns were developed via polymerization of a precursor (monomer) directly inside a capillary or the precipitation of silica from various functionalized alkoxysilanes. This preparation method is much more effective than filling a capillary with microparticles (packed capillary columns) since it avoids producing gas bubbles, which are harmful to the proper functioning of the column. Furthermore, adding a porogenic reagent to the solution also ensures good homogeneity of the porosity of the stationary phase.

As for OT-CEC, innovation targets the monolithic material, explaining the rise in publications. It focuses on developing polymer or hybrid materials and their application in the selective separation of analyte mixtures with varied functionalities. In the 2000s, many studies were already devoted to developing and using monolithic stationary phases (Svec 2005), and today, this trend is still as strong as ever (Hu et al. 2020). The difference between the two periods lies in the use of new materials, such as functionalized nanomaterials (metal NPs, carbon nanotubes, graphene oxide NPs), molecularly imprinted materials, organometallic compounds such as metal-organic frameworks (MOFs), the objective being to obtain increased selectivity and controlled porosity with separation techniques that can be adapted to portable microfluidic devices.

Optimizing the monolithic stationary phase is not always obvious and sometimes requires an empirical search for weighting between the different porogenic compounds involved in the structure of the polymer monolith (Murauer 2019).

Of the many examples of product mixtures separated by CEC, the enantiomeric separation of chiral drugs – a major pharmaceutical problem – can be solved by CEC using monolithic chiral stationary phases. The general idea is to obtain a monolithic stationary phase with a controlled porosity in which a chiral molecule responsible for distinct interactions between the two enantiomers of a racemic mixture has been included.

Recently, significant separation results were obtained by grafting pepsin (chiral protein) onto a MOF (MOF-5), itself grafted onto a porous polymer (Miao et al. 2022)[16]. The choice of MOF is critical because these organometallic compounds, with a quasi-crystalline structure, are typically characterized by a porosity that can be predefined by certain ligands, optimizing selectivity (see Chapter 1).

The analysis carried out on six pairs of enantiomeric drugs, identified as racemic hydroxychloroquine-HCQ (anti-inflammatory), chloroquine-CHQ (antimalarial), nefopam-NEF (non-opioid analgesic), chlorpheniramine-CHL (antihistamine), clenbuterol-CLE (anabolic) and hydroxyzine-HXY (antihistamine), indeed shows a significant improvement in enantiomeric resolution (Rs) when MOF-5 is part of the column packing[17] (Figure 2.19).

16 The preparation of the column is relatively complex and involves several steps. It is first silanized by γ-methacryl-oxypropyltrimethoxysilane (γ-MAPS – $CH_2{=}CH(CH_3)COO(CH_2)_3Si(OCH_3)_3$), which releases the reactive methacrylate groups on the surface. In a second step, co-polymerization occurs inside the column when reacting the glycidylmethacrylate monomers (GMA-$CH_2{=}CH(CH_3)COOCH_2(CHOCH_2)$ and ethylenedimethacrylate (EDMA-($CH_2{=}C(CH_3)COO{-}CH_2CH_2{-}OCOC(CH_3){=}CH_2$) in the presence of a polymerization initiator (AIBN, 2,2-azo bis isobutyronitrile) and a mixture of porogenic compounds (1,4-butanediol and 1-propanol). After the polymer obtained (GMA-co-EDMA) is reacted with N-3 aminopropylimidazole, the addition of Zn^{2+} and 1,4 benzene dicarboxylic acid (H_2BDC) forms MOF-5 or MOF-5@ GMA-co-EDMA due to the addition of successive layers. Finally, pepsin is grafted onto MOF-5 by a condensation reaction between the COOH of MOF-5 and the NH_2 groups of pepsin, leading to the structure Pepsine@MOF-5@GMA-co-EDMA.

17 The resolution Rs between two enantiomers 1 and 2 is defined by the ratio $2(t_2 - t_1)/(W_1 + W_2)$, where t_1 and t_2 are the migration times of the two enantiomers and W_1 and W_2 are their peak width at half height.

Other, simpler solutions have also been proposed to control the porosity of the monolith. Xu et al. (2017) developed a hybrid monolith column combining a polymer structure and mesoporous silica NPs (MSN), to which a chiral molecule (pepsin) is added; porosity is optimized by controlling the quantity of silica, which leads to enantiomeric separations that are just as efficient. For comparison, an Rs resolution close to 3, obtained in the case of nefopam, proves to be greater than that obtained with the MOF-based monolith.

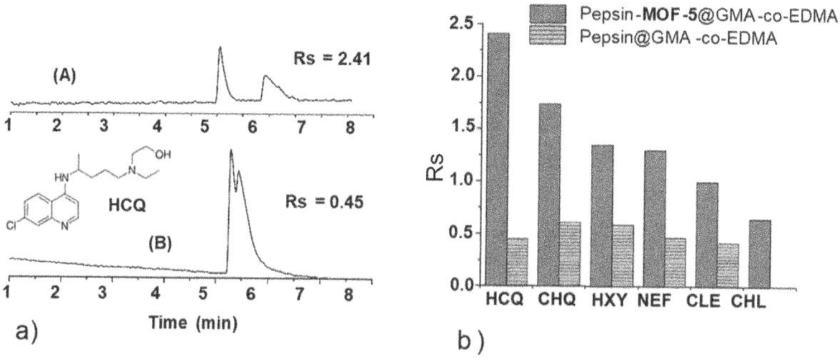

Figure 2.19. *CEC separation of enantiomer pairs with a monolithic stationary phase. (a) Electropherogram of racemic hydroxychloroquine (HCQ) in the case of (A) a pepsin-MOF-5@GMA-co-EDMA column and (B) without the presence of MOF-5 (pepsin@GMA-co-EDMA). (b) Resolution characteristics (Rs) for the six pairs of enantiomers obtained with the two kinds of monolithic materials demonstrating the separation improvement in the presence of MOF-5. Experimental conditions: acetic acid/ammonium acetate buffer (pH 4 for CHL; 4.5 for HCQ, CHQ, HXY, NEF; 5.5 for CLE); electrophoresis voltage 8 kV; electrokinetic injection, 1 kV × 2 s; 215 nm UV detection for CHL; 210 nm for CLE; 205 nm for NEF, HXY and 220 nm for HCQ, CHQ; temperature, 20°C. We note the very short analysis time, a characteristic of CEC. Adapted from Miao et al. (2022)*

2.4. Conclusions

All these concentration separation and amplification methods for analyte mixtures have a very ancient origin and have since undergone multiple improvements, which, to date, mean that their use has become widespread in many fields relating to biology, health, agri-food and the environment. Electrophoresis is a remarkable example of this. Since its discovery in the 1950s, it has been a fundamental technique for separating mixtures containing various compounds and has many advantages over traditional chromatography methods. The distinction between the two techniques depends on the mode of separation. In the case of chromatography, the separation is based on an analyte retention phenomenon in a stationary phase, resulting from an equilibrium partition of the analyte between the mobile carrier phase and the stationary phase, whereas in the case of electrophoresis, the separation is due to the electromigration of charged species. The advent of CE in the 1980s, following plate electrophoresis, has considerably improved the separation possibilities by allowing the analysis of very small amounts of analyte with much shorter analysis times. However, identifying the analytes using UV spectroscopy as they pass through the detection window limits the method. Its very short optical path due to the small thickness of the capillary tube and the very low analyte concentrations, the detection signal remains very weak, whatever capillary technique is used. This required the development of different amplification modes; amplification using an electric field (FASS) – the simplest and most common – which involves carrying out localized stacking of the analyte by amplification of the electric field, proves to be easily adaptable to commercial electrophoretic devices and can give rise to high-performance detection systems, particularly when UV absorption is replaced by LIF detection. On the other hand, the isotachophoric technique (ITP), which gives the highest amplification rates, requires specific electrolyte mobility conditions with respect to the analyte, making its use less general. The same is seen for the dynamic pH junction technique, whose advantage is that it can be applied to analytes in a saline environment but requires specific electrolyte pH conditions adapted to the analytes. Electrochromatography, with its two stationary phase options (open or monolithic tubes) combining the concepts of chromatography and electrophoresis, is also a very general technique and offers many possibilities, including the more recent use of nanomaterials, described in the review article by Jozanovic et al. (2023). Finally, the coupling of CE to MS is another important improvement, which requires heavier investments in equipment but allows, in addition to the separation properties of the analytes, to add those of a very precise analyte identification specific to the MS and its databases. This technology has developed considerably after designing and producing analyte vaporization-sputtering-ionization devices, such as ESI, which can be directly connected to a CE. Such equipment is available from major manufacturers and corresponds to CE-ESI-MS or CE-ESI-MS/MS assemblies.

3

Recognition Principles

This chapter describes the different identification methods that allow the selective detection[1] of a target molecule (analyte) relying on its affinity for a capture probe. The analyte is generally free in solution or in gaseous form, contained in a mixture known as the matrix. The probe, however, is generally attached to the surface of a porous and nanostructured material. It can also be fixed to nanoparticles, which are free to diffuse in solution. Natural or synthetic molecules, macromolecules of biological origin made of a matrix that acts as a trap, or even compounds with catalytic properties, are the main types of natural or synthetic probes.

Furthermore, given the considerable advancement in manufacturing and computing capabilities, a paradigm shift is to be expected. New analysis techniques are likely to emerge, different from designing a specific probe for a given target[2]. One new technique is machine learning[3]. It can perform a detailed analysis of weak signals emanating from a network of several probes with low selectivities and compare the signals with databases, making it possible to link these signals to the composition of the mixture analyzed.

1 According to the definition of IUPAC (International Union of Pure and Applied Chemistry), a capture probe is said to be selective if it is capable, in contact with several analytes, of binding with one of them without affecting the others.

2 According to IUPAC, specificity is the ultimate degree of selectivity. In other words, qualifying a probe as specific means that it is strictly selective for a given target.

3 Machine learning is the use of computer systems capable of learning and adapting without following explicit instructions, using algorithms and statistical models to analyze and draw conclusions from sufficiently large databases.

3.1. The different molecular or chemical identification techniques

A chemical sensor is different from a physical sensor in that it is designed to report the presence of an analyte, which can be in a molecular, ionic or gaseous form or a solubilized compound in water or other liquid media.

Thus, the chemical sensor detects ions in solution (e.g. an anion or a cation present in a biological fluid such as blood, sweat or urine), molecules in solution (a neurotransmitter such as dopamine, a pollutant such as glyphosate, etc.) or gaseous species (dihydrogen for monitoring fuel cells, nitrogen oxides for monitoring air pollution in cities, various digestive gases allowing the identification of gastric or intestinal pathology).

All cases involve fixing or capturing a target on an active surface carrying the capture probe. This can take very diverse forms. Among the most typical are ion complexing compounds (ionophore ligands), antibodies to capture molecular or macromolecular antigens, short strands of DNA or RNA used to capture complementary strands, aptamers made of nucleic acids or amino acids to recognize molecules or macromolecules, molecularly imprinted polymers (MIPs) and matrices well suited to capture small molecules. The probe can also be made up of enzymes, which catalyze a reaction with the captured molecule or a catalytic surface for gas recognition, functioning as an enzyme in the case of an aqueous analysis medium.

3.1.1. *Ion detection using synthetic ligands*

Ion sensors (organic or inorganic) are based on ionophore ligands, which complex a family of ions or a specific ion (Bühlmann et al. 1998; Bobacka 2006; Bobacka et al. 2008). Most ionophores have been designed for inorganic cations because they generally have spherical symmetry, and it is much easier to develop a ligand with such symmetry (typically cyclic or hemispherical). These are crown- or calixarene-type macrocycles. The cations interact with these cavities through Lewis acid–base interactions between a basic element in the cycle (oxygen, nitrogen and sulfur) and the cation, which has an acidic character. Typically, ligands carrying "hard" basic atoms (N, O, Cl, etc.) have a stronger affinity for "hard" acidic atoms, such as alkaline or alkaline earth ions. In contrast, sulfur atoms tend to interact with transition metal ions. The other factor contributing to the ligand selectivity is the diameter of their cavity in relation to that of the target ion. For example, for the K^+ ion (diameter of 1.33 Å), the most suitable crown ether is the ether 18-Crown-6

(Figure 3.1). In comparison, 12-crown-4 ether, which is smaller, excludes potassium and has a good affinity for the Li^+ ion (0.68 Å).

Even if used less often, acyclic ionophores have satisfactory affinities, although lower than those of cyclic ethers. This is because, although linear, they have a folded secondary structure, which requires rearrangement to be able to complex the cation, resulting in an additional energy cost that decreases the affinity for the target. These are, therefore, flexible molecules; the ionophore ETH1001, specific for calcium, is an example (Johnson and Bachas 2003).

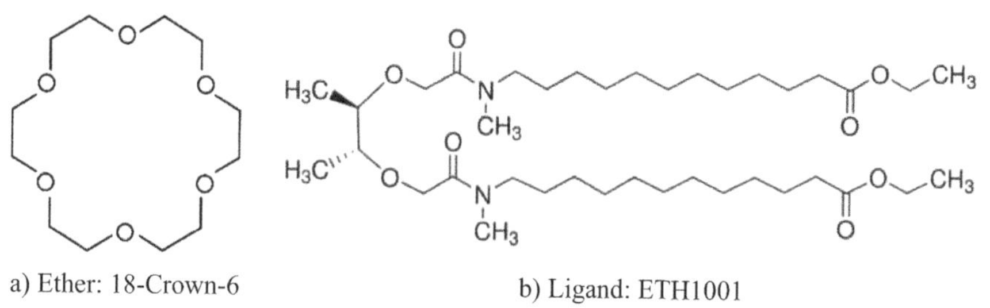

Figure 3.1. *Examples of cation-specific ionophores. (a) Simple cations such as K^+ and Na^+ have a natural spherical symmetry. The ligands with a high affinity for these cations are generally cyclic, such as 18-Crown-6 ether. (b) Acyclic ligands can also bind to this type of cation if the groups that constitute them are flexible enough to form a secondary structure folded around the cation; this is the case of the ligand ETH1001*

The case of anions is different. As indicated above, due to their highly diverse geometry, their ligands, much less common than those dedicated to cations, are often inspired by biomimetism (Sanchez et al. 2005). Among them, we find derivatives of cobyric acid (inspired by vitamin B12), which have a particular affinity for chloride or nitrite anions, and phthalocyanines and porphyrins, respective ligands of thiocyanates and nitrites. All three molecules have a central cobalt atom, considered as a hard Lewis acid, giving them great affinity for these anions. Cationic ligands carry a Lewis acid-type center (Figure 3.2), contributing to the complexation of anions carrying Lewis bases such as Cl, P, O, N and S atoms found in chlorides, nitrates, sulfates, phosphates, etc.

For both cation-selective and anion-selective ionophores, these macromolecules must be grafted onto a surface via a spacer (generally a carbon chain); this ensures the ionophore's flexibility and promotes ion complexation.

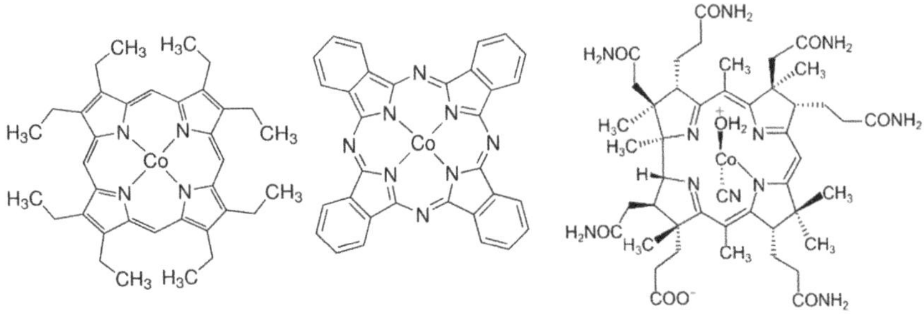

a) 2,3,7,8,12,13,17,18-Octaethyl-21H,23H-porphyrine cobalt(II)

b) Cobalt(II) phthalocyanine

c) Cobyric acid

Figure 3.2. *Examples of ionophores corresponding to cobalt complexes, specific for anions. The metallic center acts as a Lewis acid, acceptor of a non-binding electron doublet from the anion. Ligands (a) and (b) are selective for thiocyanate anions SCN^- and nitrite NO_2^-, and (c) is selective for chlorides and cyanides (CN^-). The cyanide anion CN^- and a water molecule interact in the latter case*

3.1.2. *Immunological recognition: antibodies, nanobodies and peptides*

When identifying a molecule, a macromolecule, a peptide or a protein, isolated or as part of a larger entity such as a virus, antibodies are one of the most frequently chosen solutions. They are generally specific for their target ("the antigen"), even in complex matrices containing many other molecules or macromolecules, and offer high complexation constants, allowing the detection of very low concentrations. They are particularly suitable for identification in biological media (sweat, urine, serum or blood). Still, they can also be used in other media provided they do not denature the protein structure of the antibody (i.e. pH around 7 and non-zero ionic strength).

Immunoglobulins (Ig) are a particular class of antibodies produced by mammals as part of their immune system (Figure 3.3).

These natural proteins bind with organisms, proteins or intruding molecules, considered as antigens, when recognized by the immune system. There are several types of immunoglobulins, the most common being IgG of sizeable molecular mass (typically 150,000 $g \cdot mol^{-1}$). They have a typical Y-shaped, symmetrical structure, composed of two heavy chains responsible for approximately 75% of the molecular weight of the antibody and two light chains representing the remaining weight. The two light chains are each composed of a constant part (standard to all IgG) and a variable part specific to the antigen, located at the two ends of the "Y".

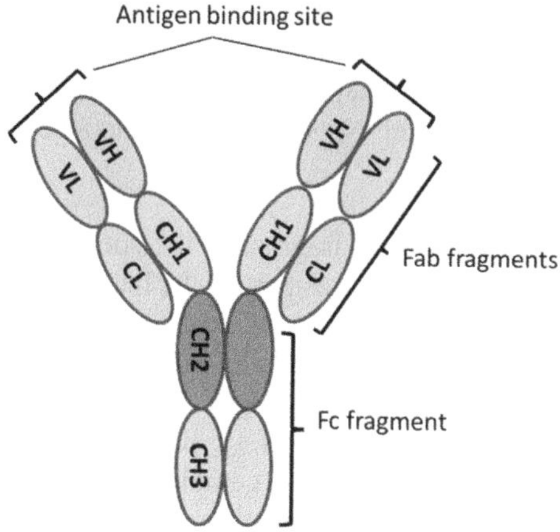

Figure 3.3. *The structure of an IgG antibody, with "heavy" and "light" chains, forms the antigen's constant non-specific domains, and the variable domains which are specific and responsible for the antibody-antigen binding. For a color version of this figure, see www.iste.co.uk/lacaze/nanotechnologies.zip*

COMMENT ON FIGURE 3.3.– *Immunoglobulin G (IgG) is made up of several domains: constant (C) variable (V), heavy (H) and light (L). Each domain is made up of approximately 110 amino acids. The heavy constant domains (CH1, CH2 and CH3) form the heavy chain, which also carries a variable domain (VH). The light chain has two domains: light (CL) and variable (VL). Specificity for a given antigen results from variable domains, more precisely, part of the VL domain and part of the VH domain. Each heavy chain has a mass of 50,000 g·mol*$^{-1}$ *(50 kDa) and each light chain has a mass of 25 kDa, a total of 150 kDa for the IgG.*

Differentiation is made between so-called polyclonal and monoclonal antibodies. Polyclonal antibodies are mixtures of different antibodies, recognizing the same antigen but attaching to different areas (called "epitopes") of the antigen. Coming from different types of immune cells, they are produced by injecting the antigen into an animal's body (mouse, rabbit, goat, etc.) and then, after a few weeks, obtained by chromatography from the animal blood sample. On the contrary, monoclonal antibodies are all identical for a given antigen and only bind to it at a single epitope. They come from the same type of immune cell. For this reason, they are much more specific than polyclonal antibodies. These antibodies are most often produced using hybridoma technology, in which a lymphocyte line produced in a bioreactor is used instead of an animal.

Whether polyclonal or monoclonal, these antibodies are heavy (150 kDa) and bulky (around 10 nm), which makes arranging them sufficiently packed on the surface of a biosensor, particularly on nanostructured surfaces. For this reason, they are advantageously replaced by smaller antibody fragments, produced by cleavage of traditional antibodies (Fab fragments, 50 kDa), by genetic engineering of single-chain variable fragments (scFvs, 30 kDa), or nanobodies, around 15 kDa (Figure 3.4).

In conventional antibodies or their fragments, the specific part responsible for binding with the antigen epitope comprises part of the heavy chain (VH) and part of the light chain (VL). However, different antibodies have been discovered in the camelid family (camel, dromedary, llama and alpaca). The variable part is carried by only a single chain and is denoted VHH (hcIgG, Figure 3.4). Nanobodies (which should be called "single domain antibodies") are precisely this VHH part isolated from the heavy chains. They contain three complementarity-determining regions (CDRs) called CDR1, CDR2 and CDR3, directly responsible for the affinity for the target, composed of a sequence of 10–20 amino acids. They are now produced in bioreactors from yeasts, fungi or bacteria.

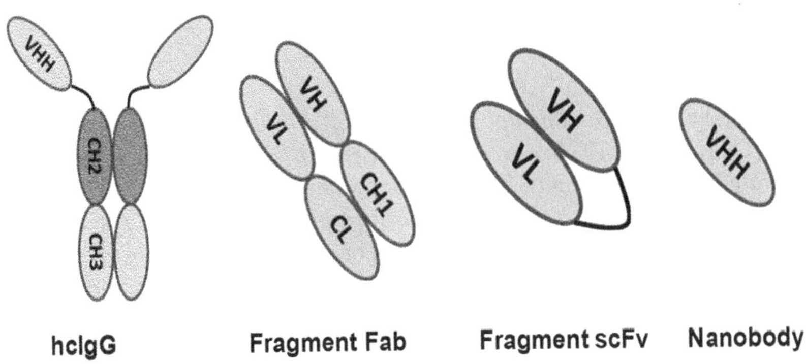

Figure 3.4. *Structure of a camelid hcIgG antibody (heavy chain IgG, symmetrical Y structure), Fab and scFv fragments from an IgG antibody, and a nanobody from an hcIgG antibody (the color codes of the different fragments are the same as those in Figure 3.3). For a color version of this figure, see www.iste.co.uk/lacaze/nanotechnologies.zip*

COMMENT ON FIGURE 3.4.– *The Fab fragment is based on the IgG structure, considering that most of the heavy chain does not improve its target specificity. Fab is lighter (50 kDa) than IgG. The scFv fragment is even lighter since it retains only the variable domains. Beyond the scFv structure, any further reduction in mass results in a loss of affinity. To maintain this affinity, researchers took inspiration from hcIgG (antibodies produced by camelids); they differ significantly from the IgG of other mammalian species because the variable specific part is carried by only a single domain, known as the VHH, which retains excellent antigen-hapten affinity. The mass of a nanobody is thus reduced by up to 15 kDa without loss of affinity.*

Figure 3.5 illustrates the different possible implementations of single-domain antibodies on nanoparticles. Their small size allows them to be combined with quantum dots, magnetic nanoparticles, enzymes or carbon nanomaterials. In extreme cases, the recognition domain of single-domain antibodies can be reduced to a peptide, taking up all or part of the CDRs described above. Interest in peptides relates to their more straightforward synthesis than nanobodies, automatically from a library of available amino acids. On the other hand, they have lower specificity and affinity than single-domain antibodies.

Antibodies form a vast subject; additional details can be found in several review articles or book chapters (e.g. Karunakaran et al. 2015). The field of nanobodies is more recent and just as broad; most of the literature focuses on their therapeutic applications (Jovčevska and Muyldermans 2020).

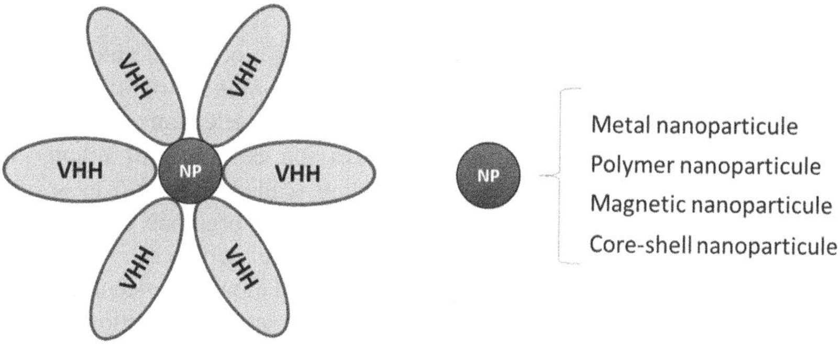

Figure 3.5. *Schematic representation of nanobodies (VHH, or single-domain antibodies) immobilized on nanoparticles (NPs). Adapted from Bastos-Soares et al. (2020). For a color version of this figure, see www.iste.co.uk/lacaze/nanotechnologies.zip*

COMMENT ON FIGURE 3.5.– *The major advantage of single-domain antibodies is their reduced size, which allows very high surface grafting densities. According to this logic and to further increase the surface-to-volume ratio, grafting these nanobodies onto NPs, used as a carrier or vector, may be advantageous. Metallic or magnetic NPs, reactive NPs, or even enzymes will make it possible to bind several antigens to an NP and ensure the transduction of probe-target recognition by modifying the reactivity of the NP or enzyme.*

3.1.3. *Identification using nucleic acids: aptamers*

Deoxyribonucleic (DNA) and ribonucleic (RNA) acids are known to code living organisms according to their sequence of bases (adenine [A], thymine [T], guanine [G], cytosine [C]), which associate in pairs (A-T, G-C) via several hydrogen bonds (see Chapter 2, Figures 2.1 and 2.2). However, these long chains can also form three-dimensional structures as complex as proteins, such as ribozymes, by folding on themselves. These natural structures do not have a coding function but an enzymatic activity.

It is also possible to define single-stranded nucleic acid sequences with a molecular recognition function rather than a coding function or a catalytic function; the sequences obtained are called aptamers. They can fold into three-dimensional structures with high affinity and selectivity for small organic molecules, large proteins or even cells. The affinity is due to a shape adjustment and an adapted distribution of weak interactions between the aptamer and specific areas on the surface of the target.

Aptamers gained status thanks to the development of a powerful automated in vitro selection method called SELEX (Systematic Evolution of Ligands by Exponential Enrichment). It is an iterative Darwinian selection process that multiplies the sequences that bind to a target to the detriment of others. To summarize this process in a few lines, a population (called a library) of several billion different nucleic acid sequences is incubated in the presence of the target. The sequences capable of binding to the target are separated by affinity chromatography and then amplified to create a new population of molecules, which all now have a minimum affinity for the target. After several repetitions of this selection and amplification process, the library contains many sequences with high affinity for the desired target. The sequences are then isolated, and their affinity is quantified by measuring the target-aptamer equilibrium's dissociation constant (K_d).

The most common targets of aptamers are proteins. The most often studied as models are α-thrombin, streptavidin, vascular endothelial growth factor (VEGF), hemagglutinin and adenosine-5′-triphosphate (ATP), rather than small organic molecules. This is because the dissociation constants for proteins are typically nanomolar (nM) or less, while those associated with small molecules are micromolar (μM), which are not as optimal. This difference in affinity can be explained by considering that there are more interaction sites between the aptamer and the target with a protein than with a small molecule. The aptamer can have hairpin-type conformations and form cavities, which bind with a specific protein. Their flexible structure forms a "Key-Lock" type complex comprising many linked sites. This association between large surfaces allows more bonds than with a small molecule, which can explain the greater stability of the aptamer–protein complex (Figure 3.6). Further details regarding the properties and applications of aptamers can be found in the review article by Dunn et al. (2017).

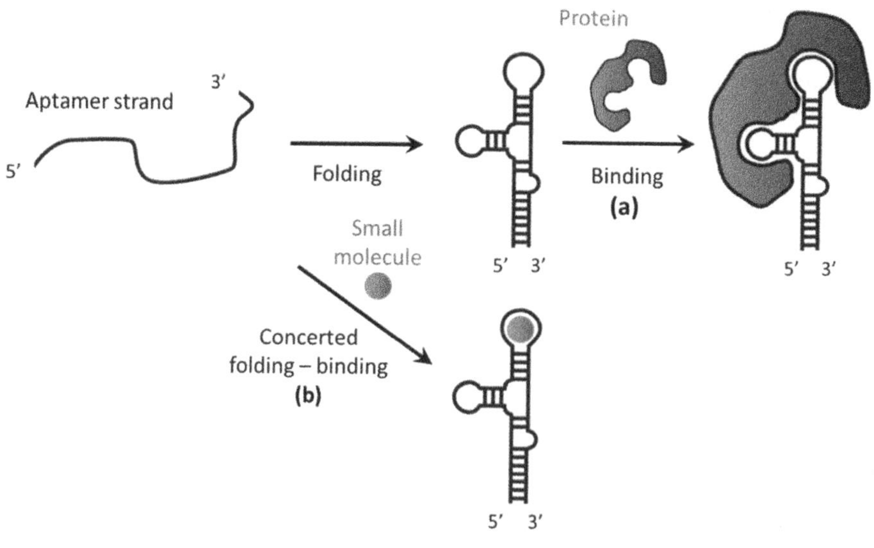

Figure 3.6. *General diagram of the aptamer-target association, according to a key-lock type mechanism. Adapted from Dunn et al. (2017). For a color version of this figure, see www.iste.co.uk/lacaze/nanotechnologies.zip*

COMMENT ON FIGURE 3.6.– *The aptamer is an unpaired strand of DNA or RNA. Therefore, it has excellent flexibility and can fold on itself according to two distinct processes. Either the strand folds into a stable secondary structure even before encountering the target, its geometry allowing it to associate with the target, according to a key-lock principle (a), or the folding occurs in concerted association with the target (b). Due to the difference in size and number of target-probe interaction sites, case (a) is typical of macromolecular targets (e.g. a protein), while case (b) is typical of molecular targets.*

3.1.4. *Recognition by molecularly imprinted polymers*

MIPs operate according to a key-lock type mechanism, extremely simple in principle, described for the first time in the 1980s (Andersson 1984). There are several methods to prepare these MIPs. All follow the same protocol: (i) mixing the monomer and the target molecule according to defined proportions (the monomer is chosen for its ability to polymerize and form weak bonds with the target molecule, ensuring that it does not inhibit polymerization); (ii) washing the formed MIP matrix with a solvent of the target molecules without being a solvent of the polymer. The result is a polymer whose cavities are the same size as the target molecule. The polymer chains are precisely arranged to maximize MIP/target interactions. When the MIP is exposed to a complex sample containing the target, the latter can enter the cavities and bind selectively. The general process is shown schematically in Figure 3.7.

Three different experimental protocols were developed based on this general principle. The first, described previously, is the bulk polymerization of a monomer in the presence of a target molecule; the second involves solubilizing a polymer in the presence of a target molecule and precipitating the mixture; and the third is a surface process known as stamping.

3.1.4.1. *Bulk polymerization*

This is the most common preparation method. It allows a functional monomer to interact in solution with the target molecule, developing a non-covalent interacting network. The monomer must be functional, meaning that besides the functional group responsible for polymerization, at least one other functional group must allow weak bond interactions. The most common functional monomers are acrylamide, methyl methacrylate (MMA), methacrylic acid (MAA) or their derivatives. We can also find monomers leading to conductive polymers such as aniline or pyrrole. An intermediate solution involves adding conductive particles (carbon nanotubes, gold nanoparticles) to an insulating polymer or copolymerizing an insulating polymer (polyacrylamine, polymethyl methacrylate, poly(methacrylic acid), polyacrylonitrile) with a conductive polymer such as polypyrrole or polyaniline.

In most cases, particularly for vinyl monomers, radical chain polymerization occurs, triggered by a small amount of initiating agent such as azobisisobutyronitrile (AIBN) or benzoyl peroxide (BPO). Furthermore, a crosslinking agent is frequently added, such as ethylene glycol dimethacrylate (EGDMA). Crosslinking stiffens the cavities, so they do not collapse during washing; however, this crosslinking means less accessibility to the cavities when the target is added. A variation of bulk polymerization is emulsion polymerization, which produces granules. These polymers in bulk or as granules can be reduced to a powder to increase the surface/volume ratio, then washed several times (e.g. with methanol) to eliminate the printer molecule and, finally, introduced into a chromatography column for use.

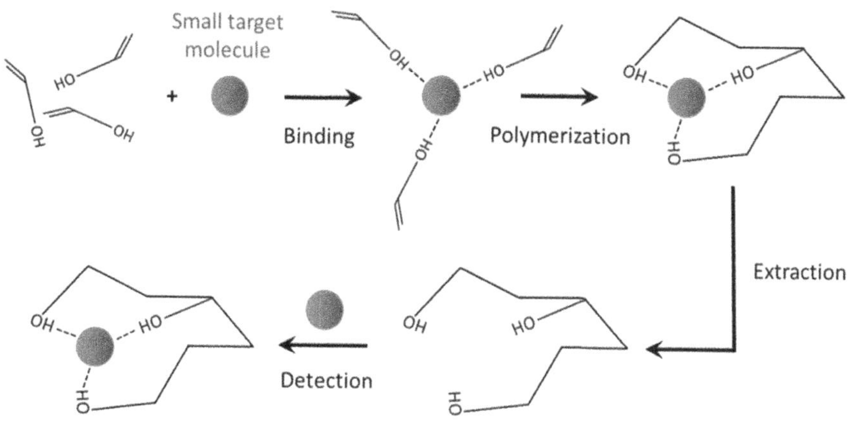

Figure 3.7. *General recognition process of molecularly imprinted polymers. For a color version of this figure, see www.iste.co.uk/lacaze/nanotechnologies.zip*

COMMENT ON FIGURE 3.7.– *A vinyl monomer carrying an -OH function (alcohol or carboxylic acid) is taken here as an example. The weak interactions between the target molecule and the functional groups of the monomer initially ensure the three-dimensional organization of the polymerization, which produces a material with cavities calibrated and functionalized by the OH groups after extracting the target molecule. Thanks to the functional groups in the cavity, the polymer formed can recognize and capture a target molecule with similar dimensions to those of the cavity.*

3.1.4.2. *Printing on a pre-existing polymer*

The idea is to avoid the tricky polymerization step and replace it with molecular imprinting on a pre-formed polymer. The technique involves choosing a polymer that can be solubilized in the same solvent as the target molecule simultaneously. This solvent is then removed, usually by evaporation, which returns the polymer to its solid state but with the target molecule trapped inside. The polymer, which must be sufficiently porous and swell as much as possible in contact with the washing solvent, is then washed in the same way to release the target molecule. The washing solvent differs from the solvent used for the imprint since it must solubilize the target, not the polymer.

3.1.4.3. *Stamping method*

Molecular recognition is reduced to the outer surface of a substrate and mainly applies to large targets (proteins, viruses, bacteria, etc.) that cannot diffuse in the polymer matrix. The stamping method uses a printing pad and a solid support to be printed. The stamp is thus modified on the surface by the target. The solid support is impregnated with a thin layer of the monomer, the polymerization of which can be initiated thermally or by UV irradiation. The stamp and the support are then brought into contact for sufficient time to polymerize the monomer. The stamp is then removed, releasing the thin printed polymer layer, which is washed as before (Poller et al. 2017) (Figure 3.8).

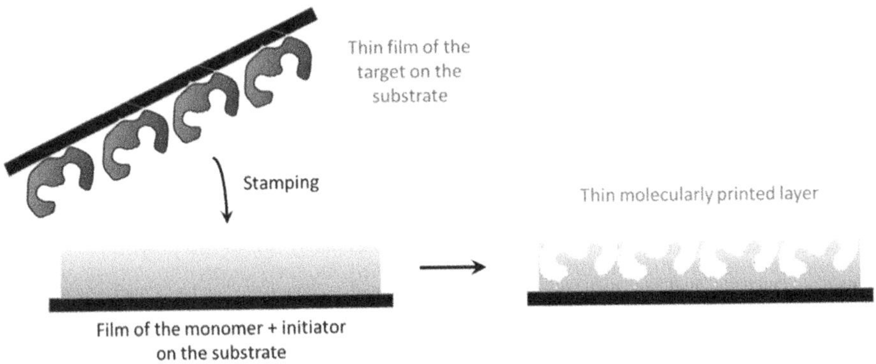

Figure 3.8. *Illustration of a stamping method for printing macromolecular targets on a thin film, whose large size would not allow diffusion toward the internal cavities of a solid polymer. For a color version of this figure, see www.iste.co.uk/lacaze/nanotechnologies.zip*

Other examples of MIP applications and exhaustive documentation can be found in several review articles, in particular that of BelBruno et al. (2019).

3.1.5. *Recognition followed by transformation: enzymes and catalysts*

3.1.5.1. *Enzymatic recognition*

Enzymes are proteins that can carry out the molecular recognition of a target (designated as the substrate) and ensure its catalytic transformation into a product. Its qualification as a catalyst means that the enzyme is constantly regenerated and is not consumed. Figure 3.9 illustrates a schematic example of how an enzyme (E) catalyzes the reaction between a substrate S and a cosubstrate CS to give a product P and a coproduct CP. Without the enzyme, the reaction S + CS = P + CP is thermodynamically favorable but kinetically slow.

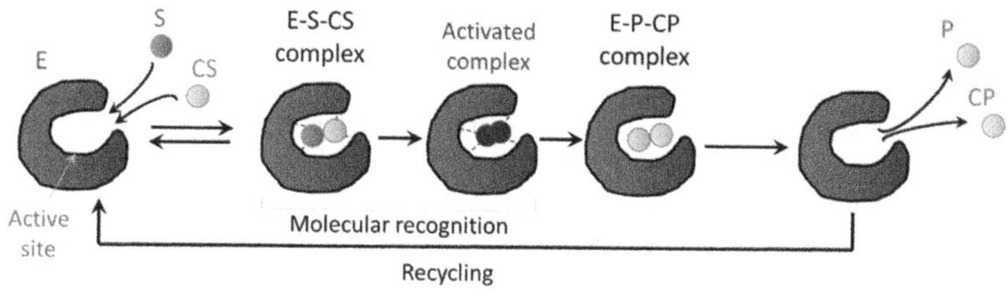

Figure 3.9. *Enzymatic recognition of a target (S) and its chemical transformation. For a color version of this figure, see www.iste.co.uk/lacaze/nanotechnologies.zip*

COMMENT ON FIGURE 3.9.– *The enzyme generally comprises a protein envelope with an active site within. This site is active because it can selectively bind the substrate-cosubstrate couple, or more precisely, the activated complex of these two reagents. It can also contain (not shown) a cofactor participating in the reaction, attached to the active site, or diffusing the same way as the substrate. The weak bonds established with the substrate S and the cosubstrate CS make it possible to modify the reactivity of these two molecules and to form an activated complex, which would not form in the absence of these interactions; it is this step that allows the transformation to take place according to rapid kinetics, giving a product P and co-product CP.*

In kinetic terms, the reaction can be simplified as follows:

$$\mathrm{E + S + CS} \underset{k_{-1}}{\overset{k_1}{\rightleftarrows}} \mathrm{E{\cdot}S{\cdot}CS} \xrightarrow{k_{cat}} \mathrm{E + P + CP}$$

where k_1 is the kinetic constant of formation of the enzyme-substrate-cosubstrate complex; k_{-1} is the kinetic constant of reformation of the enzyme, the substrate and the free cosubstrate; and k_{cat} is the kinetic constant of formation of the free enzyme and the product and co-product. Under these conditions, the so-called Michaelis–Menten equation (Michaelis and Menten 1913) applies. It gives the reaction rate V at equilibrium, corresponding to the disappearance of S, according to the relation:

$$V = \frac{V_{max}\,[S]}{[S] + K_M}$$

where V_{max} is the maximum reaction rate, given by $V_{max} = k_{cat}$ [E]. [S] is the substrate concentration, [E] is the enzyme concentration and K_M is the Michaelis–Menten constant, homogeneous to a concentration ($K_M = (k_{-1} + k_{cat})/k_1$). It corresponds to the substrate concentration for which the reaction rate is half the maximum rate ($V = V_{max}/2$) (Michaelis and Menten 1913; Sunney 2013).

In addition to the fact that enzymes are very effective catalysts for greatly accelerating reactions that would be excessively slow without catalysis, enzymes are generally highly selective catalysts. The reason is that the enzyme's active site has an architecture that better matches that of the activated complex rather than that of the substrate S (Figure 3.9) (Kraut 1988; Sunney 2013). Thus, specificity results from steric effects and weak bonds between the enzyme active center and the activated complex.

The enzyme's active site comprises peptide residues composed of amino acids whose general formula is $HOCO\text{-}CH(NH_2)\text{-}R$. R is a side group corresponding to acidic or basic, hydrophilic or hydrophobic functions, which can be electrically charged (+ or –) or neutral and can form hydrogen bonds. These characteristics of positions, sequences, 3D architectures and the capacity to form weak bonds create a particular chemical environment within the active site and give enzymes their catalytic properties. It is now known that the specificity of the enzyme active site for the activated complex is achieved through a host–guest type mechanism with induced fit. The bond formation between the enzyme and the complex occurs gradually; it begins with interaction forces, which cause slight variations in the structure of the enzyme active site, the substrate and the cosubstrate. We thus obtain an enzyme–substrate complex in which the substrate is significantly different from the free substrate: we speak of an activated complex because these deformations weaken certain bonds in the substrate and reduce the activation energy of the substrate-cosubstrate reaction.

The enzyme active site can also provide the reaction with a favorable acidic, polar or apolar environment. All this information and other examples of enzymatic reactions from the perspective of selectivity can be found in the review article by Benkovic and Hammes-Schiffer (2003).

3.1.5.2. *Catalytic activity of nanozymes*

Enzymes are often natural proteins that must be extracted and purified from living organisms, which incur high costs. Furthermore, enzymes are generally unstable once extracted from their natural environment and denature, like any protein, without the possibility of renaturation. Finally, enzymes see their activity drop if the system deviates from mild pressure and temperature conditions, which generally do not correspond to industrial conditions. Therefore, synthetic catalysts, which are more resistant and reproducible, are essential. Unfortunately, synthesizing these artificial enzymes requires considerable effort, significantly limiting their applicability. However, there are synthetic catalysts based on nanoparticles, called nanozymes, which can be obtained relatively quickly and replace a natural enzyme's activity. Their properties and applications have been the subject of several review articles (Zhou et al. 2017; Dong et al. 2019; Huang 2019; Liang et al. 2019; Liu and Zhang 2021; Yang 2021).

Similar in size to enzymes, their reactivity results from the charge distribution on their surface. They are made up of monometallic, or more often bimetallic nanoparticles (to avoid the structural reorganization of monometallic nanoparticles and loss of activity), metal oxides or carbon structures (graphene oxide sheets, in rare cases, carbon nanotubes). They have mainly been developed to replace peroxidases and oxidases; few effective nanozymes have been developed for other catalytic reactions, such as hydrolases.

Although the catalytic activity of nanozymes can reach or exceed that of natural enzymes, this is not the case for their selectivity. This property, essential for the functioning of nanozymes, has generated significant research efforts oriented toward improving the coupling between metal oxide particles or carbon compounds and an organic network, making it possible to develop weak bonds with the substrate. The use of MIPs, to which catalytic NPs have been added to reproduce the selective pockets of enzymes, has proven particularly favorable (Figure 3.10).

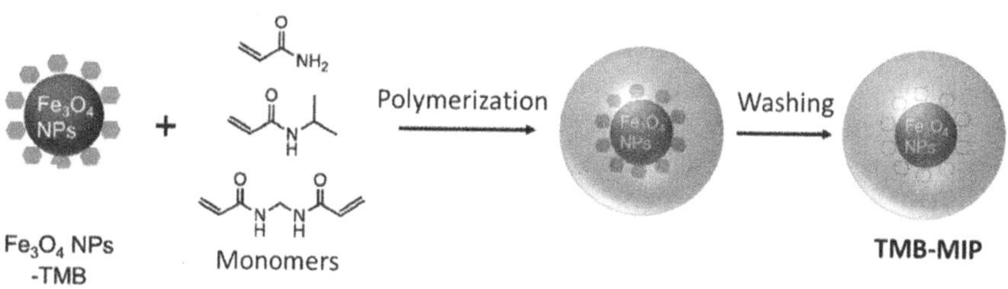

$$\text{TMB-MIP} + \text{TMB}_{red} + H_2O_2 \longrightarrow \text{TMB}_{ox} + H_20$$

$$\text{TMB-MIP} + \text{ABTS}_{red} + H_2O_2 \longrightarrow \text{no reaction}$$

Figure 3.10. *MIPs functionalized by mimicking-peroxidase catalytic nanoparticles. a)The MIP is formed by copolymerization of acrylamide, N-isopropylacrylamide and N,N-methylenebisacrylamide in presence of iron oxide NPs (Fe_3O_4 magnetite) functionalized by the chromogenic printing molecule TMB (3,3',5,5'-tetramethylbenzidine). TMB is adsorbed onto the nanoparticles before polymerization, then eliminated by washing to give TMB-MIP. In the presence of TMB and ABTS (2,2'-azino-bis(3-ethylbenzothiazoline-6-sulphonic acid), adding H_2O_2 and TMB-MIP allows the selective oxidation of TMB without oxidizing ABTS. Adapted from Zhang et al. (2017). For a color version of this figure, see www.iste.co.uk/lacaze/nanotechnologies.zip*

COMMENT ON FIGURE 3.10.– *Magnetite NPs enable the decomposition of hydrogen peroxide (H_2O_2) by oxidizing a third molecule, typically chromogens used in colorimetric tests, such as TMB or ABTS. It is used as a printing form to ensure the oxidation selectivity for a given chromogen (here, TMB).*

3.1.6. *Recognition of gaseous species*

Unlike the molecular recognition processes, applying the same mechanisms to gaseous molecules is ineffective. Also, generating a recognition signal does not involve mobilizing van der Waals interactions between the gas and a sensitive surface but instead directly using the reactivity of gases (oxidizing, reducing, acidic, or basic) such as CO, NO, NO_2, NH_4, SO_2, CO_2, CH_4 – the most common – and putting them in contact with a sensitive solid to cause, for example, a variation in the conductivity of its outer layer.

3.1.6.1. *Behavior of metal oxide semiconductors*

Materials used to detect gaseous species can be organic or inorganic. In the case of organic materials, most are semiconductor-conjugated oligomers or polymers (polyaniline, polypyrrole, polyethylene dioxythiophene [PEDOT], polyoctylthiophene [POT]) whose electronic structure will be modified by interaction with the targeted gaseous species resulting from semiconductor doping. Some of them, such as the POT, have given rise to reliable gas sensors operating at room temperature. Likewise, inorganic materials, metal oxide semiconductors (MOS), or carbonaceous materials (CNT, graphene), often combined with catalytic metals such as Pt or Pd, have proven efficient in recognizing gaseous species at high temperatures.

The n-MOS or p-MOS type metal oxide semiconductors are the most widely used, whose recognition signal is a variation in conductivity induced by the reducing or oxidizing gas. The mechanism that induces this variation in conductivity results from the depletion or enrichment of the electronic charges on the surface of the semiconductor.

For an n-MOS, whose charge carriers are electrons, its interaction with a reducing gas (electron donor) will increase the concentration of electrons in the material and, therefore, reduce its electrical resistance. Conversely, its interaction with an oxidizing gas reduces the electron concentration and increases resistance. For a p-MOS, the situation is reversed, with a reduction in resistance for oxidizing gases (or an increase for reducing gases).

In this case, no selectivity can occur at room temperature since the interactions correspond only to physisorption forces. On the other hand, by bringing the oxide to temperatures between 300 and 600°C, specific molecules develop strong interactions with the substrate, unlike others, which are only physisorbed, resulting in a specific selectivity. Moreover, operating at high temperatures prevents the formation of a water layer that would render the material inoperable, as it is more sensitive to ambient humidity than to the desired gas.

However, this direct interaction between oxidizing or reducing gases on n-MOS must be modified slightly when the analysis is carried out in air since it implies the additional presence of oxygen, which also acts on the sensitive oxide layer when brought to a high temperature. Its interaction with the oxide causes a transfer of electrons from the oxide to O_2, generating the superoxide ion O_2^- adsorbed on an electron-depleted layer, the thickness of which depends directly on the surface concentration of adsorbed oxygen.

A reducing gas, if present, will react with O_2^-, reinjecting electrons into the depleted layer by reducing its thickness, increasing the material's electrical conductivity. Conversely, an oxidizing gas will reinforce the effect of oxygen, increasing the thickness of the depleted layer, and reduce the material's conductivity. In both cases, it is a surface effect, which is why the nanostructure of the receiving material is of considerable importance, as it significantly increases the surface/volume ratio.

3.1.6.2. *Behavior of metal oxide NPs*

Substantial conductivity variations are expected when an NP surface assembly forms the sensitive layer. The thickness *L* of the depletion layer in the oxide and the size of the NPs (diameter *D*) are two criteria involved in the conductivity process (Figure 3.11).

Their use has a double effect: on the one hand, a larger specific surface increases the exchange surface with the gas and, on the other hand, a new charge distribution between NPs, decisive for the conductivity of the sensitive layer.

An n-type MOS material composed of sintered nanoparticles can be compared to a set of grains in contact. Each grain is characterized by a thin layer of depletion of electrical charges on its surface, determining the electrical contact between the grains. The smaller the diameter *D* of the nanoparticles (and, by extension, the grain size) is compared to the thickness *L* of the charge carrier depleted layer, the more this thickness affects the macroscopic conductivity. Therefore, the smaller the grains, the greater the detection sensitivity and the greater the depleted volume of grains in terms of charge carriers (Figure 3.11).

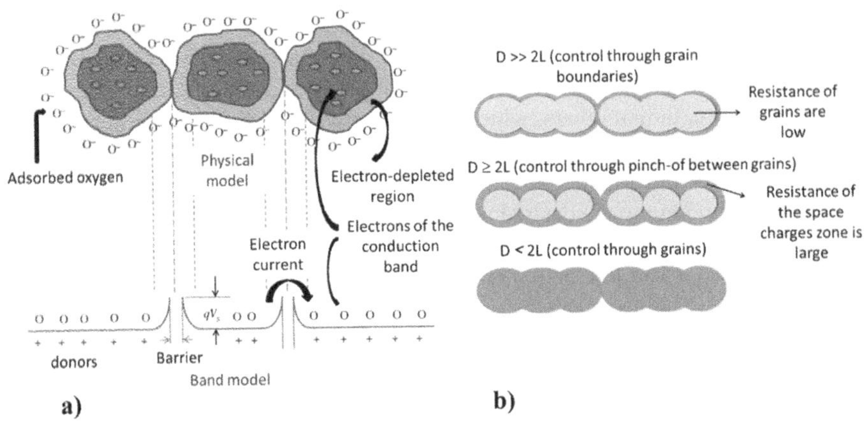

Figure 3.11. *Mechanisms of oxygen interaction on n-type semiconductor NPs. (a) Interaction of oxygen with the surface of a grain formed by the assembly of three semiconductor NPs. The surface layer of the NPs is depleted in electrons following their transfer from the oxide to the adsorbed oxygen layer. This results in a depletion layer and a potential barrier at the boundaries between NPs, which opposes the percolation of charges between particles inside a grain and increases the material's resistance. (b) The smaller the diameter D of the particles compared to the thickness L of the depletion layer, the larger the volume the depletion layer occupies in the NP. Adapted from Sharma and Madou (2012). For a color version of this figure, see www.iste.co.uk/lacaze/nanotechnologies.zip*

COMMENT ON FIGURE 3.11.– *The effect of NP size on the macroscopic properties of materials is particularly marked because it affects the surface/volume ratio and, therefore, the ratio between surface properties and volume properties. A more realistic representation of NPs is to group them into "grains", whose contact surfaces form "grain boundaries". When O_2, very electronegative, adsorbs on the NPs, it exerts a repulsion effect on the electrons of the n-type SC oxide, creating a thin, depleted layer with thickness* L *on the surface of the oxide into charge carriers. This NP superficial region is characterized by high electrical resistance, unlike the NP core, whose resistance remains low. Three structured forms of the NPs, depending on the* D/L *ratio, determine their electrical resistance. When D is much greater than* L *(*D>>L*), conductivity remains high, although charge transport between grains is limited by the charge carrier-depleted (double) layer. With smaller NPs (*D ≥ 2 L*), pinch points form between NPs, which control the passage of charges from one grain to the other and reduce the conductivity, compared to the previous case. Finally, if* D < 2L*, the entire volume of the NPs is depleted with electrons, and the conductivity is even lower. So, the smaller the NPs, the more oxygen will influence the conductivity.*

In the case of large nanoparticles, the sensitive zone is located at the grain boundaries, while in the case of smaller NPs, the entire material is sensitive to gas chemisorption. This is why, in the case of MOS combined with catalytic Pt or Pd nanoparticles, it is essential to distribute them as homogeneously as possible, as close as possible to the grain boundaries. Indeed, these metallic NPs form an activated complex with the gaseous species, dissociating the atoms so that they adsorb more easily onto the MOS surface. Under these conditions, the closer the metal nanoparticle is to a grain boundary, the more significant its effect on the resistivity variation (Figure 3.12).

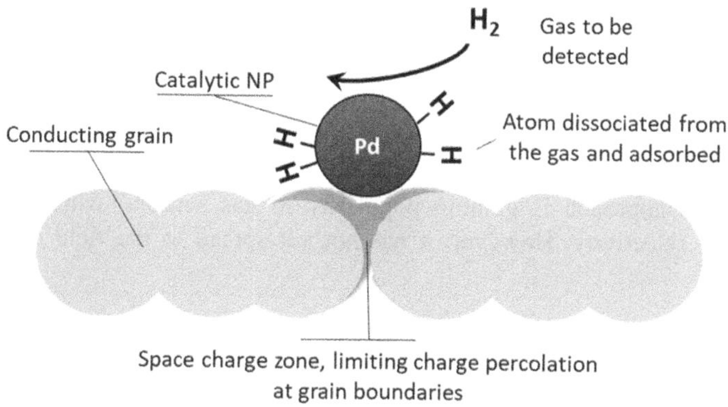

Figure 3.12. *Role of the distribution of NP catalysts at grain boundaries. Representation of two grains and a grain boundary on which an NP catalyst is adsorbed (palladium Pd). The gas to be detected (dihydrogen H_2) is known to adsorb onto palladium and form a hydride, which helps to repel the conduction electrons of the metal oxide semiconductor (MOS) and form a depletion layer. For a color version of this figure, see www.iste.co.uk/lacaze/nanotechnologies.zip*

COMMENT ON FIGURE 3.12.– *The position of the catalytic NP significantly influences the detection sensitivity. Suppose the particle is located far from a grain boundary. In that case, the effect of the dissociation of the gas atoms will be exerted on the grain without any significant change in conductivity. On the contrary, if the catalytic NP is placed at the grain boundary, the charge-depleted zone will limit percolation between boundaries, significantly increasing the macroscopic electrical resistance of the material.*

However, identifying a particular gas mixed with other gaseous molecules is still problematic when this involves a mixture, especially when we still need a mass spectrometer associated with an extensive database at our disposal and the guarantee that each compound will be identified. Except in exceptional cases, using a sensor based on oxide semiconductors is insufficient due to a lack of selectivity. However, a new technique for analyzing gas mixtures is developing and could become widespread. It involves networking many sensors whose signals can be processed according to the new machine learning concept (ML), which allows the identification of each element with the support of substantial databases. This approach is widely discussed in the review article by Milone et al. (2023).

3.2. Sensor networks and artificial intelligence

Recent advances in ML, driven by ever-increasing information storage capacities, make it possible to collect, store, analyze and interpret unprecedented information from all sizes and types of sensors. As discussed before, sensors have typically been limited by their selectivity. Machine learning now allows the fusion of the results of several sensors or even several detection modalities to deduce correlations, which can identify the composition of a mixture by comparison with a database. This approach is gaining popularity in gas sensors, which historically needed more selectivity. However, it has not yet spread to the field of biological sensors.

The strength of machine learning lies in the ability to extract information and common characteristics from a dataset without human intervention. This approach for image processing has already been applied, for example, to the automatic recognition of pedestrians in a road environment by modern cars. The versatility of machine learning algorithms allows them to adapt to many other applications, with the only requirement being to access a database or to acquire the data. This technique makes it possible to replace traditional detection approaches when the system studied is particularly complex, such as that corresponding to a mixture of numerous molecules in a matrix.

3.2.1. *Basic principles*

Consider a device with two sensors (sensor 1 and sensor 2), one sensitive to analyte A and the other sensitive to analyte B, but each suffering from cross-interference: sensor 1 can also respond to analyte B, and sensor 2 can respond to analyte A.

An operator will first record the responses of both sensors in many diverse situations (temperature, pH, presence of different interferents in known quantities, variable concentrations of A and/or B, etc.). The first dataset will be created, compiling the raw signals from both sensors. From these raw data, identifiable characteristics are extracted by machine learning, such as potential variations over time (if it is a potentiometric sensor), potential differences between the two sensors, different speed variations, etc. For each dataset, the operator will create a second dataset containing the operating conditions (temperature, pH, the concentration of each analyte, and, where applicable, the presence and identification of interferents, etc.). A second ML algorithm will then be used to associate the two datasets. The machine will be able to identify correlations between all raw data and all the experimental conditions, establish the most significant variations, and associate them with different experimental conditions (which we will call the "model"). When the model has been created (when the operator judges the model to be sufficiently reliable), the machine will be able, from raw data, to automatically associate experimental conditions (Figure 3.13).

The model can be fixed (defined during the development of the device, saved locally, and subsequently used without modification) or evolving (that is to say, dynamically updated with each use and at each validation (or non-validation) by the user to improve it over time. This considerably reinforces its relevance since the device thus gains "experience" as it is used.

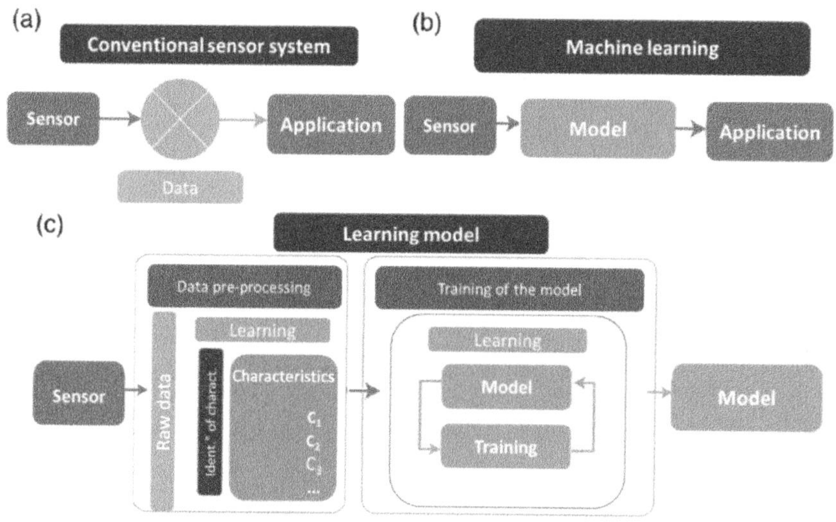

Figure 3.13. *Comparative designs of traditional and intelligent sensors. (a) Traditional method of acquisition from a sensor delivering a signal. (b) Modern method where data are acquired and processed to establish a model and deliver a signal. (c) General functional diagram of a smart sensor. Adapted from Ha et al. (2020). For a color version of this figure, see www.iste.co.uk/lacaze/nanotechnologies.zip*

COMMENT ON FIGURE 3.13.– *A reliable smart model should be trained with reliable datasets before being fed into the smart sensor system. The general block diagram is shown in Figure 3.13(c). In the initial phase, raw data are acquired. Typical characteristics are identified either by a "human" expert (e.g. a current value corresponding to a concentration value for a given analyte) or by machine learning (e.g. repeated data sequences that cannot be identified manually). These features form the training database. The model is then trained based on machine learning algorithms, and training stops when the model reaches an acceptable level of accuracy.*

3.2.2. *Application to chemical sensors*

3.2.2.1. *Gas sensors*

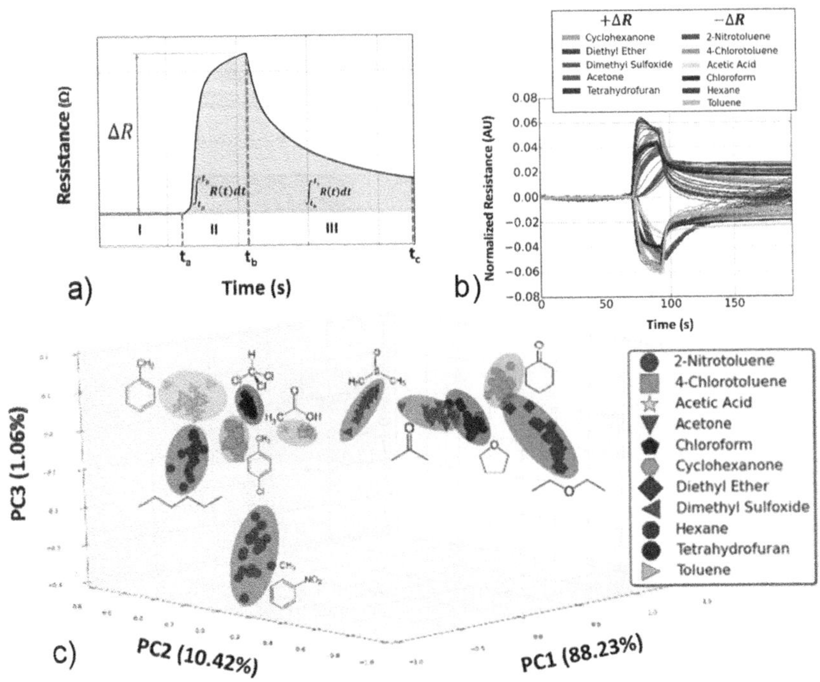

Figure 3.14. *Machine learning detection and gas mixture analysis using a single, non-specific graphene-based resistive sensor. (a) Typical sensor response with response zone, recovery zone and maximum amplitude ΔR. (b) Superposition of signals obtained for 11 different analytes, some showing increases in resistance (+ΔR) while others showing resistance decreases (-ΔR). Without detailed analysis, these signals cannot be interpreted. (c) Extraction of three main components for identification of volumes (in the component space) specific to each analyte. Adapted from Nallon et al. (2016). For a color version of this figure, see www.iste.co.uk/lacaze/nanotechnologies.zip*

COMMENT ON FIGURE 3.14.– *In the case presented here, the operator chose three relevant signal characteristics, one corresponding to the resistance variation amplitude when adding gas and the other two to the respective areas of the response and recovery curves. However, these characteristics should not be confused with the three principal components chosen (PC). These PCs result from a multivariate statistical analysis of the database, making it possible to reduce the number of system characteristics while remaining sufficiently descriptive. As a result, PCs only have statistical significance without any physical significance.*

Gas sensors are an excellent example of "smart" sensors. The results detailed in Figure 3.14 were obtained with a resistive sensor comprising graphene as a sensitive layer. Graphene does not initially show selectivity for any of the 11 analytes with which it is brought into contact.

For five of them, the resistance variation is positive; for the other six, it is negative. The following relevant signal characteristics are identified: the amplitude ΔR, the area under the response curve A_{rep} in the presence of the analyte, the area under the recovery curve (just after removal of the analyte) A_{rec} and the ratio A_{rep}/A_{rec}. The raw results recorded for the 11 analytes are uninterpretable; however, repeated 20 times for each analyte (i.e. 220 raw curves and 4 × 220 values of ΔR, A_{rep}, A_{rec}, and A_{rep}/A_{rec}), a principal component analysis (PCA) makes it possible to identify three coordinates and establish a three-dimensional plot (y-PC1, x-PC2, z-PC3) in which distinct volumes associated with each analyte appear[4].

3.2.2.2. *Biosensors*

Unlike a gas sensor, a biosensor aims to detect species in solution, and for more than 50 years, it has focused on the search for capture probes, which are as specific as possible to the analyte. However, the matrices analyzed are generally very complex (they contain numerous molecules of extremely varied masses and reactivities), which makes this selectivity very difficult to obtain unless biological probes such as enzymes or antibodies are used. Furthermore, whether for the characterization of living systems (health sensors) or environmental systems (pollution sensors), the analysis of a single molecule is insufficient, and it is necessary to characterize, at the same time, a series of molecules, each providing complementary information. For example, the health sector usually needs to analyze a series of biomarkers for which their concomitant presence at specific concentrations only allows a diagnosis to be given (Gaggin and Januzzi 2013). In the environmental field, this involves, for example, identifying "cocktails" of pollutants (pesticides, endocrine disruptors, etc.) that, individually and below regulatory doses,

---

4 The theory of PCA is well explained and detailed in the review article by Gewers et al. (2021).

do not induce metabolic disorders but which, together, constitute a major health risk (Relyea 2009).

Without an ML strategy, developing a device capable of independently analyzing all these molecules would amount to compiling data from as many sensors as analytes, generating large datasets with poorly interpretable information, and risking that the signal from each independent sensor is polluted by interference and generates interpretation errors. Furthermore, whether for the environment or health sector, a sensor must be able to provide decision support. Artificial intelligence is also relevant in this area, as shown by Cui et al. (2020) in a recent review article.

3.3. Conclusions

Recognizing a known molecular target mixed with other analytes requires a so-called "capture" probe. Generally immobilized on the surface, this probe must be sufficiently selective to bind only to its target. The recognition process is thus based on probes of biological origin (DNA, peptides, proteins, antibodies and enzymes), synthetic but bio-inspired probes (specific ligands, in particular), and those made of MIPs.

Today, biological probes (natural antibodies or nanobodies) are still the most relevant regarding selectivity. However, they are costly, and their use can only be generalized to some types of molecules, particularly low molecular mass targets. Furthermore, they are only effective under temperature, pH and ionic strength conditions close to physiological conditions. Finally, these biological probes are fragile and degrade quickly when the experimental conditions become too aggressive, meaning synthetic probes retain their importance.

When the analyte is a gas, the principle of molecular recognition becomes more complicated to implement and is based on catalytic-type mechanisms. The sensitive surfaces of the sensors are thus designed to react catalytically with gas and produce an easily measurable signal; they are nanostructured to increase their surface area and improve their interaction with the gases to be analyzed. Their fault, unfortunately, is their lack of selectivity.

The first approach to circumvent this problem was using a model from biology and taking inspiration from enzymes to recreate areas near nanostructured catalytic sites, developing weak bonds with the target in the manner of "nanozymes". The use of MIPs, as well as that of MOFs (detailed in Chapter 1), is an auspicious approach, which makes it possible to control the sensitive layer's porosity and offer calibrated sites concerning the target.

Another approach involves no longer aiming for the selectivity of a probe for a target, but of taking the animal nose as a model. Therefore, the new solution is to use networks of sensors (electronic noses) and process all the information through multivariate analysis, making it possible to identify a target not by the presence of a single sensor but by the intersection of different signals from different sensors. Currently under development for gases, this approach has yet to be applied to liquid-phase sensors and biosensors.

4

Physico-chemical Transduction Techniques

Identifying and analyzing chemicals in a particular environment has always been an essential societal objective, motivated by safety and health reasons. Chemical and biological sensors make it possible to accomplish this objective. All are developed according to the same principle, implying a cause-and-effect relationship between three entities: the compound to be analyzed (analyte), the receptor (the element that recognizes the analyte) and a transduction device that transforms the interaction between the analyte and the receptor into a measurable physical signal.

Different types of analytes can be detected and measured using biosensors or chemical sensors whose specificity is determined by the choice of recognition and transduction systems. Analytes include ions in aqueous solution or a biological fluid (blood, sweat, or urine), molecular species in solution, or gaseous species, particularly those involved in monitoring urban air pollution and various digestive gases, which help identify gastric or intestinal pathologies.

Among the different physical techniques used for the detection of analytes, some of them, such as mass spectrometry (MS), atomic absorption spectrometry (AAS), and atomic emission spectrometry coupled to an inductively coupled plasma (ICP) source, are analysis methods that must be considered as references but are challenging to apply in the field or near a patient when it comes to medical analyses.

This is why the trend has always been to replace these techniques with simpler devices that can be easily relocated, leading to the design and production of miniaturized, portable chemical and biochemical sensors that can be used by non-specialized personnel. They are manufactured from innovative assemblies with, very often, nanomaterials (NMs) and physical techniques from electrochemistry, optics and electronics. Among the electrochemical methods involved, the most common include potentiometry, amperometry and its derived techniques (linear sweep and cyclic voltammetry (CV), differential pulse voltammetry (DPV), electrochemical impedance spectroscopy (EIS), electrochemical gravimetry accessible by piezoelectricity and quartz crystal microbalance (QCM). Others, based on optical methods, involve photoluminescence (PL), colorimetry, plasmonics (surface plasmon resonance [SPR]), surface-enhanced Raman spectroscopy (SERS), but also techniques combining optics and electrochemistry such as electrochemiluminescence (ECL) and photoelectrochemistry (PEC). PEC is well suited for quantitatively analyzing biomolecules and achieves very low femtomolar detection limits (10^{-15} M). More recently, the use of field effect transistors (FETs) is also a new path that has made achieving extremely low detection limits possible, with the added advantage of easy transfer to portable systems.

In the case of mixtures of mineral, organic or biological analytes, their separation by chromatography or electrophoresis followed by physical detection is also a preferred analysis method, which, in some instances, requires amplification of the concentrations when these correspond to infinitesimal traces. Polymerase chain reaction (PCR) techniques and DNA sequencing are principal biological tools used for specific analyses.

Finally, unlike high selectivity sensors, another way to determine the composition of an analyte mixture is by using non-specific sensor networks, whose response signals can be processed by applying artificial intelligence methods. Functioning in a manner comparable to the olfactory system of mammals, they are therefore referred to as electronic noses or tongues (e-nose and e-tongue).

4.1. Electrochemical methods

The most used are potentiometry, voltammetry and impedancemetry. Voltammetry encompasses various methods involving a particular relationship between the current and voltage of an electrode, which can be subjected to different types of electrical excitations. Among these, we will mainly focus on amperometry with one or two indicator electrodes and methods derived from amperometry in which the electrode potential is programmed as a function of time. The most widely used is CV, a technique in which the measured current results from a linear sweep of cyclic voltage imposed on the indicator electrode.

4.1.1. *Potentiometric analysis*

An oxidoreducible chemical species in solution and in contact with a chemically inert metal electrode (gold, Pt, or even carbon) is characterized by an equilibrium redox potential E, resulting from a reversible exchange of n electrons between the oxidized (Ox) and reduced (Red) forms of the compound, according to the reaction:

$$\mathrm{Ox} + n\,\mathrm{e}^- \leftrightarrow \mathrm{Red}$$

Depending on the activities of a_{Ox} and a_{Red} [1], the electrode potential E is defined by the Nernst equation:

$$E = E_0 + (RT/nF)\ \mathrm{Log}\ (a_{Ox}/a_{Red})$$

(E_0 represents the standard redox potential when the activities a_{Ox} and a_{Red} are equal; R, T and F correspond, respectively, to the ideal gas constant (8.31 $\mathrm{J.mol^{-1}.K^{-1}}$), the absolute temperature (in Kelvin) and the Faraday constant (96,485 $\mathrm{Coulomb.mol^{-1}}$).

This electrode potential is deduced from the voltage V measured between two electrodes, one of which has a fixed potential and is called the reference electrode (E_{Ref}), and the other is an indicator of the potential E characteristic of the redox species in solution. Most often, the reference electrode is either a "Calomel" electrode, corresponding to the redox couple Hg/Hg_2Cl_2, or an electrode made up of the couple Ag/AgCl, each in contact with a KCl solution of fixed concentration (generally the concentration of the salt at saturation in water). Their respective potentials evaluated in relation to the couple H^+/H_2, with a voltage 0 V (by definition, under standard conditions) are, respectively, 0.241 V and 0.197 V. This also implies that the reference electrode is separated from the solution to be tested by a porous diaphragm (generally sintered glass), which slows down the diffusion of Cl^- ions toward the solution containing the redox species but maintains electrolytic contact with it (Figure 4.1(a)). The potentiometric dosage of a redox species in solution can thus be carried out by measuring the potential difference between an indicator electrode (Platinum) and a reference electrode. Such a device should rather be considered as a system for measuring the concentration of an analyte, the selective aspect being due to the redox potential of the analyte (Figures 4.1(b) and (c)). Numerous examples of potentiometric analysis applications are described in Chapter 5, including ion-sensitive electrodes (ISE).

---

1 Activity is a dimensionless number, defined as the ratio between the concentration C_i of the species i and a reference unit equal to 1 $\mathrm{mol.L^{-1}}$, such that $a_i = \Upsilon \cdot C_i/(1\ \mathrm{mol.L^{-1}})$, Υ being the activity coefficient ($0 < \Upsilon < 1$), a function of environment and concentration.

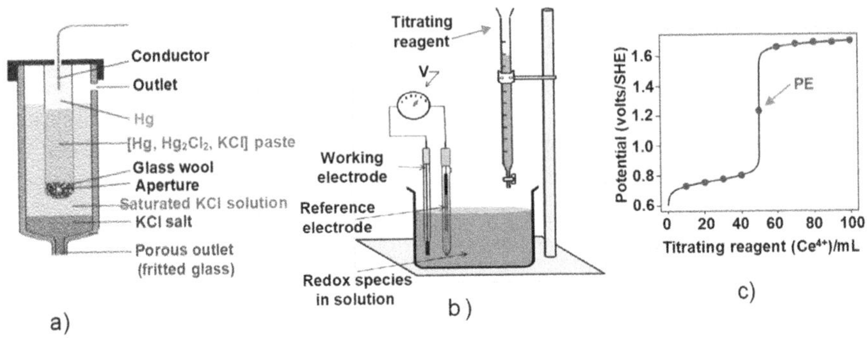

Figure 4.1. *Typical diagram of an electrochemical cell and a potentiometric analysis curve obtained by measuring the voltage between an indicator electrode and a reference. (a) Calomel-type reference electrode. (b) Basic device used for potentiometric analysis comprising the beaker with the compound for analysis (e.g.* Fe^{2+}*), the two electrodes and the titration burette containing the titrating reagent (*Ce^{4+}*). (c) Example of a potentiometric titration curve for the ferrous ion* Fe^{2+} *using* Ce^{4+}*. The neutralization can be detected more accurately by plotting the derivative curve (dE/dv). (v: volume of titrant reagent poured). (a–c) Adapted from the document https://chimieanalytique.com/potentiometrie/. For a color version of this figure, see www.iste.co.uk/lacaze/nanotechnologies.zip*

Furthermore, one of the current trends is to produce miniaturized electrochemical systems on chromatographic paper suitable for POC (point of care) analyses, known under the name of e-PAD (electrochemical paper-based analytical devices), produced for the first time more than 10 years ago by Dungchai et al. (2009) and increasingly recommended for clinical POC biological compound analyses (Maduraiveeran 2022).

4.1.2. *Voltametric analysis*

Unlike potentiometric analysis that involves voltage measurements at zero current, voltammetry concerns electrochemical methods for which reaction kinetics are involved. It mainly concerns linear voltage sweep voltammetry (CV) and DPV and, to a lesser degree, amperometry[2].

2 Amperometry is based on the measurement at a constant voltage of the electrolysis current resulting from the oxidation or reduction of a chemical species in solution, characterized by the exchange reaction Ox + n e^- ↔ Red. The applied potential is chosen sufficiently far from the equilibrium potential such that the electronic exchange rate is faster than the diffusion rate of the redox species, which is equivalent to an electrolysis current i controlled solely by the diffusion, or according to Fick's first law: $i = n\mathrm{F}AD(\partial C/\partial x)_{x=0}$, where $(\partial C/\partial x)_{x=0}$ represents

4.1.2.1. *Cyclic voltammetry*

In the presence of a redox species in solution, the application to the working electrode of a cyclic potential sweep leads to a voltammogram, characterized by two current peaks of opposite signs.

Thus, in the case of an Ox species, reversibly reducible according to the reaction Ox + $ne^- \rightarrow$ Red, the voltammogram obtained – with no solution agitation – reveals two current peaks noted $I_{p,C}$ and $I_{p,A}$ corresponding, respectively, to the reduction of Ox and the oxidation of Red, the latter forming and accumulating on the surface of the electrode during the reduction of Ox (Figure 4.2).

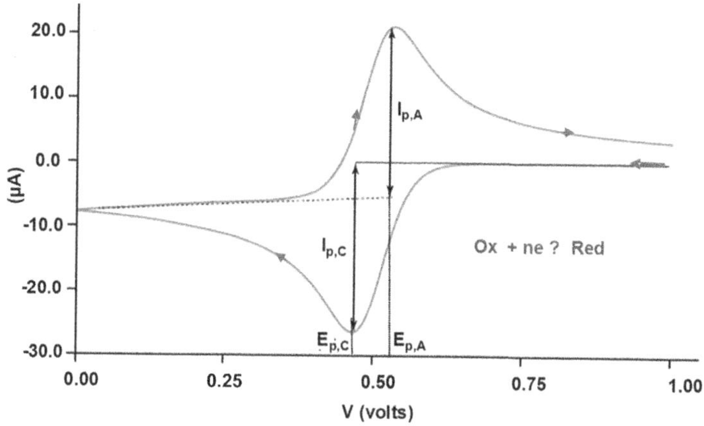

Figure 4.2. *Cyclic voltammogram model of an Ox species in solution. For a color version of this figure, see www.iste.co.uk/lacaze/nanotechnologies.zip*

COMMENT ON FIGURE 4.3.– *i/V is obtained during a cyclic voltage scan of the working electrode between 1 and 0 V (forward sweep), then 0 and 1 V (reverse*

---

the value of the concentration gradient of the Redox species on the electrode, C is its concentration (mol cm^{-3}), D is its diffusion coefficient (cm^2 s^{-1}), A is the flat surface area of the indicator electrode (cm^2), F is the Faraday constant (96,480 Coulombs), n is the number of electrons exchanged and i is the current intensity in amperes. In practice, an amperometric analysis is carried out in a stirred solution. Under these conditions, the mass transfer resulting from the diffusion of the Redox species from the core of the solution toward the surface of the electrode becomes stationary and the concentration gradient can be deduced from a simplified model by considering that the variation of concentration of the Redox species is confined inside a diffusion layer of thickness δ in contact with the electrode, $i = nFAD\,(C_0 - C_{x=0})/\delta$ (C_0 concentration in the solution, $C_{x=0}$ concentration on the electrode surface).

sweep) for a reversible redox system corresponding to Ox + e → Red reaction. The two peak currents $I_{p,C}$ and $I_{p,A}$, with voltages $E_{p,C}$ and $E_{p,A}$ are arranged symmetrically concerning the equilibrium potential of the Ox/Red couple located for the example chosen at 0.5 V. Peak current $I_{p,C}$ is calculated for an Ox concentration of 1 mmol/L, a diffusion coefficient D equal to 10^{-5} cm²/s, an electrode surface area of 0.1 cm², and a polarization rate of 100 mV/s. Adapted from Khalafi and Rafiee (2017).

Peak intensity $I_{p,C}$ is given by the relation of Randles and Sevcik, $I_p = 0.446 (F^3/RT)^{1/2} A\, n^{3/2} D^{1/2} C^* v^{1/2}$ (I_p in A, A in cm², D in cm²/s, v in V/s, C^* concentration of Ox within the solution in mol/L). This relationship shows that the intensity of the reduction peak increases with the scan rate (Bard and Faulkner 2001).

This technique is particularly useful for studying and characterizing the electrochemical properties of a compound to determine its reversibility or irreversibility and characteristics of fast or slow electronic transfer processes compared to mass transfer rates. It is also, in the case of a mixture of redox compounds, and depending on the separation of the different peaks, a means of providing the possibility of selective detection of each compound, which can be further improved by replacing it with DPV.

4.1.2.2. *Differential pulse voltammetry*

Very effective for separately evaluating the concentrations of a mixture of compounds, the technique involves recurrently applying to the working electrode a double potential pulse, per the diagram in Figure 4.3(a), corresponding to a particular form called square wave voltammetry (SWV) (Laborda et al. 2014).

The measured current corresponds to the intensity difference Δi between the intensities measured before the end of the pulse ΔE_p, that is, $i(\tau')$, and that just before the impulse, that is, $i(\tau)$. Like a Gaussian curve, the curve obtained is characterized by a differential current peak centered approximately on the equilibrium potential of the redox system (Bard and Faulkner 2001; Laborda et al. 2014). The current peak $(\Delta i)_p$ is given by:

$$(\Delta i)_p = \frac{nFA\sqrt{D}\,C*}{\sqrt{\pi\,(\tau-\tau')}}\left(\frac{1-\sigma}{1+\sigma}\right) \text{ with } \sigma = \frac{nF}{RT}\,\frac{\Delta Ep}{2}$$

The advantage of the technique compared to conventional CV is to eliminate the component due to the capacitive current and the residual background current of the supporting electrolyte, which makes it possible to improve the detection sensitivity and to achieve LODs very weak (around a few fM) in the case of biological compounds such as disease biomarkers (Tran et al. 2013).

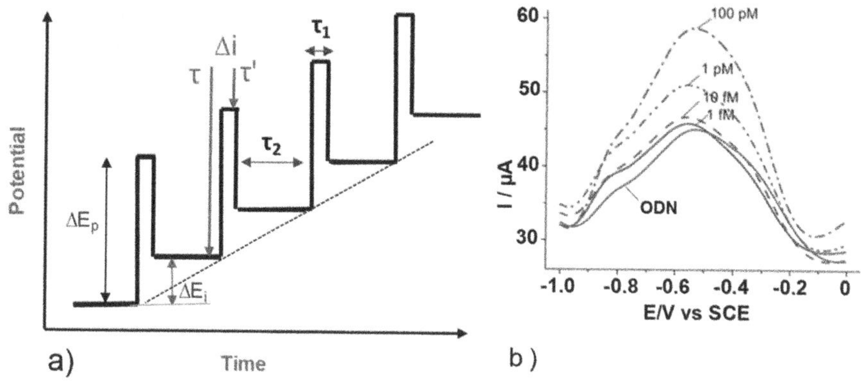

Figure 4.3. *Differential pulse voltammetry (SWV). (a) Potential variation curve as a function of time. A very short first impulse ΔEp (100 mV, τ_1 = 5–10 ms) is followed by a second pulse of the opposite sign with a greater duration (τ_2 = 50–100 ms) and amplitude (ΔEp − ΔEi) weaker than the first. Repetition of this cycle of pulses amounts to following a potential ramp represented by the dotted line. (b) Example of SWV analysis (between 1 fM and 100 pM) of a prostate cancer biomarker (miRNA-141) on an electrode modified by a composite containing SWCNTs mixed with an electroactive polymer and oligodeoxynucleotide (ODN) complementary to the biomarker. (a) Adapted from Laborda et al. (2014); (b) adapted from Tran et al. (2013). For a color version of this figure, see www.iste.co.uk/lacaze/nanotechnologies.zip*

4.1.3. *Impedancemetry*

Known since the end of the 19th century, thanks to the work of Heaviside and Warburg, EIS only really developed in the 1970s after the appearance of the first potentiostats on the market. Their use makes it possible to control the potential E of a working electrode versus a reference electrode and to superimpose an alternative voltage ΔE of low amplitude, of around 10 mV, which results in direct current I_0 modulated by a low-amplitude AC ΔI current.

Voltage E_t and intensity I_t to the working electrode are thus in the form: $E_t = E_0 + \Delta E \cos \omega t$ and $I_t = I_0 + \Delta I \cos (\omega t - \phi)$, where ω represents the radial frequency (or pulsation) of the alternating voltage and ϕ the phase shift of the current with respect to the voltage. Due to the low amplitude of the voltage, the variation in the intensity can be considered linear with respect to ΔE, from which we deduce the value of the impedance $Z = \Delta E \cos \omega t / \Delta I \cos(\omega t - \phi)$.

Using complex notation, we can write $Z = Z' + \mathrm{j}\, Z''$, and so $Z = \Delta E/\Delta I\ e^{\mathrm{j}\phi} = |Z|\, e^{\mathrm{j}\phi} = |Z|\ (\cos \phi + \mathrm{j} \sin \phi)$. Plotting impedance against ω in the complex plane

(Nyquist diagram) provides information on the different components of impedance (Figure 4.4). Another method is the Bode representation of impedance, which involves plotting log Z as a function of log ω and the phase shift ϕ as a function of log ω.

The most frequently used electrochemical system corresponds to an electrolyte solution containing a reagent that exchanges electrons at the solution/electrode interface and whose transport in solution toward the electrode is controlled by diffusion. Such a system is represented by the so-called Randles electric circuit illustrated in Figure 4.5(a), where R_e represents the resistance of the solution, R_{ct} is the charge transfer resistance (determined by the reaction kinetics at the solution-electrode interface)[3] and R_W is the Warburg resistance (corresponding to the reactant diffusion rate) both in parallel with the double layer capacitance C_{dl}.

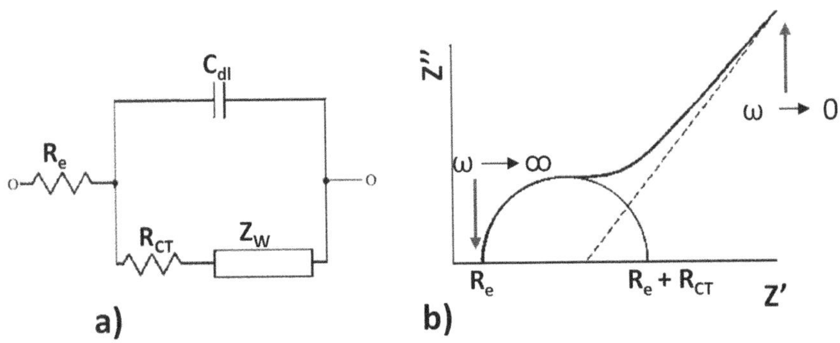

Figure 4.4. *Nyquist diagram corresponding to an electrode reaction controlled by diffusion and electron transfer. (a) Equivalent diagram of a Randles circuit indicating the resistance of the solution R_e, charge transfer resistance R_{CT}, Warburg impedance Z_W and double layer capability C_{DL}. (b) Nyquist diagram corresponding to the Randles circuit and showing the superposition between a semicircle due to the charge transfer resistance R_{CT}, the capacity C_{DL} and a straight line forming a 45° angle with the Z' axis observed for low frequencies and due to Warburg impedance Z_W [4]. For each point of the semicircle, the ordinate verifies the relationship $1 = \omega R_{CT} C_{DL}$. Adapted from MacDonald (2006). For a color version of this figure, see www.iste.co.uk/lacaze/nanotechnologies.zip*

3 The charge transfer resistance is given by the relation $R_{ct} = RT/nFi_0$, where i_0 denotes the exchange current density of an oxidation-reduction reaction Ox + ne → Red, R, T, and F representing, respectively, the ideal gas constant, the absolute temperature and the Faraday constant.

4 The Warburg resistance corresponds to the diffusion rate of the reagent and is expressed by the relation: $Z_W = \sigma\,(\omega)^{-1/2}\,(1 - j)$ with $\sigma = (RT/n^2F^2A\;2^{1/2}) \times (1/C_{Ox}D_{Ox}^{1/2} + 1/C_R\,D_R^{1/2})$. A represents the area of the electrode, and C_{Ox}, C_R, D_{Ox} and D_R correspond, respectively, to the concentrations of Ox and Red in the solution and to their diffusion constants.

The Nyquist diagram then shows two distinct impedance zones corresponding to two time constants. At high frequencies, the complex impedance Z is represented by a semi-circle, the diameter of which has the abscissa R_e and $R_e + R_{ct}$, due to kinetically rapid phenomena. This semi-circle is interrupted at low frequencies by a straight line with a slope of unity characteristic of the diffusion of the reagent in the solution, a slower phenomenon (Figure 4.4(b)).

Many sensors are thus based on impedance measurements of metal electrodes, often functionalized by chemical compounds that interact with the analytes. Typically, the device is composed of a network of interdigitated metal electrodes, of micrometric lengths and thicknesses, which makes it possible to increase the sensitivity of impedancemetry measurements (Figure 4.5).

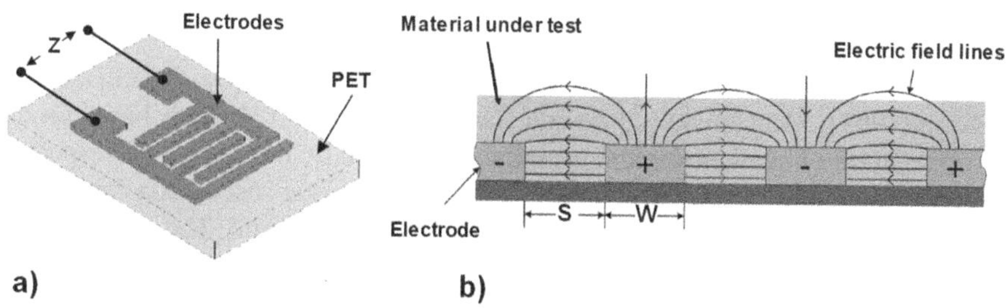

Figure 4.5. *Diagram of a sensor with interdigitated electrodes. (a) Example of a device comprising two interdigitated electrodes deposited on a flexible insulating polymer support by combined vacuum evaporation and electrolysis operations. (b) Representation of electric field lines between interdigitated electrodes. Impedance Z is measured between the two electrode terminals covered by a material that recognizes the analyte. Impedance variations are directly correlated to the adsorption of the analyte on the sensor surface. Adapted from Khan and Kang (2015). For a color version of this figure, see www.iste.co.uk/lacaze/nanotechnologies.zip*

Electrochemical impedance spectrometry (EIS) has been widely used to detect gaseous analytes and biomolecules whose adsorption on the sensor surface induces variations in resistance and capacity.

A more exhaustive description of the method and its analytical applications can be found in the recent review article by Magar et al. (2021).

4.2. Piezoelectricity for gravimetric analysis

Discovered in 1880 by the Curie brothers, piezoelectricity[5] has become a very widely used gravimetric analysis technique, for the detection of gases and many biological compounds. Two distinct techniques are in fact used to create sensors based on the use of piezoelectric materials subjected to the action of acoustic waves. The first, the most used, with the QCM as the main device is based on the variation in the transverse resonance oscillation frequency of the crystal, which is directly correlated to the mass of analyte deposited on its surface. The second, particularly for gases, uses surface acoustic waves (SAWs), whose frequency is also related to the species adsorbed on its surface.

4.2.1. *Quartz crystal microbalance*

The technique, described in detail in the review articles by O'Sullivan and Guibault (1999), MacDonald (2006) and Bard and Faulkner (2001), is based on the use of a natural or synthetic quartz crystal cut in a specific direction with respect to the crystallographic axes; this gives it a characteristic oscillation frequency depending on the thickness, mass and shear stress of the crystal. The oscillation of the quartz (5 MHz in the first years, then later much more) can be produced by applying an alternating electric field of the same frequency, provided by a broadband oscillator, capable of matching the frequency variations, themselves detected using a frequency counter.

Sauerbrey showed that a change in crystal mass caused a linear change in frequency according to the relationship:

$$\Delta f = -2 f_0^2 \Delta m\, n/(\rho\mu)^{1/2}$$

where f_0 is the resonant frequency of quartz, Δf is the frequency variation caused by mass deposition Δm, n is the number of the oscillation harmonic ($n = 1$ for the preferential frequency used), ρ is the density of the crystal and μ is the shear modulus. With the previous parameters constant for a given quartz, the Sauerbrey relation can be written more simply in the form $\Delta f = -C_f \Delta m$, where C_f represents the sensitivity coefficient of the crystal, which is higher the higher the frequency f_0. This relationship implies that the compound deposited on the crystal is rigid and firmly bonded to the crystal surface.

5 Piezoelectricity is the property of a crystal to deform under the effect of an electric field and, conversely, to generate a potential difference under the effect of pressure. The most common piezoelectric materials are quartz and some ceramics.

Currently, standard quartz microbalances are designed to operate with a resonant frequency of 10 MHz. The crystal thickness is approximately 160 μm and the diameter of the reactive surface is 15.5 mm. For a 1 cm^2 quartz crystal operating at a frequency of 5 MHz in air, the sensitivity coefficient C_f is 56.6 Hz per μg; it is four times higher for a resonance frequency of 10 MHz, or 226 Hz per μg (Bard and Faulkner 2001). Various chemical compounds (proteins, pollen grains, etc.) can be detected by modifying the surface of the crystal such that the reagent/crystal interaction is sufficiently rigid so that the Sauerbrey relation can be applied (Latif et al. 2013; Länge 2019) (Figure 4.6).

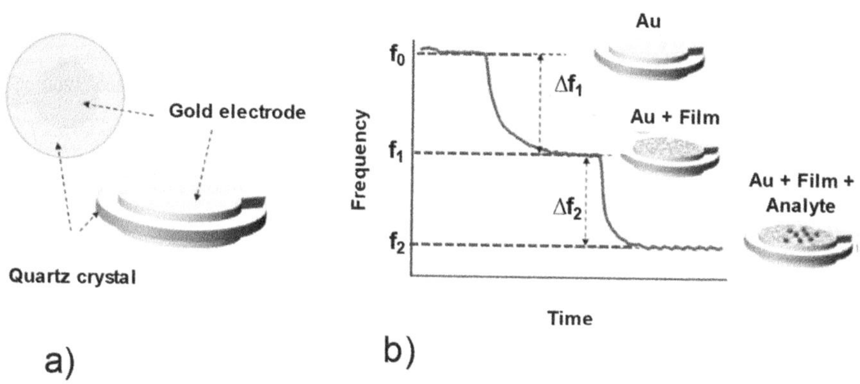

Figure 4.6. *Principle of selective detection of an analyte by the quartz crystal microbalance (QCM). (a) Representation of the quartz crystal sandwiched between two gold electrodes. (b) Variations in the oscillation frequency of quartz after attachment to a film (Δf_1) and then after sequestration of an analyte (Δf_2). Adapted from Fauzi et al. (2021). For a color version of this figure, see www.iste.co.uk/lacaze/nanotechnologies.zip*

A very wide variety of compounds can be detected and measured using this method. A typical application of QCMs is measuring air moisture. Hygroscopic composite films based on graphene oxide (GO) and polyethylene imine (PEI), firmly attached to the metalized quartz surface, can absorb moisture from the air. These modified quartz surfaces are characterized by relatively fast response and recovery times of less than 30 s for air moisture levels between 10% and 97% (Tai et al. 2016) (Figure 4.7).

Films made of molecular imprinted polymers (MIPs, see Chapter 3) have also been used with QCMs for their very strong selectivity properties since they can be synthesized with perfectly calibrated cavities to the size of the molecule to detect. Many examples of compounds important in the environment, biology, food and beverage monitoring sectors have been tested using this technique and are described in several review articles (Diltemiz et al. 2017; Fauzi et al. 2021; Liu and Zhang 2021).

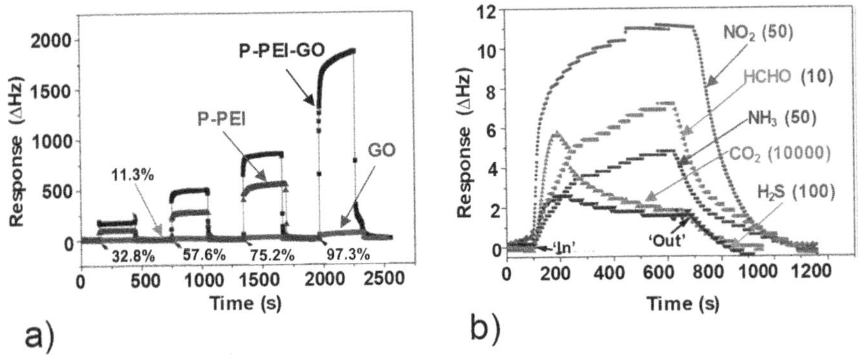

Figure 4.7. *Determination of air moisture levels using various functionalized QCMs. (a) Curves showing the variation in resonance frequency of the QCM functionalized on the surface by different coatings and in interaction with air-water vapor mixtures; P-PEI (protonated polyethyleneimine), GO (graphene oxide), P-PEI-GO (P-PEI grafted onto a layer of GO attached to the quartz surface. (b) Low-frequency variations obtained with high concentrations of interfering gases demonstrating the selectivity of the P-PEI-GO coating for water vapor. (Numbers in brackets correspond to ppm). We note that for high water vapor contents (97.3%), the P-PEI coating is ineffective (adhesion loss), unlike P-PEI-GO, which remains operational. Adapted from Tai et al. (2016). For a color version of this figure, see www.iste.co.uk/lacaze/nanotechnologies.zip*

In terms of sensitivity, March et al. (2015) showed, in the case of pesticide (carbaryl[6]) analysis, that a QCM operating at a frequency of 100 MHz achieves LODs comparable to those of standard ELISA-type immunosensors (see Chapter 6) and appear better than those of plasmon resonance sensors (PRS).

4.2.2. *Surface acoustic wave sensors*

The technology is also based on the use of a piezoelectric material on which interdigitated electrodes are printed, on which SAWs are generated with frequencies between a few MHz and a few GHz. These surface waves are then recorded by a

---

6 Carbaryl is a formula pesticide $C_{12}H_{11}NO_2$ from the carbamate family, used as a phytosanitary product. According to March et al., the LOD obtained with a frequency f_0 of 100 MHz is 0.14 ng/L (standard ELISA test 0.13 µg/L, PRS test 1.4 ng/L).

second network of interdigitated electrodes, which give an electrical signal of the same frequency as that of the acoustic wave. Any surface disturbance produces a frequency variation which, as in the quartz microbalance, varies linearly as a function of the mass adsorbed on the sensitive recognition layer, placed between the two networks of interdigitated transducers (IDT) (Figure 4.8).

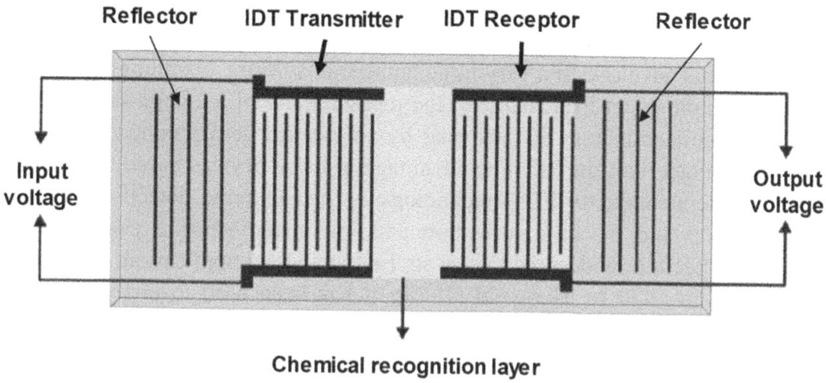

Figure 4.8. *Representative diagram of a two-port SAW sensor. Adapted from Mujahid and Dickert (2017). For a color version of this figure, see www.iste.co.uk/lacaze/nanotechnologies.zip*

COMMENT ON FIGURE 4.8.– *The device is made up of a piezoelectric material on which two networks of interdigitated transducing metallic electrodes (interdigitated transducers [IDT]) are printed, one being an emitter of an alternating surface acoustic wave by application of an alternating voltage and the other is the receiver of the wave produced. The sensitive layer (green) is where the observed frequency variation originates. The piezoelectric material is usually made up of quartz (the least expensive). Still, it can be replaced by other materials such as* $LiNbO_3$ *and* $LiTaO_3$ *for which higher propagation speeds of acoustic waves are obtained, which makes it possible to have a greater spacing between electrodes, equal to* $\lambda/2$ *of around 1* μm.

The difference with the quartz microbalance is technical. It comes from the fact that, unlike the characteristic transverse mode of excitation of a QCM crystal, limited to a few tens of MHz, the surface propagation of an acoustic wave can be carried out with frequencies ranging from several hundred MHz up to a few GHz, which makes it possible to increase the sensitivity of the SAW sensor considerably. This justifies its use for detecting very small, adsorbed quantities and explains its widespread use for determining gases and various biological compounds, a detailed list of which can be found in several review articles (Länge 2019; Mandal and Banerjee 2022).

4.3. Field effect transistors

Very widely used for the design of electronic memories and digital devices, FETs have also been adapted for the analysis and detection of various chemical compounds. This gave rise to a very large number of sensors, operating according to the basic principle of an FET, namely, the measurement of a charge current on the surface of a semiconductor (SC) induced by an electric field produced through an insulator in contact with the SC. The advantage of these devices based on MOSFET (metal oxide semiconductor-FET) technology is their extreme detection sensitivity (see Chapters 5 and 6), miniaturization and low manufacturing cost. The FET technique has become widespread since its first analytical application, dedicated to measuring pH in 1970 by Bergveld (1970). It has been used to analyze ions and various biomolecules, giving rise to various types of sensors.

4.3.1. *Operating principle of an FET*

The operation of an FET is based on a three-electrode architecture comprising: (i) a pair of source (S) and drain (D) electrodes in contact with a SC, between which is applied a low voltage V_{DS}; (ii) a gate electrode (G) itself polarized relative to the source at a voltage V_{GS} of the same order of magnitude as V_{DS}, exerting an electric field on the SC through an insulating layer. In the case of an n-type SC, where the minority charge carriers are positive charges (holes), applying a negative gate voltage repels the electrons toward the interior of the SC. It attracts the positive charges at the insulator/SC interface in a thin zone, designated as the transistor's channel, itself qualified as a p-type channel transistor (Figure 4.9(a)).

Any modification to the electric field causes a variation in the density of charges induced in the FET channel and, therefore, a variation in the current I_{DS} between S and D, identifiable as a variation in conductivity. The advantage of such a device is that it controls an output current with an almost infinite signal input impedance on the grid. The output characteristics, corresponding to the variation in the source/drain current (I_{DS}) as a function of source/drain voltage (V_{DS}) and for a fixed gate potential V_{GS}, reveal a linearity zone and a saturation zone (Figure 4.9(b)).

The linearity zone ($I_{DS,lin}$) is limited to values of V_{DS} much lower than V_{GS} and is characterized by a linear variation of I_{DS} as a function of V_{GS} according to:

$$I_{DS,lin} = (W/L)\ \mu_{lin}\ C_i\ (V_{GS} - V_{Th})\ V_{DS}$$

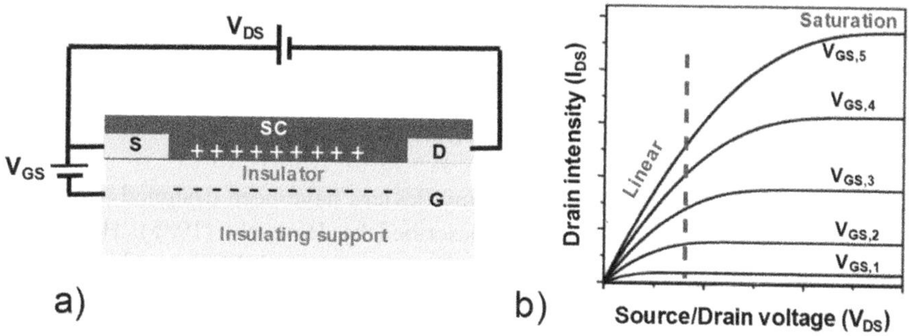

Figure 4.9. *Field effect transistors. (a) Configuration of a field effect transistor in the 'back gate' configuration. (b) Transfer characteristics $I_{DS} = f(V_{GS})$ of an FET subjected to gate voltages of increasing amplitude of $V_{GS,1}$ to $V_{GS,5}$. Adapted from Zaumseil and Sirringhaus (2007). For a color version of this figure, see www.iste.co.uk/lacaze/nanotechnologies.zip*

COMMENT ON FIGURE 4.9.– *The positive charges + located at the insulator/SC interfaces (channel) correspond to a negative V_{GS} gate potential and an n-type SC. It is the positive minority charges that determine the conductivity of the channel and, therefore, the intensity of the I_{DS} current, which is the output signal. With a p-doped SC, the minority charges being electrons, creating an n-type channel implies a positive polarization of the gate, which repels the positive charges toward the interior of the SC and localizes the electrons at the insulator/SC interface.*

When V_{DS} is much higher than V_{GS}, we reach a so-called saturation zone ($I_{DS,sat}$) where $I_{DS,sat}$ varies quadratically as a function of V_{GS}, according to:

$$I_{DS,sat} = (W/2L)\mu_{sat}\, C_i\, (V_{GS} - V_{Th})^2$$

(μ_{lin} and μ_{sat} represent the respective linear and quadratic electron mobilities, C_i is the capacitance of the insulating layer located between the gate and the SC, W and L are the respective width and length of the channel and V_{Th} is the threshold voltage of the grid (potential for which I_{DS} becomes measurable).

Considering this property, we can see that for a determined gate potential, the adsorption of molecules on the surface of the SC is likely to modify the charge density in the FET channel and therefore cause a variation in the current I_{DS}, the measurement of which can then be correlated to the presence of chemical compounds in solution. The effect is similar if adsorption occurs on the grid surface. The variations of the induced signal will be greater as the SC channel becomes thinner, which is why many one-dimensional and two-dimensional NMs arranged as monolayers have been used to make FETs for analytical chemistry.

4.3.2. *Main FET devices applied to chemical analysis*

Inspired by MOSFET technology, many devices have been designed to produce sensors, giving rise to various names described by Bergveld (1985). He himself developed the first ion sensitive field effect transistor (ISFET), with which he carried out the first pH measurement in the early 1970s, covering a range of around 10 units ($2 < pH < 12$). Also used to analyze biomolecules, they are typically known as BioFETs. Subsequently, many FET devices were proposed, differing by the choice of the SC in the FET channel. Furthermore, different types of SC NMs have been used, such as silicon nanowires (Zheng et al. 2005; Kim et al. 2016; Ghosh and Giri 2017), two-dimensional SCs such as MoS_2 (Liu et al. 2019), and more recently MXenes (Khan and Andreescu 2020; Alwarappan et al. 2022; Chen et al. 2023) and organic SC films (organic field effect transistor [OFET]). OFETs comprise two categories of sensors: OECTs (organic electrochemical transistors) and EGOFETs (electrolyte-gated organic field effect transistors). These devices have many advantages, such as the possibility of device miniaturization, allowing their operation on microdroplets of solution. Also noteworthy is that labeling the molecules to be analyzed (label-free detection) is unnecessary. It is a simplification in the experimental procedure that differentiates them from most traditional immunoassays for which fluorophore labeling is usually necessary.

4.3.2.1. *Field effect transistors sensitive to ionic and biological species (ISFETs and BioFETs)*

ISFETs are the oldest chemical sensors incorporating MOSFET technology for analytical purposes. Their structure differs from that of a conventional FET in that the FET metal grid, which is no longer in contact with the insulator, is replaced by a reference electrode immersed in the solution to be analyzed, itself in contact with the SC channel. Initially designed for analyzing ionic species, they were subsequently used for biomolecules and designated as BioFETs (Schöning and Poghossian 2002; Sung and Koo 2021).

However, the interest in ISFETs is rather focused on the environment, for which the use of ISFET sensors may present an advantage over ISE (ion selective electrodes) and glass electrodes (see Chapter 5), due to their better adaptability to miniaturization and also, for their greater robustness and ease of design, inspired by commercial CMOS (complementary metal oxide semiconductor) technology (Moser et al. 2016). Their use for ion detection, however, involves choosing a particular gate insulator. Thus, to measure the acidity of a solution, grid insulators (type SiO_2) are modified by thin layers of alumina (Al_2O_3) or tantalum oxide (Ta_2O_5) sensitive to H^+ ions. These oxides react reversibly with the H^+ ions in the solution, which leads to a variation in the threshold voltage of the transistor (V_{Th}) proportional to the solution pH (Bergveld 1985). With V_{DS} and I_{DS} constant, V_{GS} is then proportional to the pH of the solution. This type of sensor is commercially produced under the name pH-ISFET.

The same type of FET can be used to determine biomolecules using a conventional insulator in contact with a solution, in which a reference electrode plays the role of the grid, as shown in Figure 4.10.

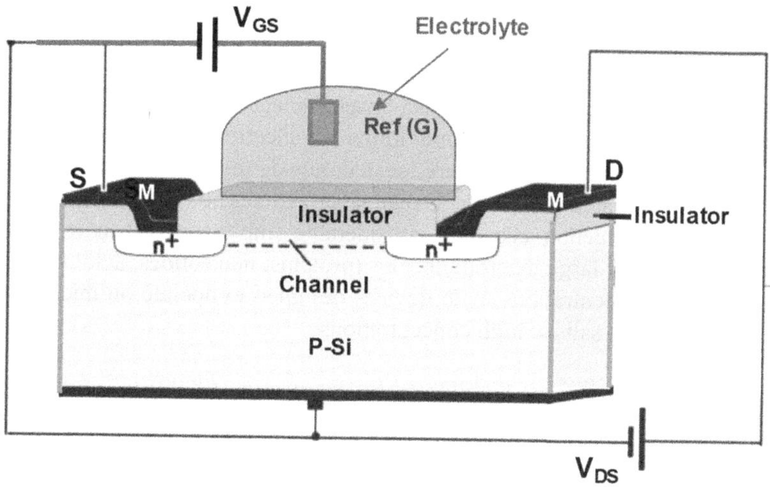

Figure 4.10. *Schematic diagram of an n-type channel BioFET biosensor. We note that a positive polarization is applied between the reference electrode and the source S, which induces a charge inversion in the channel with a predominance of minority negative charges (electrons) of the p-type semiconductor. Adapted from Schöning et al. (2002). For a color version of this figure, see www.iste.co.uk/lacaze/nanotechnologies.zip*

Due to their high detection sensitivity, their ease of design, their suitability for miniaturization and their ability to produce low-cost portable biosensors, which can be mass manufactured using industrial techniques derived from the microelectronics sector, BioFETs now occupy a privileged place in the family of biosensors.

Their operation is analogous to that of a FET, and the analysis of a biomolecule is thus based on an electrostatic interaction with the SC, which is added to or subtracted from that of the gate potential. This causes a variation in the channel conductivity and, consequently, in the drain current, which can then be correlated to the concentration of the biological compound. The molecule's electrostatic effect is essentially due to the presence of electric charges resulting from anionic or cationic groups connected to the body of the molecule, whose ionization is a function of the solution pH.

For this mode of operation and knowing that the biological solutions analyzed (blood, blood plasma, saliva, etc.) contain electrolytes in sometimes high concentrations, we must then consider the screening effect exerted by the ions of the solution. These accumulate in a layer, known as the Debye layer, in contact with the grid insulator, whose thickness (λ_D)[7] is inversely proportional to the square root of the ionic strength of the solution. This thickness is less than 1 nm for typical physiological solutions, while biological macromolecules easily reach lengths of around 10 nm[8]. This ionization zone forms an electrostatic screen against the charges of the biomolecules, which are located outside this layer, meaning that part of the molecule is located outside the Debye layer and has no action on the transistor channel. This phenomenon takes on considerable importance when it comes to carrying out assays of large macromolecules (proteins, nucleotides, aptamers, DNA, etc.) at very low concentrations, with devices designed to operate on microdroplets of solution containing salts at high concentrations.

This screening effect was highlighted by Stern et al. (2007) by measuring the variation in current I_{DS} of an FET during the addition in a buffer medium of streptavidin on biotin molecules adsorbed on p-Silicon nanowires, themselves attached to the gate insulator (Figure 4.11(a)).

Streptavidin is a tetrameric protein of very high molecular mass (42.8 kDa), with a very high affinity for biotin with which it forms very stable complexes. Characterized by an isoelectric point of around 5, this implies that in a slightly basic buffer medium (pH 7.4 with the phosphate buffer [PB]), it is ionized as a polyanion. We understand that in this ionization state it exerts a significant electrostatic effect on the FET channel and, therefore, can induce significant variations in the

7 The thickness λ_D of the Debye layer is given by $\lambda_D = (\varepsilon_0\varepsilon_r k_B T/I_s)^{1/2}$, where ε_0, ε_r and k_B represent, respectively, the dielectric constant in a vacuum, the relative dielectric constant of the solution and the Boltzmann constant, and I_S is the ionic strength of the solution: $I_S = ½ \Sigma C_i z_i^2$ (C_i and z_i are the concentration mol/L and charge of i). In numerical form, $\lambda_D = (2.15 \times 10^{-10})/I_S^{1/2}$.

8 An antibody, an aptamer comprising around 30 bases as well as high-mass proteins can vary between 5 and 15 nm in size.

source-drain current, provided that the molecule is inside the Debye layer (Figure 4.11(b)).

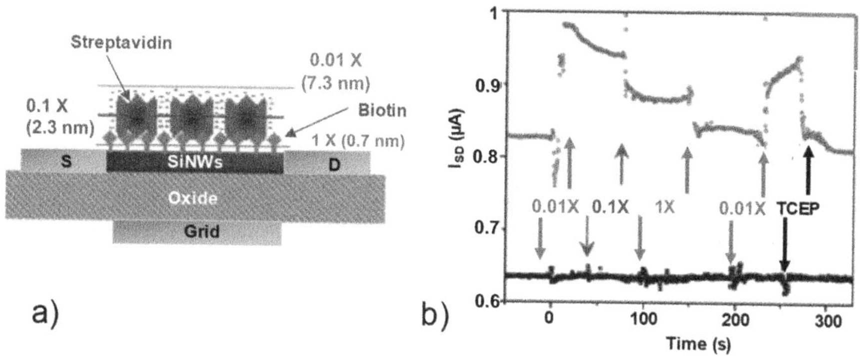

Figure 4.11. *Variation in source-drain current I_{DS} of an FET attenuated by the electrostatic shielding effect caused by the electrolyte. (a) Variation in the Debye length (λ_D) as a function of the dilution of the phosphate buffer (PB) solution and position of the Debye boundary layer in relation to the biotin/streptavidin complex adsorbed on p-Si nanowires (SiNWs). (b) Change in I_{DS} current intensity as a function of the dilution of the buffer solution with (red curve) or without (black curve) addition of streptavidin. Adapted from Stern et al. (2007). For a color version of this figure, see www.iste.co.uk/lacaze/nanotechnologies.zip*

COMMENT ON FIGURE 4.11.– *The thickness of the Debye layer varies between 0.7 and 7.3 nm when the solution ion concentration changes from 1X (0.15 M/L) to 0.01X (a). The black curve (b) relating to FETs functionalized only with biotin shows no change in the I_{DS} current (0.63 μA) when buffer solutions diluted from 1 to 100 times (1X, 0.1X, and 0.01X) are added, which indicates that the molecule is entirely inside the Debye layer. However, the red curves relating to injections of streptavidin (10 nM) show a significant decrease in I_{DS} the more diluted the buffer solution is. The highest I_{DS} values (~0.95 μA) are obtained with the highest dilution of PB (0.01X), corresponding to a Debye length of 7.3 nm, longer than the biotin-streptavidin unit. The charges specific to streptavidin are then fully seen by the FET, which explains the maximum value of the I_{DS} current. The addition of TCEP (tris(2-carboxyethyl)phosphine hydrochloride) releases streptavidin into the solution, which restores the initial I_{DS} current value, characteristic of the biotin-modified FET (I_{DS} ~ 0.8 μA). (PB: phosphate buffer solution with pH 7.4 and ionic concentration of around 0.15 M; V_{DS} = 2 V, V_{GD} = –35 V; the arrows pointing up and down in the figure indicate, respectively, the moment when PB solutions of different dilutions with and without streptavidin are injected). Biotin and biotin/streptavidin curves are from different devices.*

In conclusion, using FETs for direct analysis in biological environments with high saline concentration and without the need to label macromolecules (label-free detection) is hampered by an electrostatic screen. The risk is a limitation of the transduction signal and, consequently, poorer LODs.

This shielding phenomenon, which determines the effectiveness of electronic biosensors, has recently been the subject of research. Different experimental solutions for biosensors based on using FETs have been proposed. One of them, recommended by Lieber and team, attractive for its simplicity, involves adsorbing polyethylene glycol (PEG) onto the surface of silicon nanowires, which artificially increases the thickness λ_D of the Debye layer and thus allows the direct analysis of biological molecules in their original physiological environment (Gao et al. 2015) (Figure 4.12).

Si nanowires functionalized with APTES molecules (3-aminopropyl) triethoxysilane) and PEG, included in an FET configuration, made it possible to directly detect the prostate-specific antigen (PSA) in a physiological environment. PSA is approximately 2 nm in size, much greater than the Debye length in the case of buffer solutions and physiological media, for which the electrolyte concentrations are around 0.1 M.

The effect due to PEG is spectacular. With a PET functionalized only with APTES, no transduction signal appears after the addition of the buffer solution containing PSA (Figure 3.14(c)). On the contrary, in the case where FET SiNWs are modified by APTES and PEG, we observe a significant transduction signal, and a good linear correlation of the FET signal with the logarithm of the PSA concentration for a concentration range between 10 and 1000 nmol/L[9]. The signal latency time is quite long (around 8–10 min) caused by the slowing down of the diffusion of PSA through the PEG layer. According to the authors, the effect of an increased λ_D could be due to a decrease in the dielectric constant in the area where the polymer is located.

A similar effect would also be obtained by placing multilayer films of polyelectrolytes in an FET-type biosensor with graphene as a sensitive layer (Piccinini et al. 2018). Based on a thermodynamic model, the authors predict that the effective thickness of the Debye layer can be increased by an order of magnitude and reach lengths of 10 nm for physiological solutions for which the length of

9 This value is very high. PSA has a molecular mass of around 25 kDa, which corresponds to approximately 250 ng/mL for a concentration of 10 nM. The average value of PSA in the blood is around 3 ng/L and the sensitivity of the technique appears insufficient for such measurements. It should be compared to techniques that use an OECT, for which a PSA detection threshold of 1 pg/mL has been achieved (Kim et al. 2010).

Debye is generally around 0.8 nm. However, the interpretation is controversial and other hypotheses have been proposed, as discussed in an article by Kesler et al. (2020).

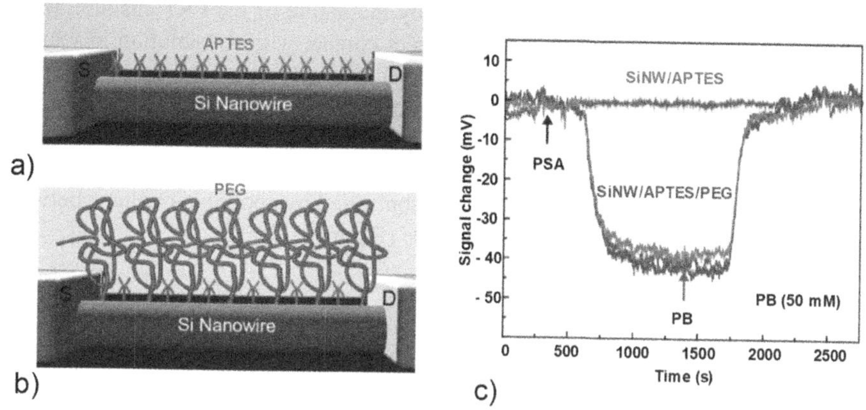

Figure 4.12. *Attenuation by PEG of the electrostatic screen effect experienced by an FET. (a) Diagram of the FET comprising SiNWs functionalized with APTES. (b) Diagram of the FET after attachment of APTES and PEG. (c) Variation of the transduction signal of three variously functionalized FETs after injection of PSA. Adapted from Gao et al. (2015). For a color version of this figure, see www.iste.co.uk/lacaze/nanotechnologies.zip*

COMMENT ON FIGURE 4.12.– *The purple, red and blue curves (diagram c) are obtained after injection of PSA (100 nM, black arrow), where APTES and PEG are grafted onto SiNWs. After washing the FET (green arrow) with the PB buffer solution (pH 6), the transduction signal returns to its initial value. We note the absence of a transduction signal when the SiNWs are modified only with APTES (green curve); the I_{DS} transduction signal is transformed into mV.*

It is also worth mentioning that experiments have been performed with local desalting of the solution near the FET sensitive layer to recreate low ionic strength conditions that increase the Debye length. This type of device has been proposed in the case of micro-sensors where the solution to be analyzed is a microdroplet deposited on the gate insulator. By placing two electrodes with opposite polarizations (–V/2 and +V/2) at both ends of the droplet and outside the gate insulator, cations and anions relocate around these electrodes and considerably lower the ionic concentration in the center of the microdroplet, redefining a Debye length characteristic of a solution with high ionic dilution (Dak et al. 2016). This option, however, carries a risk in the case where the analysis involves biological molecules, which can denature in a desalinated environment.

4.3.2.2. *Electrochemical organic transistors (OECTs)*

Early OECTs date back to the mid-1980s, with the creation of the first electrochemical transistor by Wrighton et al. (1984). The general diagram of an OECT differs from that of a classic OFET by the nature of the FET channel, which is a thin layer of conductive polymer in direct contact with the solution in which a reference electrode (Ag/AgCl type) is immersed and acts as a metal gate (G). Unlike the case of a bio-FET, where the channel conductivity variation is induced by an electrostatic field across an insulator, an OECT has no gate insulator. The OECT channel comprises a thin layer of conductive polymer, whose conductivity variations depend on its oxidation state, which can be controlled by applying a voltage between the source electrode and a reference electrode (Figure 4.13(a)).

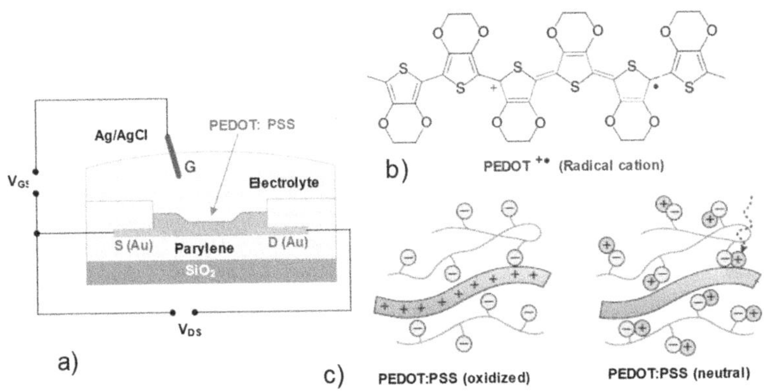

Figure 4.13. *Typical diagram of an OECT operating with a conductive polymer as the channel. (a) Structure of the transistor showing the PEDOT:PSS film deposited on a support (parylene-SiO$_2$) in direct contact with the electrolyte and the electrodes S and D. (b) Structure of PEDOT in the oxidized state of radical cation. (c) Representation of the two states of PEDOT:PSS. (a–c) Adapted respectively from Khodagholy et al. (2013) and Rivnay et al. (2018). For a color version of this figure, see www.iste.co.uk/lacaze/nanotechnologies.zip*

COMMENT ON FIGURE 4.13.– *The oxidation or reduction of PEDOT is carried out by applying a polarization between S and the reference electrode (Ag/AgCl), which acts as a gate. With the PSS chains being entangled with those of PEDOT, when PEDOT is reduced to the neutral state, the negative charges carried by the PSS polymer are then compensated by the cations of the solution (orange-colored charges) to maintain the overall neutrality of PEDOT:PSS. Parylene is deposited to serve as a mechanical support and isolate S and D from the solution.*

The best results are obtained with poly(3,4-ethylenedioxythiophene)-PEDOT, combined with poly(styrenesulfonate)-PSS. The PEDOT:PSS couple is quasi-insulating in the neutral reduced state and conductive in the oxidized state and can reach a conductivity of more than 1,300 S/cm after treatment with an alcohol (Alemu et al. 2012).

Such a system has a very high transconductance (g_m)[10], which sets it apart from other field effect devices. Khodagholy et al. (2013) showed that an FET of this type, with a structure corresponding to that of Figure 4.14, can reach a g_m value of 4 mS (milliSiemens) for V_G values between 0 and 0.6 V in the case of a channel with a thickness of 400 nm and length and width of 10 μm.

These high transconductance values together with high gain (I_{ON}/I_{OFF}) values, of around 10^5, have enormously contributed to the widespread use of this type of sensor, which subsequently gave rise to flexible systems that can be used for the detection of certain physiological functions such as the activity of the brain, heart or living cells. Its use has also become widespread for assaying biomolecules (DNA, enzymes, proteins, glucose, etc.) after the adequate functionalization of the channel or the gate of the transistor (Rivnay et al. 2018; Liang et al. 2018; Liao et al. 2019). In the case of a PEDOT channel, its direct functionalization, however, requires a more or less complex synthesis preparation; a copolymerization between simple EDOT and an EDOT functionalized by a chain, capable of reacting with an antibody, is a frequently used technique, allowing an immunological type of assay.

4.3.2.3. *Organic electrolyte-gated OFETs (EGOFETs)*

With this type of biosensor, the gate insulator, which is typical in FETs, is also replaced by the electrolyte in direct contact with the SC, like in an OECT. Unlike OECTs, which are characterized by a channel made of a conductive polymer whose conductivity depends on its oxidation state, the conductivity of the EGOFET channel is modulated by the electric field resulting from the accumulation of ions at the [electrolytic solution/SC channel] interface (Figure 4.14).

10 The transconductance g_m of a transistor, measured by the ratio $\Delta I_D/\Delta V_G$, represents the gain of the transistor for a given value of the voltage V_{SD}. It is a function of the size and geometry of the channel and characterized by the ratio g_m/W, where W is the width of the channel. For the PEDOT/PSS channel cited in the text, its maximum value is 402 S/m, while transistors manufactured with a channel made of Si nanowires and in contact with a buffer solution only reach 0.25 S/m. The presence of the electrolyte is in fact responsible for the low value of g_m which, in the absence of electrolyte, is much greater around 200 S/m. The highest g_m/W values are obtained for all-solid transistors and in the case of SC III/V, for which g_m/W reached 1950 S/m with n-In/As nanowires (Khodagholy et al. 2013).

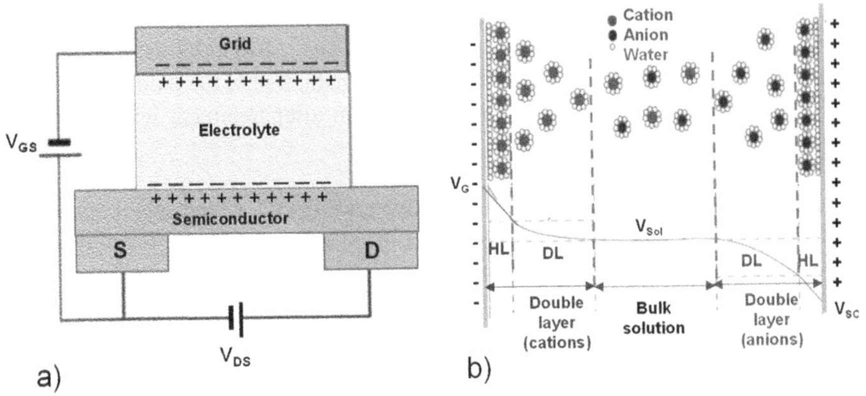

Figure 4.14. *Representative diagram of an EGOFET in the "Top-Gate" configuration. (a) Charge distribution at the gate/electrolyte and electrolyte/SC interfaces for a negative gate polarization. (b) Potential profile between the grid (V_G) and SC (V_{SC}). Adapted from Wang et al. (2016b) and Kergoat et al. (2012). For a color version of this figure, see www.iste.co.uk/lacaze/nanotechnologies.zip*

COMMENT ON FIGURE 4.14.– *A double layer of ions of opposite signs characterizes each interface. This double layer comprises a thin, compact layer corresponding to an ion adsorption layer (Helmholtz layer [HL]), followed by a diffusion layer (Debye layer [DL]) whose thickness depends on the ionic concentration. The inner solution is electrically neutral, and the potential variation appears to be entirely localized at the interfaces.*

Compared to OFETs comprising an insulating gate dielectric, the presence of the electrolyte means that the overall capacitance of an EGOFET is much higher (1,000 times higher) and results from the strong double-layer capacitances at the gate/electrolyte and electrolyte/SC interfaces (Wang et al. 2016b). This means that for a fixed gate potential, any change in the overall capacitance of the [gate/electrolyte/SC] assembly can modify the charges in the channel and, consequently, significantly vary the drain current.

The sensitivity of the device is based on the variations in capacitance induced by the presence of a target molecule adsorbed either onto the metal grid or the SC. The device was tested by Kergoat et al. (2012b) for human DNA strands of the HIV (human immunodeficiency viruses) virus using a composite consisting of two polythiophenes as the channel, one substituted by a hexyl chain (P3HT), the other by a 5-carboxypentyl chain (P3PT-COOH), onto which an oligonucleotide (comprising around 30 purine bases) complementary to HIV DNA can be grafted.

The absence of a gate insulator makes it possible to obtain variations in the drain current for very low gate bias values, <1 V. The device can also operate without an electrolyte, for example, when depositing a drop of water containing the HIV target onto the polythiophene film. The response is better than a saline solution, and the sensitivity, defined from the grafting density, can reach 25 pM/cm^2.

4.4. Optical and optoelectrochemical detection methods

Exploiting the optical properties of molecules has been the source of a very large number of assay techniques for a long time, the oldest of which are based solely on the exploitation of their absorption and emission properties and, more recently, on properties relating to a combination of electrochemistry and photochemistry. Fluorescence and surface-enhanced Raman scattering (SERS), SPR, as well as combined systems based on the interaction between plasmons and excitons, obtained with metallic NPs and quantum dots (QDs), are purely photonic processes used for the detection of compounds in very low concentrations.

A new trend has emerged combining electrochemistry and luminescence techniques for analyzing chemical compounds. PEC and ECL are the two representative techniques that differ by the nature of the signal detected. With PEC, the measured signal is an electric current caused by light excitation, while for ECL, the transduction signal is light emission induced by an electrochemical electrode reaction. These methods have very high detection sensitivities, with essential applications in the biomedical field, particularly for carrying out immunological assays.

4.4.1. *Luminescence of molecules*

The excitation of the electrons in a molecule (or atom) by a photon leads to an excited state of the molecule, followed immediately by light emission, which can lead to fluorescence or phosphorescence, two different types of luminescence. However, it is fluorescence that is most used for analytical purposes.

4.4.1.1. *Energy diagram of fluorescence and phosphorescence emissions*

In the case of fluorescence, the emission is spontaneous and stops as soon as the excitation is interrupted. On the contrary, with phosphorescence, light emission occurs after an electron conversion from the singlet excited state $^1S^*$ toward the lower energy triplet $^3T^*$ state. This conversion is followed by the emission of phosphorescence and return of $^3T^*$ to the initial singlet state $^1S_0$. Taking into account the selection rules that prohibit transitions of different electron multiplicities, spin-orbit coupling nevertheless allows inter-system crossing (ISC) $^1S^* \rightarrow {}^3T^*$, followed by the very slow return of $^3T^*$ to $^1S_0$, which explains why the emission of phosphorescence is characterized by long emissions that persist even after light excitation has stopped (Figure 4.15).

Not all molecules are either fluorescent or phosphorescent. They are generally molecules with strongly conjugated rings, which confers rigidity to their excited state, and favors radiative deactivation to the detriment of thermal vibrational deactivation. Such molecules are called fluorophores.

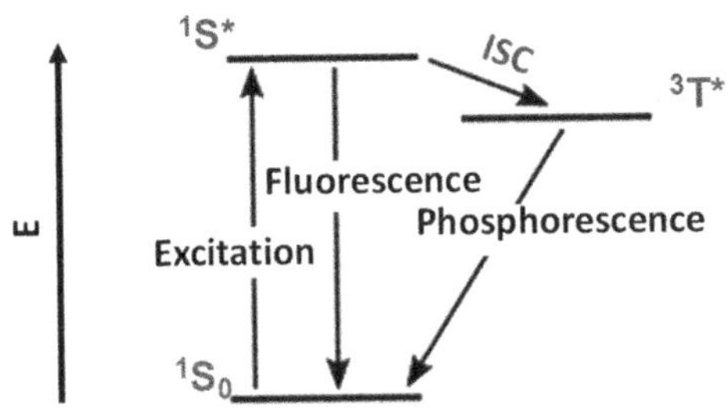

Figure 4.15. *Jablonski diagram indicating the energetic origin of fluorescence and phosphorescence. $^1S_0$ represents the energy state of the fundamental singlet state (S = 0), $^1S^*$ that of the first excited state (S = 0), and $^3T^*$ that of the triplet state (S = 1), S designating the electron multiplicity of the molecule. For a color version of this figure, see www.iste.co.uk/lacaze/nanotechnologies.zip*

Fluorescence spectroscopy is an extremely powerful analysis technique, given its very high detection sensitivity due to the emission of photons on a dark background. A photomultiplier (PM) converts this photon emission into an easily measurable flow of electrons. This technique has been widely used from the study of electron transfers in photosynthetic systems (Kondo et al. 2017), to food monitoring (Hassoun et al. 2019), to the detection of specific reactive oxygen species (ROS) or reactive nitrogen species (RNS) (Chen et al. 2011), but also to that of biomarkers, proteins and ions (Dou et al. 2022).

4.4.1.2. *Fluorescence exploitation*

In addition to the direct emission of a fluorophore, fluorescence emission can also occur after an energy transfer between molecules, corresponding to a FRET (fluorescence resonance energy transfer [RET]) process. Such an effect can occur when two fluorophore groups (or a fluorophore and a non-fluorophore) interact, one being an electron donor (D) and the other an electron acceptor (A). An energy transfer can then occur between D and A, which makes the molecule A fluoresce after the excitation of D, which transfers the acquired energy to A. This energy transfer originates from a dipolar interaction between D and A and can occur up to distances of 3–6 nm when the donor emission spectrum overlaps 30% of the acceptor emission spectrum (Sekar and Periasamy 2003).

This phenomenon has been widely used for the in vivo study of proteins or nucleic acids using gold NPs coupled to a fluorophore, according to the OFF-ON fluorescence principle. In this device, the fluorophore is in very close proximity to the non-fluorescent acceptor group (gold NP), which prevents photon emission from the fluorophore due to deactivation and energy transfer from the fluorophore to the gold NP. The addition of a compound to the medium which, by molecular recognition, moves the fluorophore away from the NP, prevents the transfer of energy toward the NP and restores fluorescence, hence the name OFF-ON fluorescence, used to detect a DNA target (Figure 4.16).

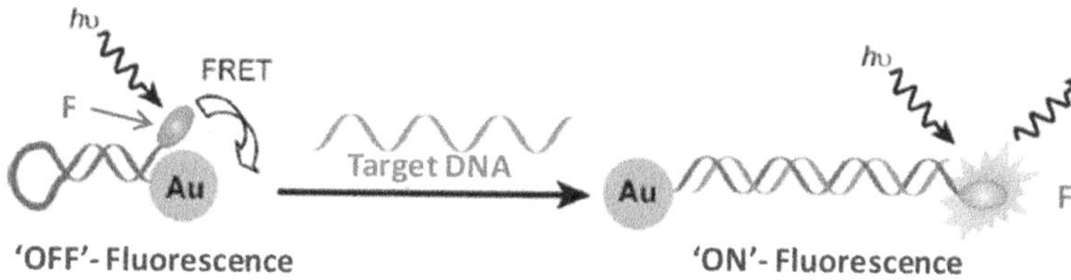

Figure 4.16. *Detection of a nucleic acid (DNA) by OFF-ON fluorescence. Adapted from Saha et al. (2012). For a color version of this figure, see www.iste.co.uk/lacaze/nanotechnologies.zip*

COMMENT ON FIGURE 4.16.– *Fluorescent F-labeled nucleic acid (green) with complementary sequences can self-hybridize, giving rise to a hairpin structure. The hairpin structure, with a gold NP attached at one end and a fluorescent group F at the other, means that the fluorophore F and the gold NP are very close to each other. This favors energy transfer between the fluorophore and the gold NP. The fluorophore's excitation is then turned off by the gold NP ("OFF" fluorescence). A target DNA strand, complementary to the nucleic acid F, causes, after hybridization, the fluorophore F to move away from the NP by undoing the "hairpin" structure. This suppresses the FRET energy transfer and restores the fluorescence of F ("ON" fluorescence).*

Other combinations, also involving energy transfers between molecules, have been carried out to design ratiometric type sensors, the principle of which involves exploiting two fluorescence signals, correlated with each other by the presence of an analyte. The method, known as ratiometric fluorescence (RF), was recently used by

Liang et al. (2022) to measure alpha-fetoprotein (AFP)[11], by adopting a procedure inspired by ELISA-type immunoassays.

Various interactions between the analyte and the fluorophores can be considered, with the principle of an energy transfer induced by the analyte, which quenches the fluorescence of one of the sensor components without modifying the fluorescence of the other.

A biosensor of this type was produced by Liang et al. (2022), by developing a nanostructure made up of SiO_2 NPs, comprising two fluorophores, carbon dots (CDs) and CdTe quantum dots (CdTe-QDs). The CDs (4–5 nm in size) are dispersed inside the SiO_2 NPs (around 200 nm), while the CdTe-QDs are distributed on their surface. Subjected to UV light excitation, the composite emits double fluorescence, one at 450 nm due to the CDs and the other at 624 nm due to CdTe. The trick with this setup is that, unlike the CdTe emitter on the surface of the SiO_2 NPs, the CDs are not exposed to any external energy transfer. The CdTe-QDs can therefore be quenched directly by an analyte, while the CDs inside SiO_2 NPs are not affected by the analyte and, therefore, continue to fluoresce at a constant intensity, emitting a reference signal.

4.4.2. *Electrochemiluminescence*

Unlike the general phenomenon of PL, which requires a molecule (luminophore) to be excited by a light source to be able to emit light, chemiluminescence (CL) and ECL proceed according to a different mechanism called annihilation, which involves reacting the oxidized and reduced forms of one or two reagents (organic or inorganic) in solution; their recombination can lead to an excited state, followed by light emission. ECL differs from CL in that the reduced and oxidized forms are produced electrochemically on an electrode. They can be produced either by using a bipolar electrode (BPE; Wu et al. 2013), or by polarizing the electrode alternately in anodic and cathodic mode, or, as we will see later, by applying a fixed polarization to the electrode in the presence of a co-reactant. It has several advantages: absence of a light source unlike PL, simplicity of the electrochemical equipment, speed of execution and high sensitivity due to luminescence. This luminescence excitation is also better than traditional fluorescence, sometimes polluted by environmental background emissions.

11 Alpha-fetoprotein (AFP) is a protein found in the blood at very low concentrations, around 10 ng/mL, in the case of healthy individuals. It is mainly produced in the liver. High concentrations of AFP can be indicative of cancer and so AFP is a marker for liver cancer. In pregnant women, it can also be an indicator of poor fetal development.

Observed since the 1960s by Hercules in the case of 9,10-diphenyl anthracene (DPA) (Hercules 1964), then taken up by Bard et al. with a ruthenium salt (Ru^{2+}) complexed by 2,2′-bipyridine ($Ru(bpy)_3^{2+}$) (Tokel and Bard 1972), the technique then developed very quickly and gave rise to many systems, described in several review articles (Richter 2004; Miao 2008; Deng and Ju 2013; Liu et al. 2015; Li et al. 2017; Liu et al. 2022a). Labeling biological molecules with emitters that give rise to ECL has made it possible to perform assays of many biological compounds, with a detection sensitivity of around 10 pmol/L (10^{-11} mol/L). This can be further improved using hybrid combinations of emitters with various NMs. This set of qualities has justified its use for the determination of biomolecules, especially proteins and nucleic acids, bacterial pathogens and ionic species (Shen et al. 2019).

With nucleic acids, the possibility of direct analysis is an advantage in terms of speed and simplicity, making it possible to avoid the preliminary and lengthy PCR-type amplification steps, which can only be carried out with specialized equipment. Another advantage of this technique is the straightforward design of miniaturized devices, such as micro- or electro-paper-based devices (μ and e-PADs) (Chinnadayyala et al. 2019). Since its initial implementation, numerous technical improvements have been proposed, including the use of new emitting materials adapted to POC-type devices, but also due to the widespread involvement of NMs.

4.4.2.1. *Basic principle of ECL*

ECL light emission was initially demonstrated by applying an alternating voltage to an electrode, in such a way as to successively carry out the electrochemical oxidation and reduction of an emitter E, thus producing radical-type species at the same site resulting from the loss or gain of an electron ($E^{\bullet +}$ or $E^{\bullet -}$). Exergonic recombination (negative free enthalpy variation ΔH) between $E^{\cdot +}$ and $E^{\cdot -}$ gives back two neutral forms E but in an excited state E* at the origin of the emission. Depending on the magnitude of the annihilation reaction enthalpy of $E^{\cdot +}$ and $E^{\cdot -}$, the excited state may be a singlet state $^1E^*$ when the enthalpy ΔH of reaction is higher than the energy of $^1E^*$, or a triplet state $^3E^*$ for lower ΔH values but still higher than that of the triplet state. In the first case, emission corresponds to direct deactivation of $^1E^*$ toward the fundamental state $^1E_0$, whereas in the other case, deactivation occurs transiently via an annihilation reaction of two triplet states, leading to $^1E_0$ and $^1E^*$.

Such a process was carried out for the first time using aromatic derivatives as emitters, such as DPA. This type of compound unusually produces relatively stable cation and anion radicals in an organic aprotic medium by oxidation and reduction, with their annihilation causing luminescence (Hercules 1964).

Another widely studied example of ECL is that of the luminophore $[Ru(bpy)_3]^{2+}$ corresponding to the Ru salt complexed with 2,2' bipyridine. Its drawback is that the oxidation and reduction potentials (1.2 and –1.4 V/ECS, respectively) are too high to be applied in an aqueous medium, which implies the use of an aprotic organic solvent such as dimethylformamide (DMF) or acetonitrile (CH_3CN), unsuitable for biological applications.

One way to overcome this drawback is to use a co-reactant whose oxidation or reduction in an aqueous medium contributes to the formation of excited-emitting species. Peroxydisulfate $S_2O_8^{2-}$, tripropylamine (TPA; $CH_3CH_2CH_2)_3N$) and 2-(dibutylamino)ethanol (DBAE) are appropriate co-reactants. The first can be reduced and give the anion radical $SO_4^{\bullet-}$, whereas TPA and DBAE are oxidized to cation radicals $TPA^{\bullet+}$ and $DBAE^{\bullet+}$, both strong oxidants producing ECL with Ru^{2+} salt.

Thus, in the case of the TPA co-reactant associated with the luminophore $Ru(bpy)_3^{2+}$, a series of reactions lead to the ECL emission:

$$Ru(bpy)_3^{2+} - e \rightarrow Ru(bpy)_3^{3+}$$

$$TPA - e \rightarrow TPA^{\bullet+} \rightarrow TPA^{\bullet} + H^+$$

$$Ru(bpy)_3^{3+} + TPA^{\bullet} \rightarrow Ru(bpy)_3^{2+*}$$

$$Ru(bpy)_3^{2+*} \rightarrow Ru(bpy)_3^{2+} + h\upsilon$$

Note that for such a process, the light emission only involves oxidation reactions with simultaneous regeneration of the luminophore in its initial state, which avoids the alternation of anodic and cathodic potentials at the electrode provided for a direct annihilation reaction in the case of the Ru^{2+} luminophore. Furthermore, the example of TPA – widely used – can be generalized to other amino compounds, which has thus enabled the direct determination by ECL of many biological nitrogen compounds behaving as ruthenium co-reactants.

Luminol (5-amino-2,3-dihydrophtalazine-1,4-dione) is also a widely used luminophore, initially known for its chemiluminescence properties[12], which can also produce ECL, but according to a different mechanism, resulting from the electrochemical transformation of luminol into a derivative of orthophthalic acid, which is responsible for the ECL emission. Note, however, that a high emission

12 Luminol emits a blue glow by chemiluminescence when mixed in a basic medium with an oxidant (H_2O_2) and a catalyst such as ferrous ion (Fe^{2+}). It is used during forensic investigations due to its ability to react on surfaces impregnated with traces of blood, and to emit a blue glow, visible in the dark, due to the catalytic action of ferrous ion in hemoglobin.

intensity is only obtained in an alkaline environment, which excludes all applications from biological compounds. Furthermore, the reproducibility of the emission seems to depend on many external factors, which makes it unreliable. Nevertheless, a clear improvement was observed when the metallic surface of the electrode was replaced by an NP assembly of the same metal (Cui et al. 2004).

The quantum yields of the ECL emission[13] for all these systems are comparable to those obtained by PL. In the case of the chelate $Ru(bpy)_3^{2+}$ without a co-reagent, it is 5% and varies depending on the nature of the ruthenium complexing agents. So, with 4,4′-diphenyl-2,2′ bipyridine [$Ru(dp\text{-}bpy)^{2+}$] or 4,7-diphenyl-1,10 phenanthroline [$Ru(dp\text{-}phen)^{2+}$], the quantum yields are higher and are 14% and 24%, respectively (McCord and Bard 1991).

Subsequently, new emitters based on heavy metals, such as iridium (Barbante et al. 2014) as well as innovative assemblies of carbon and inorganic nanocomposites, were also tested and led to many ECL devices being listed in several review articles (Miao 2008; Liu et al. 2015, 2022a).

4.4.2.2. *Impact of NMs on ECL*

Since the 2000s, the replacement of a conventional electrode with an NP assembly of the same metal has brought considerable improvements in ECL, which is attributed to an increase in the specific surface area of the electrode on which ECL occurs (Cui et al. 2004).

Subsequently, the incorporation of various NMs became widespread with the formation of nanocomposites. The use of carbon nanodots (NDs), gold nanoparticles (NPs) and inorganic QDs, such as CdSe, CdS, CdTe, PbS, ZnSe and Ag_2S (Wu et al. 2014; Chen et al. 2018; Yang et al. 2022), led to the production of new immunobiosensors with very low detection limits, compatible with the usual concentrations of nucleic acids or those of biomarkers of specific diseases. In the same way, carbon nanocomposites based on graphene (G), carbon nitride ($g\text{-}C_3N_4$) (Zou et al. 2020) and two-dimensional metal carbides (MXene) (Liu et al. 2022a) have also been used to produce ECL biosensors.

Two modes of operation were considered; a device for which the emission intensity increases or decreases with the concentration of the analyte, a process which, in the latter case, involves the extinction of the emission due to a transfer of energy between a donor and an acceptor-emitter (Liu et al. 2015).

---

13 The quantum efficiency of an ECL process is defined as the ratio between the number of photons emitted and the number of electrons consumed by electrolysis.

4.4.2.2.1. ECL based on inorganic QDs and carbon nanocomposites

Using QDs as fluorescent markers to detect biological compounds was first done by Bruchez et al. (1998) at the end of the 1990s. It was only a few years later that the chemiluminescence of QDs induced by electrode polarization was observed by Poznyak et al. (2004).

ECL is typically carried out by applying a cathode or anode potential in the presence of a reducible or oxidizable co-reagent C to create an electron-hole pair in the QD responsible for ECL emission (Figure 4.17).

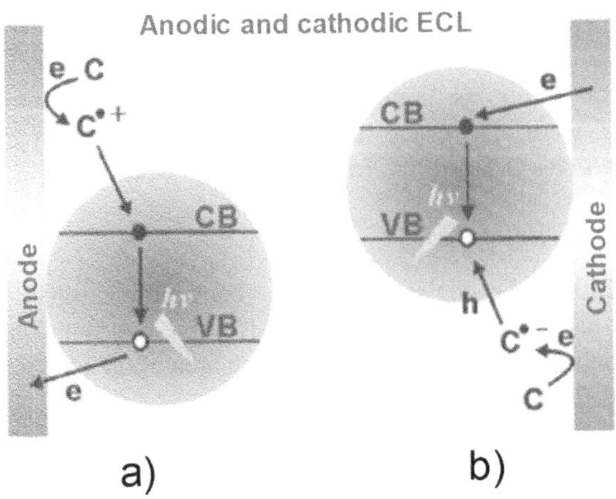

Figure 4.17. *Principle of the ECL produced by a QD and a co-reagent (C) in contact with an electrode operating as an anode (a) or cathode (b). Adapted from Zhao et al. (2015). For a color version of this figure, see www.iste.co.uk/lacaze/nanotechnologies.zip*

COMMENT ON FIGURE 4.17.– *In both cases, the emission results from the formation and recombination of an electron-hole pair in the QD; the electron (red dot) and the hole (red circle) are located, respectively, in the lowest conduction band (CB) level and the highest valence band (VB) level. The electron–hole pair formation requires the simultaneous injection of an electron and a hole into the QD. In the case of anodic polarization (diagram a), the emission of an electron from the VB to the electrode produces a hole in the VB of the QD. In contrast, the electron released by the oxidation of C is injected into the CB, which leads to the formation of the electron–hole pair and hν emission. The roles of the electrode and the co-reagent are reversed in the case of cathode polarization (diagram b).*

Many biomolecule assays using QDs are carried out using the ON-OFF technique, for which the non-luminescent compound for assay (C) intervenes by quenching the emission (quencher) due to an energy transfer from the excited fluorophore to the C. Such an effect was observed with CdTe QDs in the analysis of dopamine and L-adrenaline (Liu et al. 2007). The Stern–Volmer law[14], characteristic of an extinction effect of the analyte, is well followed for a dopamine concentration range between 0 and 5 μM.

Numerous nanocomposites made of graphene, carbon nanotubes, carbon nitrides or even MXenes mixed with metal NPs have also been used to produce ECL devices. Among these, a new graphitic SC material, with the generic formula g-C_3N_4[15] prepared as nanosheets (like QDs), has recently proven to be an excellent base material to produce ECL-emitting nanocomposites, for reduction and oxidation, and in the presence of reducible or oxidizable co-reagents. It can be adsorbed as a solid onto a glassy carbon electrode (GCE), then easily functionalized by metallic NPs, in particular gold and silver (Au-NPs and Ag-NPs).

The addition of gold NPs to C_3N_4 has a beneficial effect on its conductivity and on the quality of the ECL emission, whose lifetime is significantly improved compared to the material alone, due to the suppression of C_3N_4 passivation phenomena responsible for weakening of the emission intensity. The functionalization of g-C_3N_4 by gold NPs is also a means of facilitating the grafting of numerous antibodies involved in immunological-type assays.

A device combining nanosheets of g-C_3N_4 and gold NPs (AuNPs/g-C_3N_4) was made by Chen et al. (2014) for the ECL assay of a cancer biomarker, carcinoembryonic antigen (CEA) (Figure 4.18).

The biosensor is made of 4-nm thick g-C_3N_4 nanosheets with attached gold NPs (5 nm), directly deposited on the nanosheets by chemical reduction of a chloroauric acid solution ($HAuCl_4$). After adsorption of the AuNPs/g-C_3N_4 nanocomposites onto GCE, the CEA antibody (Ab_{CEA}) is then attached to the gold NPs, the free surface of the electrode is blocked by bovine serum albumin (BSA), which completes the construction of the biosensor. This can be considered a model for an

14 The Stern–Volmer law is represented by the equation $I_0/I = 1 + Ksv\,[Q]$, where [Q] represents the concentration of the quencher, I_0 and I are the respective intensities of luminescence emission in the absence or presence of Q, and Ksv is the Stern–Volmer constant.

15 g-C_3N_4 is a low-cost material whose preparation is easy; its water solubility (considerably improved when it is in nanometric form) and its biocompatibility generate many assay possibilities, applying to both inorganic and biological compounds, described and listed exhaustively in recent reviews (Zou et al. 2020; Liu et al. 2022a).

immunobiosensor. The ECL emission is maximum when applying a voltage of –1.17 V/Ag-AgCl in a slightly basic aqueous medium (pH 7.4) and in the presence of the co-reagent $S_2O_8^{2-}$, which corresponds to the simultaneous formation of radical anionic species $SO_4^{\bullet -}$ and g-$C_3N_4^{\bullet -}$. The anion $SO_4^{\bullet -}$, with strong oxidizing power reacts with $C_3N_4^{\bullet -}$ to give SO_4^{2-} and the excited state g-C_3N_4* causing a blue ECL emission at 450 nm.

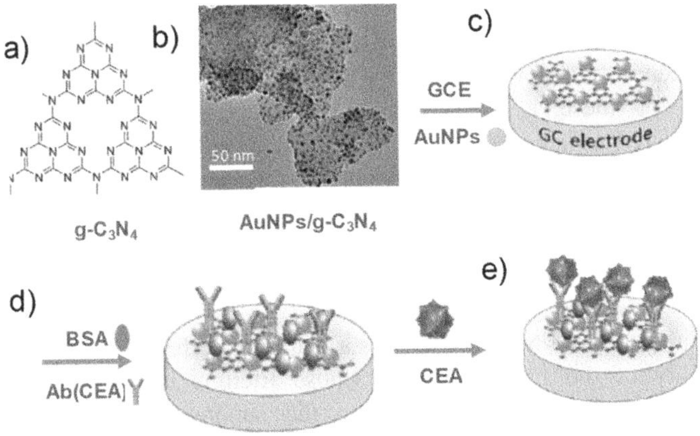

Figure 4.18. *Different development stages of the immunobiosensor ECL g-C_3N_4-AuNPs-BSA-Ab$_{CEA}$. (a) g-C_3N_4 sheet. (b) TEM image of a g-C_3N_4 sheet with attached AuNPs. (c) Adsorption of the g-C_3N_4-AuNPs sheet onto a glassy carbon electrode (GCE). (d) Blocking of the free surface by BSA and anchoring of the Ab(CEA) antibody to gold NPs. (e) Capture of the CEA antigen by the Ab antibody (CEA). Adapted from Chen et al. (2014). For a color version of this figure, see www.iste.co.uk/lacaze/nanotechnologies.zip*

This ECL emission detected by a PM can be visualized by applying to the electrode several potential sweep cycles covering the emission domain. With a sufficiently high scan rate (100 mV/s), a series of periodically spaced peaks representative of the ECL emission are obtained (Figures 4.19(a) and(i)).

We note that the addition of CEA has an inhibitory effect on the emission which, according to the authors, is attributed to an increase in the biosensor impedance, resulting from the accumulation of proteins on the surface of the electrode. This ECL intensity inhibition effect causes a linear variation in $(I_0 - I)/I_0$ as a function of the logarithm of the CEA concentration (Figure 4.19(b)), where I_0 and I are the respective intensities measured in the absence and presence of CEA.

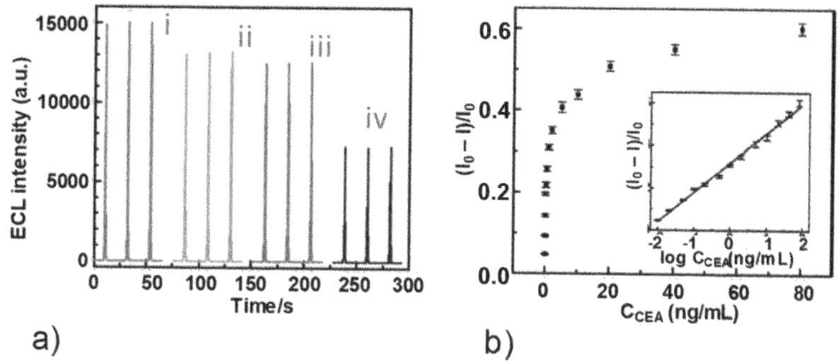

Figure 4.19. *ECL characteristics of the biosensor [AuNPs/g-C_3N_4/GCE] for detecting the CEA biomarker. (a) (i) ECL of [AuNPs/g-C_3N_4/GCE] nanosheets; (ii and iii) respective ECL of the nanosheets after adsorption of the antibody Ab_{CEA} and then BSA. (iv) ECL of the [CEA/BSA/Ab_{CEA}/AuNPs/g-C_3N_4/GCE biosensor after adding 10 ng/mL of CEA. The ECL intensities measured in (i–iv) are obtained during a cyclic voltage sweep between 0 and –1.1 V versus Ag/AgCl at 100 mV/s. (b) Relative variation in ECL inhibition as a function of CEA concentration. Adapted from Chen et al. (2014). For a color version of this figure, see www.iste.co.uk/lacaze/nanotechnologies.zip*

The range of linearity between the ECL emission and the CEA concentration is between 0.02 and 80 ng/mL, which perfectly corresponds to the expected CEA concentrations in cancer pathologies. The device's simplicity and detection specificity, proven when the biomarker is in the presence of other biological molecules, is an important asset which, according to the authors, is competitive with ELISA-type immunoassays and is already used in analytical laboratories. Such a device establishes a direct relationship between a concentration and a signal. This method, which has the merit of simplicity, does not avoid possible interferences from the environment that could disrupt the signal and distort the measurements. Hence, a different trend, applied in different fields of analytical chemistry, is used; it involves designing a ratiometric biosensor that can emit two signals directly correlated to the target concentration, notably reducing the risks of detecting false positives or false negatives.

4.4.2.2.2. Ratiometric ECL biosensors

The design of ECL biosensors that meet these criteria has therefore recently evolved, leading to the development of systems that generally include two emitters that operate according to different mechanisms. Some are identified with energy transfer mechanisms involving coupling of the two emitters, while others are based on the responses of two independent emitters, each generating an ECL at different potentials. In the latter case, biosensor operation can be achieved either by programming the two potential values by potential scanning on the same electrode (Zhang et al. 2018a), or by using a BPE that delimits two zones on its surface, anodic and cathodic, respectively (Gao et al. 2018).

a) Dual-wavelength ratiometric ECL biosensor

A dual-wavelength biosensor responding to RET was described by Feng et al. (2016a), more recently by Qin et al. (2021), who assembled two nanocomposites, one consisting of g-C_3N_4, (comparable to that described by Chen et al.; Figure 4.18) acting as a transmitter, and the other, based on $Ru(bpy)_3^{2+}$, acting as an acceptor-transmitter. The transmitter g-C_3N_4 is deposited on a GCE; it emits at 460 nm in the presence of $S_2O_8^{2-}$ for a negative electrode voltage. $Ru(bpy)_3^{2+}$ interacts with g-C_3N_4 and emits at 620 nm after an RET between g-C_3N_4 and $Ru(bpy)_3^{2+}$.

The biosensor, shown schematically in Figure 4.21(a), is developed in two stages. The first stage is devoted to the creation of the transmitter [g-C_3N_4/AuNPs/Ab1/BSA], like that in Figure 4.20, and the second stage is devoted to the development of the [TiO_2/Ru/AuNPs/BSA/Ab2] acceptor-transmitter.[16]

Applied to the determination of amyloid-β_{42} – a biomarker of Alzheimer's disease[17] – this biosensor establishes a linear relationship between the logarithm of the target molecule concentration and the logarithm of the ratio of light intensities emitted, one at 460 nm (characteristic of the emission of g-C_3N_4 in the presence of the co-reagent $S_2O_8^{2-}$) and the other at 620 nm due to $Ru(bpy)_3^{2+}$, resulting from a transfer of resonance energy (Figures 4.20 (b) and (c)).

16 The transmitter [g-C_3N_4/AuNPs/Ab1/BSA] comprises a primary antibody Ab1 with a high affinity for the target molecule; BSA and the gold NPs have the same functions as those described for the ECL transmitter in Figure 4.20. The acceptor-transmitter [TiO_2/Ru/ AuNPs/BSA/Ab2] is obtained by successively grafting the ruthenium chelate $Ru(bpy)_3^{2+}$ onto TiO_2 NPs and by depositing AuNPs by adsorption onto TiO_2. The target capture Ab2 antibody is then grafted onto the AuNPs and, lastly, BSA is added to block the surface of TiO_2 not covered by the previous reagents. The two elements of the biosensor are linked together via the target molecule (TM), sandwiched by the two antibodies Ab1 and Ab2, forming the [TiO_2/Ru/BSA/AuNPs-Ab2/TM/Ab1/BSA/AuNPs/g-C_3N_4] assembly (Figure 4.20(a)).

17 Amyloid β42 (A-β42) is a small protein (around 40 amino acids), located mainly in the brain. It appears to be involved in the development of Alzheimer's disease and is a marker of this disease.

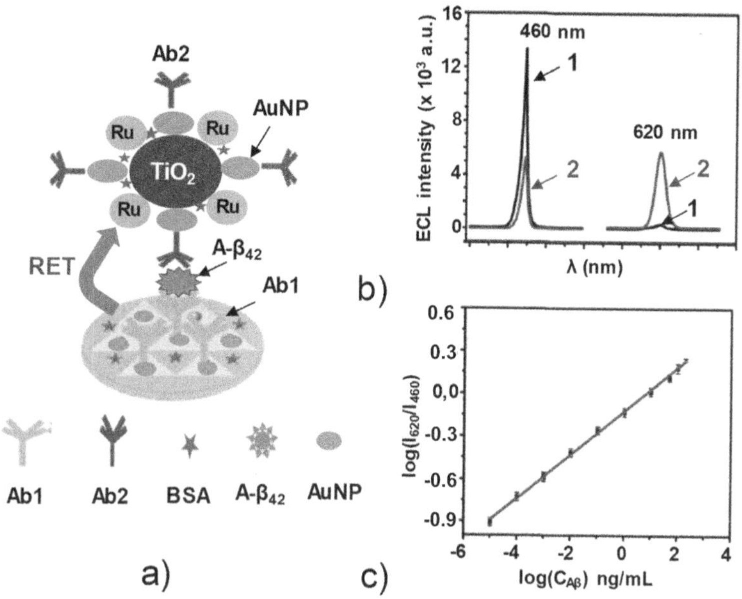

Figure 4.20. *Schematic of the dual emission ECL Immunosensor. (a) Representation of the composite assembly between the transmitters g-C_3N_4/AuNPs/BSA/Ab1 and TiO_2/Ru/AuNPs/BSA/Ab2 linked together by the target molecule A-β_{42}. Ab1 and Ab2 are two antibodies adsorbed onto gold NPs (AuNPs) and chosen for their affinity with amyloid-β_{42}. (b) The emission peaks denoted 1 and 2 (black and red), obtained as previously by cyclic sweeps of the potential correspond to the signals emitted, respectively, by g-C_3N_4/AuNPs/BSA at 460 nm and by the biosensor (TiO_2/Ru/ AuNPs/BSA/Ab2/A-β_{42}/Ab1/BSA/AuNPs/ g-C_3N_4), which emits at 460 and 620 nm. The two emissions take place simultaneously when the electrode potential is close to −1 V versus Ag/AgCl. The red peaks (2) are obtained for an amyloid-β concentration of 10 ng/mL. (c) Linear relationship between the logarithms of the A-β_{42} concentration and the ratio of emission intensities at 460 and 620 nm. Adapted from Qin et al. (2021). For a color version of this figure, see www.iste.co.uk/lacaze/ nanotechnologies.zip*

The origin of the double emission, due to the energy transfer (RET) between the transmitter and the receiver, is confirmed by the significant overlap of the emission spectrum of the donor transmitter (g-C_3N_4) and that of absorption of the acceptor transmitter AuNPs/ $Ru(bpy)_3^{2+}$. The AuNPs play an amplifying role due to the plasmon resonance, absorbing the emission of g-C_3N_4 and emitting electromagnetic radiation at a short distance from their surface, contributing to the excitation of the luminophore $Ru(bpy)_3^{2+}$. Optimizing the preparation conditions of the immunobiosensor extends the linearity range from 10^{-5} to 200 ng/mL and the limit of detection (LOD) to 2.6 fg/mL[18].

b) *Dual potential ratiometric ECL biosensor*

One of the devices uses two distinct emitters that can be excited at two different voltages in the presence of a common co-reagent. A typical example is the assembly of CdS SC nanocrystals (CdS-NCs) and luminol, emitting at negative and positive potentials, respectively, in the presence of the coreagent H_2O_2. Such a system was used by Zhang et al. (2013) to carry out the mp53 oncogene assay[19] based on the extinction of CdS emission caused by the proximity of Pt NPs (PtNPs) and the enhancement of that of luminol (Figure 4.21).

The procedure chosen, first, involves functionalizing the CdS NCs with hairpin-shaped (HP) DNA molecules, whose function is to capture the target mp53 oncogene molecule by hybridization. The composite thus formed is adsorbed onto the surface of a GCE, which gives the modified electrode [HP-CdS/GCE], characterized by a strong emission in the blue due to CdS (between −1 and −1.3 V), and an emission of luminol (emission in the red) of a lower intensity at around 0.3 V (Figure 4.21(a)).

Along with this electrode surface treatment, the luminol (Lu) and the mp-53 oncogene reactivities are also modified by grafting them onto Pt NPs. The presence of Pt NPs attached to one end of the oncogene and their functionalization by luminol has a double effect. On the one hand, the Pt NPs fixed onto the oncogene will be found close to the CdS NCs and will contribute to quenching the emission of CdS and, on the other hand, the Pt NPs covered by luminol will catalyze its light emission reaction[20].

---

18 The linearity relationship established between the logarithms of concentration of the target molecule and the logarithm of the ratio of the intensities emitted at 460 and 620 nm verifies the equation: $(I_{620}/I_{460}) = -0.1244 + 0.1534 \log C_{A\beta 42}$ with a linear correction coefficient close to unity (0.9978).

19 The mp53 oncogene is a nucleotide comprising the sequence GGCACAAACATGCAC CTCAA-(CH2)6-SH, which is frequently found in most malignant tumors and which is believed to be at the origin of the development of cancer cells. It is a mutant form (allele) of the p-53 gene, which, on the contrary, is considered a tumor suppressor and inducer of apoptosis.

20 This functionalization of Pt NPs is carried out in solution, applying the classic method of reduction using sodium citrate of a solution of $PtCl_6^-$ in the presence of luminol or oncogene. The respective sizes of the Pt NPs and the CdS NCs are 2 and 5 nm. The oncogene mp-53 is linked to Pt NPs via the SH group, obtained by cutting the S-S bond of a disulfide -SS- group contained in the molecule with the reductant TCEP (tris (2-carboxyethyl)phosphine). BSA is also added to the colloidal solution of the mp53-modified Pt NPs to block free sites on the Pt surface.

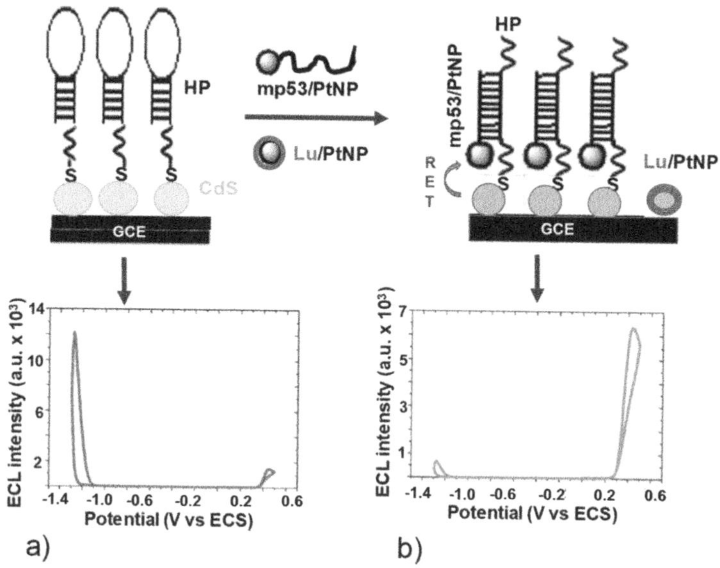

Figure 4.21. *Operating principle of the ratiometric immunosensor with two excitation potentials. (a) Before the addition of the mp-53 oncogene, a strong emission from the HP/CdS/GCE immunosensor occurs between –1 and –1.3 V while luminol gives a weak emission at 0.3 V. (b) After the addition of the mp-53 oncogene grafted onto the Pt NPs and hybridization with HP, the emission due to CdS disappeared while the emission from luminol at 0.3 V intensified. Adapted from Zhang et al. (2013). For a color version of this figure, see www.iste.co.uk/lacaze/nanotechnologies.zip*

The ECL curves obtained as a function of a cyclic voltage scan and after hybridization of mp53 with HP clearly show the quenching phenomenon, attributable to the energy transfer between the CdS nanocrystals and the Pt NPs carried by the oncogene mp53 (Figure 4.22(a)), whereas the H_2O_2-luminol couple normally emits at positive voltages, with an increase in emission due to the presence of Pt NPs. The energy transfer between the CdS NCs and the Pt NPs is also corroborated by the overlap of the Pt NP absorption curve (between 250 and 600 nm) and the CdS NC emission spectrum.

The sensitivity of the device is very high and, as for the double emission system, a linear variation is obtained between the logarithms of the mp53 concentration (between 5 fM and 1 pM) and the ratio of the light intensities of luminol and NCs of CdS (log (I_{Lu}/I_{CdS}) (Figure 4.22(b)).

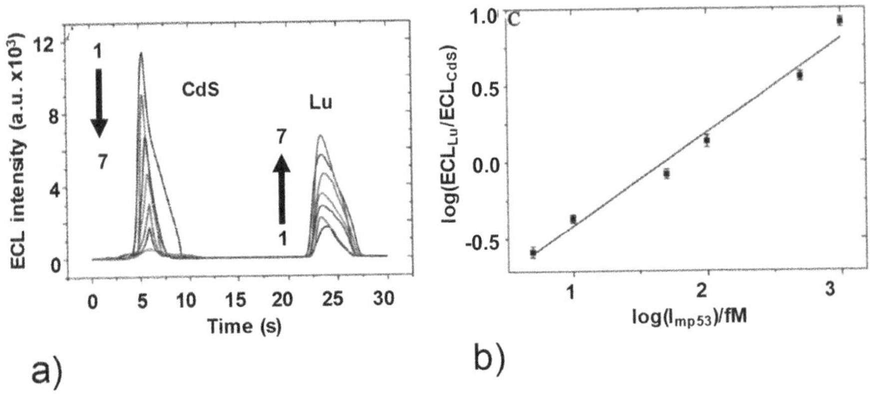

Figure 4.22. *ECL characteristics of the ratiometric device (mp53-PtNP/Lu/CdS/GCE) with double excitation potential. (a) ECL of CdS and Luminol (Lu) NCs in the presence of the coreagent H_2O_2 observed during voltage sweeps between –1.2 V and 0.5 V and for increasing concentrations of mp53 (1: 0 fM, 2: 5 fM, 3: 10 fM, 4: 50 fM, 5: 100 fM, 6: 0.5 pM, 7 :1 pM). Sweep rate of 100 mV/s. (b) Linear relationship between the logarithm of the ratio of ECL intensities and the mp53 concentration logarithm (in fM). Adapted from Zhang et al. (2013). For a color version of this figure, see www.iste.co.uk/lacaze/nanotechnologies.zip*

Another way to consider double excitation potential ECL is to use a closed BPE,[21] which locates two distinct zones on the surface of the electrode: the cathode and anode. The cathode and anode poles are physically separated, making it possible to have two reservoirs, each containing a particular electrolytic solution adapted to a specific ECL.

Such an assembly is particularly well suited to produce portable biosensors, and numerous examples of assays based on the use of BPEs have recently been described in the literature, particularly with closed BPEs (Gamero-Quijano 2019). Wang et al. (2016c) thus proposed such a device to carry out the ratiometric determination of PSA, the primary biomarker of prostate cancer (*prostate-specific antigen [PSA]*) (Figure 4.23).

21 A closed bipolar electrode can be made by placing an electrical conductor devoid of any wired connections in a channel containing an electrolyte solution and establishing a potential difference at both ends of the channel via two electrodes. Bipolar electrodes are particularly applied in the manufacture of electrolyzers and a detailed description of closed and open BPEs is given by Mavré et al. (2010) and Gamero-Quijano et al. (2019).

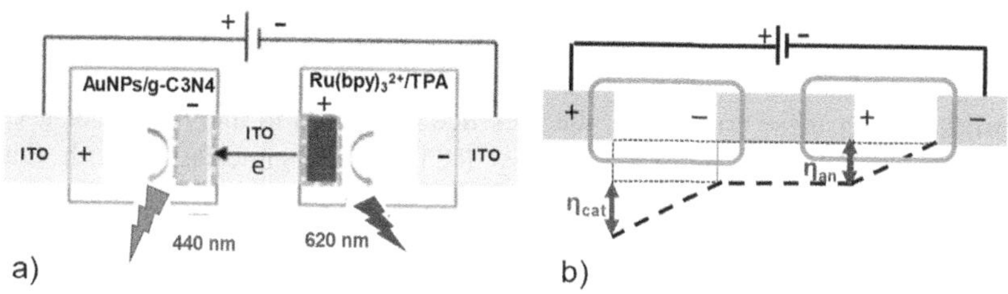

Figure 4.23. *Schematic representation of the ratiometric electrochemiluminescence device comprising a closed bipolar ITO electrode used for the determination of PSA. (a) Representation of the cathodic and anodic isopotential parts of the BPE, each in contact with a distinct electrolytic solution. (b) Potential variation (in dashed lines) applied between the two outer terminals + and – of the external electrodes. (a) Adapted from Wang et al. (2016c). For a color version of this figure, see www.iste.co.uk/lacaze/nanotechnologies.zip*

COMMENT ON FIGURE 4.23.– *The biosensor comprises three thin ITO conductive layers, one of which, in the middle, is the BPE. Insulating walls forming two volume spaces partially cover each BPE exterior zone as well as the exterior electrodes. They form "reservoirs" in which different electrolytic solutions can be placed. By applying a potential difference between the two outer electrodes, a cathode zone and an anode zone are induced at the two ends of the middle zone, although the latter is almost equipotential. The left part of the BPE is cathodic, while the right part is anodic.*

After modification of the cathode pole by the g-C_3N_4 composite, decorated with gold NPs (AuNPs/g-C_3N_4) to improve the conductivity of g-C_3N_4, and bringing the anodic zone into contact with a solution containing the luminophore $Ru(bpy)_3^{2+}$ and tri-n-propylamine (TPA), co-reagent of ruthenium luminescence, a polarization between the two external zones causes electroluminescence, blue at the cathode and red at the anode.

To perform a ratiometric PSA assay using this device, further modifications of the cathode zone are necessary. g-C_3N_4 is thus previously modified by a strand of DNA (c-DNA) complementary to an aptamer, the latter being chosen for its very high affinity with PSA (Figure 4.24).

The strategy envisaged for the PSA dosage then consists of fixing on g-C_3N_4, modified by cDNA, a composite that quenches the cathode's emission and reactivates it when it detaches from g-C_3N_4 under the influence of PSA. Such a composite is prepared by linking together a DNAzyme[22], a dendrimer (PAMAM [poly-amidoamine]) containing Pt NPs and the aptamer that hybridizes with both DNAzyme and cDNA, which determines the anchoring of the composite to the electrode. Under these conditions, the presence of Pt NPs in the vicinity of g-C3N4 causes its emission to be quenched by RET without interfering with that of ruthenium, located in the anodic part of the BPE (Figure 4.25).

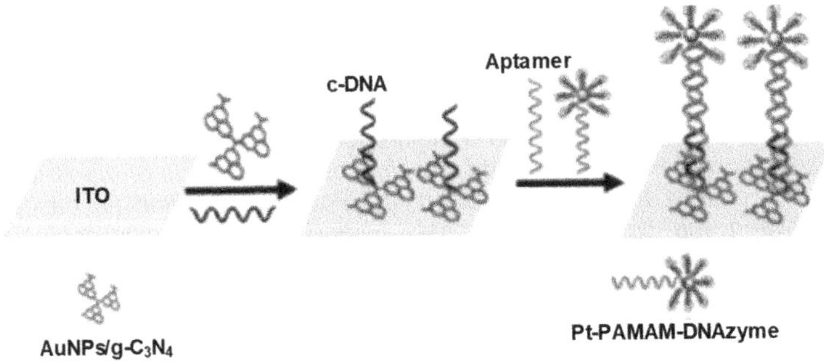

Figure 4.24. *Schematic representation of the different stages of grafting on ITO carried out in the cathode compartment of the BPE. Adapted from Wang et al. (2016c). For a color version of this figure, see www.iste.co.uk/lacaze/nanotechnologies.zip*

COMMENT ON FIGURE 4.24.– *The complementary DNA of the aptamer (c-DNA) is grafted onto g-C_3N_4 and can also hybridize with a DNAzyme, which is a DNA strand with a small number of bases. PAMAM (polyamidoamine) is a hyperbranched dendrimer obtained by condensing ethylene diamine with vinyl methyl ester. It reacts easily, via its amine functions, on Pt NPs as well as on DNAzyme (green), to give the Pt/PAMAM/DNAzyme composite.*

The two emissions produced by the BPE correlated with each other, because it is the same current intensity that circulates in the cathode and anode compartments. In the absence of PSA, the blue emission due to g-C_3N_4 is partially quenched due to a charge transfer between g-C_3N_4 and Pt NPs carried by PAMAM and the DNAzyme hybridized with the aptamer (Figure 4.25(b)).

Any increase in the electrolysis current has little effect on the blue emission and, in fact, increases the red emission of the Ru^{2+} due to more intense oxidation. The phenomenon is reversed as soon as the PSA is in solution in the cathode compartment. Its strong affinity for the aptamer destroys the previous composite and

22 A DNAzyme is a strand of synthetic DNA made up of a few nucleotides and endowed with a particular catalytic activity.

releases the PSA-aptamer couple into the solution (Figure 4.25(c)). The blue emission returns to its initial state, as represented in Figure 4.25(a).

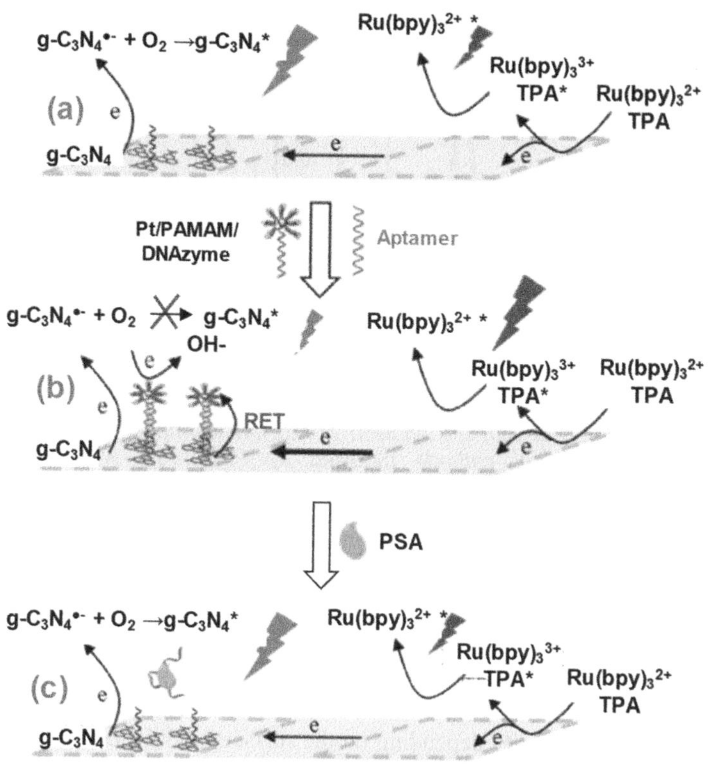

Figure 4.25. *Principle of a ratiometric ECL assay of PSA on BPE. (a) The cathode compartment is modified by AuNPs/g-C3N4 onto which c-DNA is grafted. (b) The addition of the Pt/PAMAM/DNAzyme/aptamer composite, which hybridizes with c-DNA, attenuates the blue emission from the cathode due to the energy transfer (RET) between g-C3N4 and the Pt NPs, while that of the anode compartment is determined solely by the intensity of the current, common to both compartments. (c) The addition of PSA and its high affinity for the aptamer releases the composite and restores part of the cathode to its initial state. Adapted from Wang et al. (2016c). For a color version of this figure, see www.iste.co.uk/lacaze/nanotechnologies.zip*

From a practical point of view, these variations in emission intensity at the BPE cathode and anode are perfectly correlated with the PSA concentration. A linear relationship between the logarithm of the ratio of light intensities and that of the PSA concentration is observed, between 0.3 and 200 ng/L (Figure 4.26).

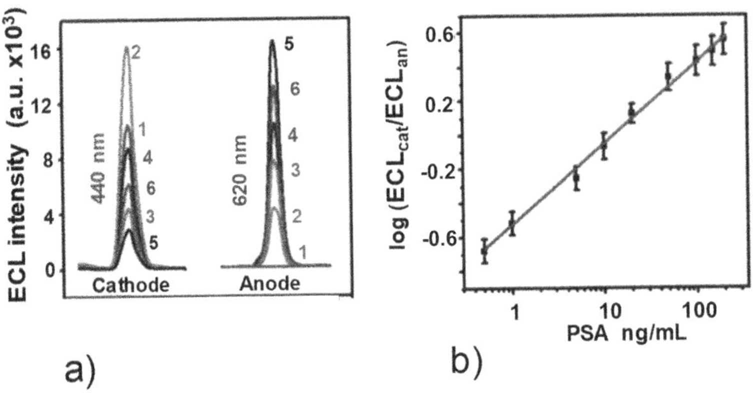

Figure 4.26. *Evolution of ECL emissions on the cathode (ECL_{cat}) and anode (ECL_{an}) according to the different surface modifications of the cathode. (a) Curve 1 (dark green): ECL of AuNPs/C_3N_4/c-ADN in the absence of $Ru(bpy)_3^{2+}$; 2 (light green): same cathode configuration, but with the addition of 0.5 mM $Ru(bpy)_3^{2+}$ and 25 mM TPA at the anode; 3 (purple): cathode modified by Pt/PAMAM with the structure Pt/PAMAM/AuNPs/C_3N_4/c-ADN; 4 (blue): cathode modified by DNAzyme leading to DNAzyme/AuNPs/C_3N_4/c-ADN; 5: further modification Pt-PAMAM-DNAzyme/AuNPs/g-C_3N_4/c-ADN; 6: addition of 1 ng/mL of PSA in the cathode modified as per 5. (b) Linear relationship between the logarithm of the ratio of intensities ECL_{cat}/ECL_{an} and the PSA concentration. Adapted from Wang et al. (2016c). For a color version of this figure, see www.iste.co.uk/lacaze/nanotechnologies.zip*

The ECL emissions at the cathode (440 nm) and anode (620 nm) (Figure 4.26(a)) vary significantly in intensity depending on the surface modifications applied to the cathode for the same concentration of Ru^{2+} and TPA at the anode. Higher emissions are observed at the anode when the Pt/PAMAM (curve 3), DNAzyme (curve 4) and Pt/PAMAM/DNAzyme (curve 5) groups are grafted onto AuNPs/g-C_3N_4, while emissions in the cathode compartment decrease sharply between 4 and 5. Such an increase in the ECL at the anode and its decrease in the cathodic compartment can only be explained by an increase in the current intensity, part of which is diverted with a non-emissive reduction reaction occurring at the cathode.

4.4.2.3. *Conclusions and perspectives*

ECL appears to be a quasi-universal assay method combining the advantages of electrochemistry and luminescence. The recent application of the ratiometric technique to ECL has considerably extended its field of investigation, which applies to biology and environmental science. Two systems with similar performances, dual-wavelength ECL and dual-potential ECL, recognized for their analytical reliability, find their place in very low-concentration detection and analysis. Detecting concentrations up to femtomolar (fM) is possible and inherent to luminescence, combined with immunological type procedures, with, in addition, the possibility of detecting and transforming an emission of extremely low electrical signal easily measurable thanks to the use of PMs. Using BPEs also simplifies the dual-potential emission method, allowing easier access to portable devices and POC assays. ECL is a method that has established itself in most pharmaceutical analysis laboratories. Commercial devices are available and very active research in the field is developing to offer new, simpler and ever more efficient systems, adapted to specific compounds.

4.4.3. *Photoelectrochemistry*

Unlike ECL, which produces light emission from an electrode reaction, PEC does the opposite and generates a photocurrent from the light excitation of an SC (generally n- or p-doped Si), on which a thin layer of insulator is deposited, itself functionalized by a receptor material sensitive to a particular analyte. As with ECL, the physical difference between signal excitation and detection means that residual background noise signals are eliminated, leading to a highly sensitive device. It is a method that has become very widespread since the 2000s and is typically used to detect various nucleotides, with very low detection limits reaching fM (10^{-15} mol L^{-1}). They are referred to as light-addressed potentiometric sensors (LAPSs). Recent review articles list and describe many application examples (Liang et al. 2019; Yoshinobu et al. 2021; Liu et al. 2022b).

4.4.3.1. *Basic principles of a light-addressed potentiometric sensor*

A photoactive SC material that absorbs photons and produces an electron–hole pair induces a photocurrent or photopotential, when the electron–hole pair is separated and transferred between two electrodes. This photocurrent is only observable when a depletion layer is present at the insulator/SC interface. This depletion layer, responsible for the space charge, whose width depends on the voltage applied between the SC and the insulator, is formed when the surface potential created by the external polarization is positive or negative for a type p or n SC, respectively. Such a device can be used either by measuring the variations in photocurrent intensity for a determined polarization or by measuring the variation in voltage at a constant photocurrent, hence its name in that case, LAPS (Hafeman et al. 1988). In practice, the device works by applying alternating irradiation to the rear face of the sensor with a frequency between 5 and 10 kHz, which creates an alternating photocurrent (AC). This is then rectified, amplified and converted into a direct potential signal (AC/DC converter), which is transformed into a digital signal (A/D converter) to be finally processed on a computer (Poghossian et al. 2001) (Figure 4.27(c)).

Although the structure of the sensor is comparable to that of an ISFET (ion-sensitive field effect transistor) (Figure 4.27(b)), or to that of a capacitive EIS (electrolyte insulator semiconductor) sensor (Figure 4.27(a)), with the same three insulator-SC-receiver layer arrangement, the electrical signal correlated to an analyte concentration is very different. Although an EIS sensor measures capacitance, an ISFET and a LAPS measure a drain current and a photocurrent, respectively. The drain current, characteristic of an FET, is located at the extreme surface of the SC (SC channel). In contrast, with a LAPS, the measured photocurrent crosses the different layers and therefore the insulator, leading to a very low intensity (a few hundred picoamperes) compared to a drain current (around a mA). This very low photocurrent value also justifies replacing the traditional three-electrode assembly controlled by a potentiostat (auxiliary and calomel-type reference SC) with a two-electrode assembly consisting only of a reference and the SC.

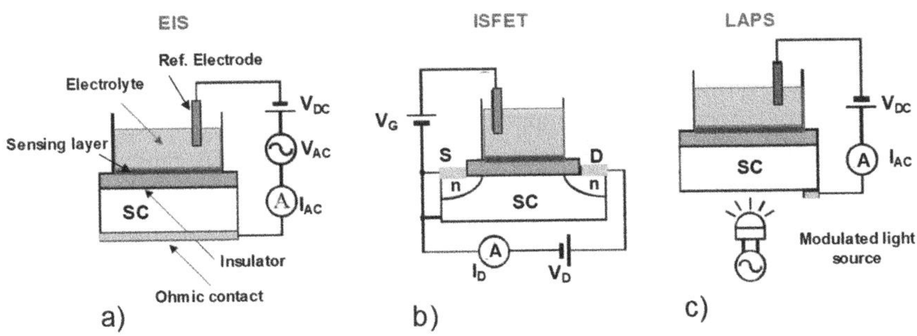

Figure 4.27. *Comparison of structures of EIS, ISFET and LAPS sensors. (a) EIS capacitive sensor (electrolyte–insulator and sensing layer–SC). (b) ISFET sensor with an oxide layer sensitive to ions (Ta_2O_5, Al_2O_3) and adjacent to the insulating layer (SiO_2). (c) LAPS sensor. Adapted from Yoshinobu et al. (2005). For a color version of this figure, see www.iste.co.uk/lacaze/nanotechnologies.zip*

COMMENT ON FIGURE 4.27.– *Variations in the capacitance of the EIS device produced by the analyte are measured by superimposing an alternating voltage V_{AC} onto a fixed voltage V_{DC}. The voltage V_{DC} is chosen in such a way as to create a depletion zone at the insulator/SC interface. In the case of an ISFET and for a determined value V_G, the variations in the SC drain current are measured as a function of the analyte concentration, which interacts with the sensing layer adsorbed onto the insulator. With LAPS, a direct voltage V_{DC} is applied between the backside of the SC and the reference electrode which, as before, creates a depletion zone between the SC and the insulator. The alternating irradiation of the SC generates an I_{AC} photocurrent of the same periodicity whose amplitude is correlated to the presence of the analyte, interacting with the sensing layer. The color codes* *are identical for the three diagrams: white (SC), gray (insulator), red (sensing layer), yellow (metal contacts) and blue (electrolyte).*

LAPS photocurrent/voltage curves have sigmoid shapes. In the presence of an analyte, there is a translational shift of the signal curve parallel to the potential axis. A typical example is that of a LAPS used to measure the pH of a solution, in which the sensing layer is tantalum oxide (Ta_2O_5) sensitive to H^+ ions, with a thickness comparable to that of the insulator (SiO_2) and around 50 nm (Figure 4.28(a)).

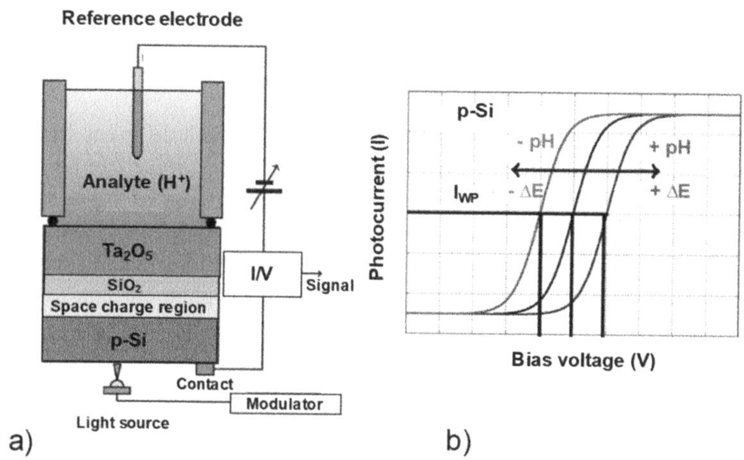

Figure 4.28. *Operating principle of a photoelectrochemical sensor to H^+ ions. (a) Diagram of the sensor subjected to a voltage V between a reference electrode and the rear face of the SC. Ta_2O_5 (50 nm) is the H^+ ion sensitive layer for which the voltage–pH relationship is of Nernstian type. (b) I/V curves of the photocurrent obtained for different solution pH values and in the case of a p-type Silicon SC. Adapted from Wagner et al. (2016). For a color version of this figure, see www.iste.co.uk/lacaze/nanotechnologies.zip*

COMMENT ON FIGURE 4.28.– *Note that when the pH of the solution decreases (higher concentration of H^+ ions), the I/V curve shifts toward lower potentials, which corresponds to the fact that the H^+ ions adsorbed on the Ta_2O_5 layer add a positive Nernstian voltage to the applied potential. The opposite is observed when the pH increases. According to this diagram, the analyte can be characterized at a constant voltage or constant current, with the reference voltage chosen at the inflection point of the I/V curve (working point [WP]). In the first case, we measure the photocurrent intensity as a function of the analyte concentration while in the second, for an average photocurrent (I_{WP}) value, we measure a voltage as a function of the analyte concentration.*

The current induced by polarization is very low, around a picoAmpere per cm^2 (pA/cm^2). Under irradiation, photoelectron production increases significantly and the current then reaches a limiting intensity of around a μA corresponding to a space charge totally reversed with respect to the SC core. The presence of H^+ ions in the solution and their adsorption onto the Ta_2O_5 layer modifies the polarization by adding to or subtracting from the applied voltage, resulting in a shift of the photocurrent I/V curve parallel to the voltage axis (Figure 4.28(b)). A linear quasi-Nernstian variation in pH as a function of the voltage (57.5 mV/pH) is obtained by selecting I_{WP}, the photocurrent value at the inflection point of the curve (Yoshinobu et al. 2005).

4.4.3.2. *Using LAPS for imaging and multi-detection*

This is one of the specificities of the technique for which different arrangements have been proposed. The idea is that the sensitive LAPS surface can be artificially divided into surface elements, each of which can be irradiated by a light beam. Several methods have been tested. One involves scanning the sensitive sensor surface by moving the light source, the other uses a network of light sources which impact specific surface regions to which specific photocurrents correspond (Figure 4.29).

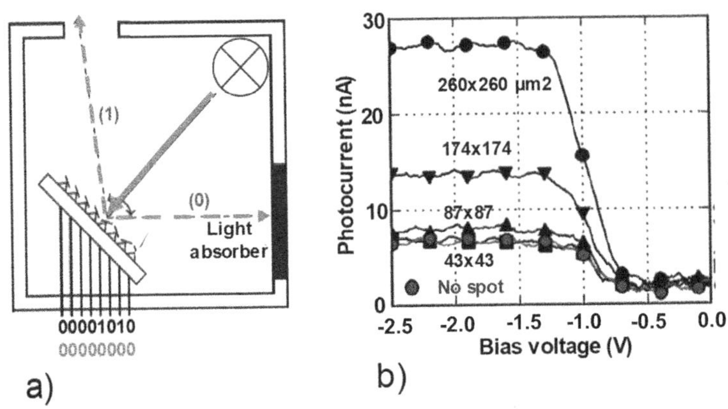

Figure 4.29. *Operating principle of a LAPS comprising a light source of constant intensity and an array of micro-mirrors with alternative orientation. (a) Diagram of the device showing the light beam irradiating a set of eight micro-mirrors, only two of which are controlled and can periodically switch between two orientations corresponding to 0 and 1. (b) Photocurrent variation curves (after amplification) as a function of the applied voltage and the size of the light spots (expressed in µm²) impacting the LAPS SC. Adapted from Wagner et al. (2012). For a color version of this figure, see www.iste.co.uk/lacaze/nanotechnologies.zip*

COMMENT ON FIGURE 4.29.– *LAPS is made up of alternating layers of n-Si/SiO_2/Si_3N_4 in contact with a pH 7 buffer solution, Si_3N_4 being the H^+ ion sensitive transducer layer. The incident light beam is reflected by a micro-mirror and directed toward the rear face of the LAPS, then toward the light absorber as it oscillates toward the respective orientations 1 and 0, which modulates the light intensity at the micro-mirror oscillation frequency. According to the diagram in Figure 3.30(a), two of the eight micro-mirrors are controlled to switch from position 1 to position 0, and the other six are held at position 0. The position corresponding to (1) sends the light beam to a precise point on the transducer surface, corresponding to a determined pixel. At the same time, the other orientation corresponding to (0) directs the light outside the sensitive surface*[23]*. The photocurrent/potential curves in Figure 3.30(b) show that the minimum size of a light spot triggering a measurable photocurrent corresponds to a surface area of 7,569 μm^2 (87 × 87 μm^2).*

An assembly of this type, designed from a light source irradiating a network of adjustable micro-mirrors (digital micromirror device [DMD]), was proposed jointly by the groups of Dantism et al. (2017), Yoshinobu et al. (2005), and Yoshinobu and Schöning (2021).

Another device is also possible and consists of illuminating the LAPS, either by a series of LEDs (light emitting diodes), which irradiate distinct parts of the sensitive surface of the sensor, or by a single laser diode, with a modulated intensity frequency, which can be moved in the x and y directions. It is this process that was used by Schöning et al. to study microorganism metabolism, for example Chinese hamster ovary cells (CHO)[24] which, after ingesting glucose, release ions H^+ (Figure 4.30).

The LAPS comprises the superposition of p-Si, SiO_2 and Ta_2O_5 layers, the thin tantalum oxide film forming the biosensor sensitive layer. The sensitive surface of the LAPS is then divided into four parts (called "chambers"), in which the reagents and the CHO cells have been arranged (Figure 4.30(a)) (Dantism et al. 2017). The photocurrents produced by laser irradiation are used to produce differential imaging and surface colorations directly correlated to the evolution of the metabolism of microorganisms, subject to the influence of various ingredients (Figure 4.30(b)).

23 Such a device is used in the manufacture of video projectors operating according to the DLP (digital light processing) technique, which involves digitally processing light. It was developed in the 1980s by Texas Instruments with commercial production by DMD systems, subsequently adopted by other manufacturing companies.

24 Epithelial cells derived from CHO are widely used in the pharmaceutical industry to produce recombinant proteins for therapeutic use, generally by microorganisms whose genome has been modified by the introduction of genes specific to the protein.

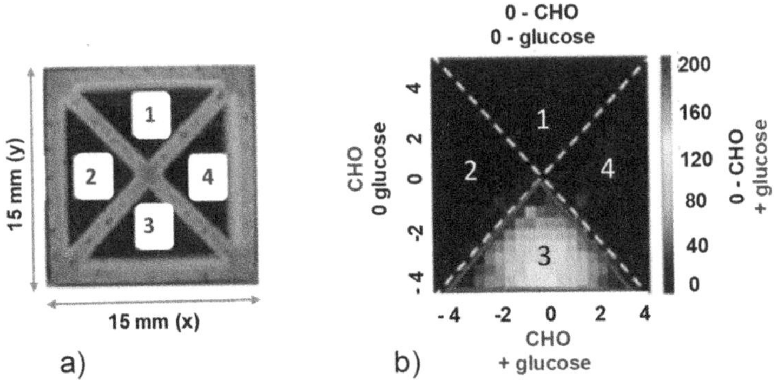

Figure 4.30. *Photocurrent imaging of a LAPS divided into four sectors (chambers) and irradiated by a mobile laser diode. (a) Dimensional aspect of triangular-shaped chambers. (b) Differential imaging representing the photocurrent intensity observed in the four chambers for CHO-glucose mixtures of different compositions. Adapted from Dantism et al. (2017). For a color version of this figure, see www.iste.co.uk/lacaze/nanotechnologies.zip*

COMMENT ON FIGURE 4.30.– *The LAPS is formed by the superposition of p-Si, SiO_2 and Ta_2O_5 layers with thicknesses of 540 μm, 30 nm and 60 nm, respectively. The photocurrent intensity is graded using a color scale ranging from blue (zero photocurrent) to red (maximum photocurrent). The experiment is carried out by first injecting the CHO culture medium into the four chambers; CHO is then added to chamber 2 (without glucose), CHO and glucose to chamber 3 and glucose without CHO to chamber 4. Coloring corresponds to differences in photocurrent intensity measured in chambers 2–4 compared to chamber 1. The yellow-green coloring of chamber 3 reveals significant metabolic activity.*

4.4.3.3. *LAPS based on nanostructured composites operating by near-infra-red irradiation (NIR)*

This is a new line of research aiming to replace the initial UV-visible irradiation step, required to create a photocurrent in most SCs, with much less energetic near-IR radiation; one advantage of it is to prevent the photo-degradation of specific biological molecules involved in the operation of a PEC sensor.

For this, it is necessary to use materials that can transform the initial IR radiation into higher frequency radiation (up-conversion), compatible with the excitation of an SC and the creation of a photocurrent. Such a property is specific to certain mixtures of rare earth metals, which, in interaction with SC nanocrystals, particularly CdTe

QDs, can then create a photocurrent. Such a device was described by Wang et al. (2016d) for the specific detection of breast cancer cells (MCF-7)[25].

MCF analysis is thus based on the decrease in the photocurrent caused by their adsorption onto the electrode surface. For this, a particular aptamer [26] capable of hybridizing MCF-7 cells is previously adsorbed onto the electrode. Under these conditions, the MCF cells present in the solution, whose size is around 10 μm, aggregate on the aptamer and partially block the surface of the electrode. This reduces electron transfers and, consequently, the photocurrent intensity (Figure 4.31(b)).

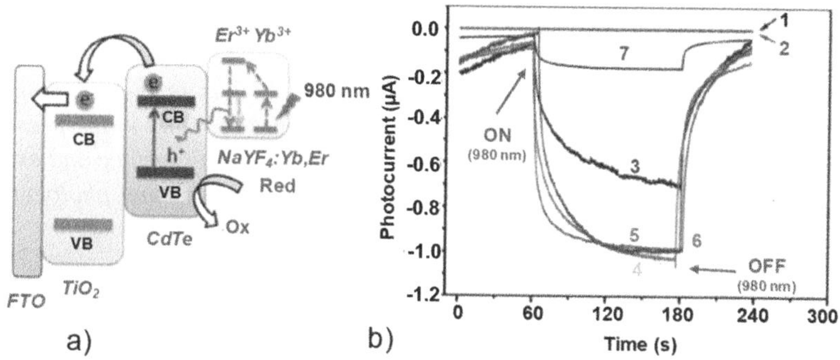

Figure 4.31. *Operating principle of a LAPS frequency up-conversion. (a) Energy diagram of the device indicating the absorption mechanism at 980 nm and the production of an electron–hole pair at the origin of the photocurrent. (b) Photocurrent variation as a function of illumination time for different LAPS structural combinations and after adsorption of MCF-7 cells. Adapted from Wang et al. (2016d). For a color version of this figure, see www.iste.co.uk/lacaze/nanotechnologies.zip*

COMMENT ON FIGURE 4.31.– *The FTO electrode is previously modified by depositing a composite layer of TiO_2 NPs (approximate size of 30 nm) mixed with the compound $NaYF_4$:Yb,Er. The latter is responsible for the IR radiation frequency up-conversion (980 nm) at 520 and 550 nm in the visible, compatible with the forbidden band width of SC CdTe (of around 1.5 eV) and generates an electron–hole*

25 MCF -7 represents a cancer cell line discovered from a pleural effusion by the Michigan Cancer Foundation in 1973. They are considered as a xenograft model best suited for the study of breast cancer.

26 The chosen aptamer (AS1411) is an oligonucleotide comprising 26 bases. It is made up of many GGT (guanine-guanine-thymine) codons and its 5' end has been functionalized by an NH2 group, which corresponds to the structure 5'-NH2-GGT GGT GGT GGT TGT GGT GGT GGT GG-3'.

pair. The positive hole formed in the valence band (VB) of CdTe is then neutralized by the oxidation of a sacrificial redox species (ascorbic acid in low concentration) in the solution. The i/t curves of the photocurrent established for different LAPS structures show that no photocurrent is detected when the electrode has the structure FTO/TiO_2 *(curve 1) or FTO/*TiO_2*/CdTe (curve 2), which proves, on the one hand, that* TiO_2 *has a band gap that is too high (3.8 eV) to generate an electron–hole pair, and, on the other hand, that CdTe without* $NaYF_4$*:Yb,Er cannot create an electron–hole pair with excitation at 980 nm. A photocurrent appears (curve 3) when the electrode is modified by the composite* $NaYF_4$*:Yb,Er and CdTe and it becomes stronger (curve 4) when the same materials are distributed on* TiO_2 *NPs that contribute to more efficient separation of the electron-hole pair, by accepting CdTe electrons into the* TiO_2 *conduction band. The photocurrent intensity practically does not change (curve 5) when the aptamer (not shown in the diagram) is added to the previous structure; the same when MCF-7 is in the solution, but without the aptamer being present on the electrode (curve 6). Only in the presence of the aptamer can MCF-7 (105 cells/mL) attach to the electrode surface and exert a blocking effect, which results in a significant reduction in the photocurrent (curve 7).*

A linear relationship between the decrease in photocurrent intensity and the logarithm of the MCP-7 concentration is observed in concentrations from 10^3 to 10^5 cells/mL, with a detection threshold of 400 cells/mL, comparable to that obtained by the plasmon resonance technique. The technique, however, appears less sensitive than ECL and resonance Rayleigh scattering (RRS), which achieve detection thresholds of between 20 and 30 cells/mL and 12 cells/mL, respectively. However, they have lower upper detection limits of 2×10^4 cells/mL for ECL and 300–2000 for RRS (Cai et al. 2015). Differential pulsed voltammetry (DPV) techniques have also been used for quantifying cancer cells and have much higher upper limits detection (around 10^8 cells/mL) but, with less good detection thresholds, around 10^5 cells/mL.

Also note that this technique, as for ECL, can be used ratiometrically, as recently shown by Qiu et al. (2019) for the detection of the CEA[27]. The device, printed on a paper substrate, consists of two photoelectrodes PE_1 and PE_2 on which the composite $NaYF_4$:Yb,Er was deposited coupled with CdTe nanocrystals (Figure 4.32).

27 CEA is a glycoprotein involved in cell adhesion. In a healthy individual, it is present in very low levels, around 2–4 ng/mL. It can be indicative of a form of cancer when its concentration increases, and, as such, is the subject of numerous studies aiming to create sensors for this type of biomarker.

The surface of each photoelectrode must be modified further to carry out CEA quantification. Photoelectrode PE_1 is modified by aptamer A1 (capable of combining with the CEA on PE_1) and PE_2 by a CEA capture DNA strand (AC) on PE_2, to which aptamer A2 combines, modified by a gold NP (AuNP-A2). The latter attenuates the photocurrent from PE_2 due to a plasmon resonance interaction between the gold NP and the CdTe emission, corresponding to an exciton–plasmon interaction (EPI) effect.

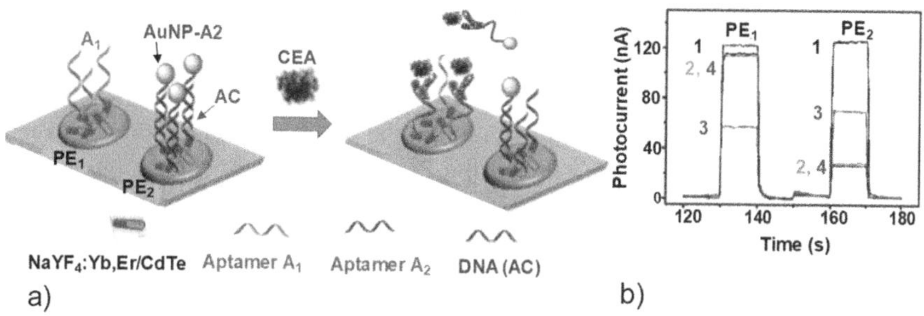

Figure 4.32. *Photoelectrochemical ratiometric sensor comprising a pair of photoelectrodes PE_1 and PE_2 modified by the composite $NaYF_4$:Yb,Er/CdTe. (a) Descriptive diagram of the heterostructures grafted onto the photoelectrodes PE_1 and PE_2 before and after adding CEA. (b) Photocurrents from photoelectrodes PE_1 and PE_2 for different heterostructures, numbered from 1 to 4. (1) PE_1 and PE_2 in the state $NaYF_4$:Yb,Er/CdTe (state "0"); 2) PE_1 and PE_2, respectively, modified by the aptamer A_1 and the composite AC/A_2-Au NP; (3) addition of 1.0 ng/mL CEA after "2"; (4) return of the sensor to the state "2" (0 ng/mL CEA). Adapted from Qiu et al. (2019). For a color version of this figure, see www.iste.co.uk/lacaze/nanotechnologies.zip*

COMMENT ON FIGURE 4.32.– *The photocurrents (2) from PE_1 and PE_2 show that the grafting of A_1 onto PE_1 initially in the "0" state has little impact on the intensity of the photocurrent unlike PE_2 that undergoes a significant drop in its intensity after grafting of the AuNP-A_2/AC composite, which highlights the photocurrent attenuation effect due to the exciton–plasmon interaction between CdTe and the gold NP. The addition of CEA (curves 3) in the biosensor causes the opposite effect. On the one hand, in PE_2, the AuNP-A_2 aptamer is separated from AC and ejected into the solution which eliminates the PE_1 attenuation effect, and therefore increases the intensity of the photocurrent I of PE_2. On the other hand, in PE_1, CEA aggregates with A_1 and contributes to the decrease in electronic transfers and the decrease in the corresponding photocurrent. The $NaYF_4$:Yb,Er/CdTe composite is obtained in the form of quasi-cylinders of sub-micron sizes (0.4 μm long and 0.2 μm in diameter) on which a thin layer of CdTe about 5 nm thick is deposited.*

In the presence of the CEA target, two reactions take place simultaneously with the photoelectrodes PE_1 and PE_2, attenuating and increasing, respectively, the photocurrents supplied by PE_1 and PE_2. In the first case, and according to the authors, the photocurrent reduction is due to an increase in steric hindrance on the PE_1 surface, resulting from the formation of the A1-CEA composite, which partially inhibits the operation of the photoelectrode. In the second case, CEA reacts with the A2 aptamer and causes its separation from the capture DNA (CA), which reduces the quantity of CA/A2-AuNP grafted to the PE_2 photoelectrode surface and, consequently, increases the photocurrent intensity by eliminating the EPI effect caused by the gold NPs.

On a practical level, such a ratiometric biosensor makes it possible to establish a linear relationship between the logarithm of the CEA concentration (between 10 pg/mL and 5 ng/mL) and the ratio of the PE_1 and PE_2 photocurrent intensities. According to the authors, the detection limit is approximately 5 pg/mL, which exceeds the values provided by commercial CEA-ELISA sensors in terms of performance, which are approximately 50 pg/mL; however, for the latter, a much higher upper concentration limit of around 50 ng/mL is obtained.

4.4.3.4. *Conclusions on PEC*

PEC, characterized by the measurement of a photocurrent, is a technique whose performance is similar to ECL. It is probably more general in application than ECL, since it only requires an SC material, for example, silicon, a material that benefits from advances in microelectronics, or any other QD SC material, offering a multitude of design possibilities for sensors. ECL is although very sensitive in terms of detection given that fluorescence emission is measured, it nevertheless requires the presence of oxidoreducible compounds, which can sometimes restrict the field of application. One of the advantages of ECL compared to PEC is its simpler operation, unlike PEC, which requires irradiation, then amplification of the photocurrent coming from the sensor.

4.4.4. *Raman scattering spectroscopy*

Raman spectroscopy originates from a process of inelastic scattering of light by a molecule, which leads to a pseudo-excitation state S' of the molecule (different from the first electronic excitation state S_1 and designated as a virtual excited state), followed by deactivation through energy exchange with the molecule vibrations (Figure 4.33).

Compared to IR spectroscopy, which is based on the direct absorption of light energy by different vibrations of a molecule in its fundamental state, one could think that both techniques obtain the same vibrational information. They are different in that the vibrational transitions observed in IR and Raman spectroscopy are often distinct due to their selection rules. In IR absorption, a vibration will be active if the variation in the dipole moment associated with the vibration is non-zero. In contrast, for Raman, it is the variation in the polarizability of the molecule linked to the vibration that determines the activity.

From a sensitivity point of view, the scattered light is very low in intensity compared to the excitatory radiation (approximately one scattered photon for 10^8 photons of the incident radiation). However, this drawback is minimized since the scattered light can be detected orthogonally to that of the incident beam. This means that the photons are scattered on a black background during detection, increasing the detection sensitivity. Furthermore, the technique used to measure the intensities is based on CCD (coupled charge device) technology, which makes it possible to integrate the electrons resulting from the photon → electron transformation over a determined period.

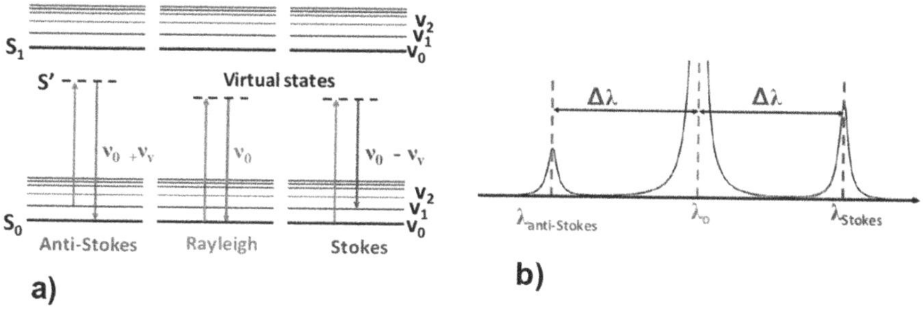

Figure 4.33. *Representative energy diagram of Raman scattering. (a) Origin of scattering (Rayleigh), Stokes and anti-Stokes. (b) Appearance of the Raman scattering spectrum defined from the wavelengths λ. Adapted from Shipp et al. (2017). For a color version of this figure, see www.iste.co.uk/lacaze/nanotechnologies.zip*

COMMENT ON FIGURE 4.33.– *S_0 and S_1 represent the fundamental state and the first excited electronic state of the molecule, respectively. S' is the virtual excited state resulting from the interaction between the monochromatic radiation of frequency ν_0 (wavelength λ_0) and the molecule. The elastic scattering of light leads to an emission with a frequency identical to that of the incident light (Rayleigh scattering line). In the case where the light gives up part of its energy to the molecule to bring it to its first vibrational level of excitation v_1, the emission is then characterized by the frequency $\nu_0 - \nu(v_1)$, where $\nu(v_1)$ being the vibrational frequency corresponding to the first vibrational level of excitation v_1. This scattering line, characterized by a wavelength higher than that of the incident light, corresponds to the Stokes line* *(so-called negative line). If, unlike the previous case, the molecule is initially in a vibrational state of excitation (level* v_1*), it is the molecule that can transfer energy to the radiation, which then diffuses at the frequency* $\nu_0 + \nu(v_1)$*, corresponding to the emission of a positive line (anti-Stokes line), lower in intensity than the Stokes line. The Raman spectrum is thus made up of a central line high in intensity, accompanied by lines lower in intensity arranged symmetrically with respect to the Rayleigh line.*

Practically, the detection limit observed for biological compounds is between 0.1 and 1 μM (Shipp et al. 2017). This sensitivity can be improved when the laser frequency is close to the excitation frequency corresponding to the absorption energy of an electron transition. The scattering is amplified by a factor of 10^3–10^4 and the technique is then known as enhanced resonance Raman scattering (ERRS) (Smith and Dent 2005).

This sensitivity can be further increased in the case where the Raman emission is associated with a surface phenomenon due to plasmon resonance. This locally enhances the electromagnetic field, which considerably amplifies the emission signal, much higher than observed by conventional Raman spectroscopy. This effect, known as surface enhanced resonance scattering (SERS), makes it possible to achieve extremely low detection thresholds for certain molecules, up to 10^{-18} mol·L^{-1} (Kudelski 2008; Shipp et al. 2017).

4.4.5. *Surface- and plasmon-enhanced resonance scattering*

The discovery of the SERS effect dates to the 1970s (Fleischmann et al. 1974). Fleischmann was the first to observe a significant increase in the Raman emission signal of pyridine adsorbed onto a silver electrode when subjected to oxidation–reduction cycles. Initially attributed to an increase in surface area, the real explanation originates from a plasmon resonance effect due to an increase in surface roughness, comparable to a deposition of silver NPs (Haynes and Van Duyne 2003). These NPs are responsible for the “hot spots”, corresponding to the formation of zones where the electromagnetic field is much more intense than on a smooth surface. The resulting phenomenon is localized surface plasmon resonance (LSPR), which intensifies the Raman signal. The two phenomena (SERS and LSPR) are closely linked and can be used to detect various compounds.

4.4.5.1. *Plasmon resonance*

SPR is due to an oscillation of the conduction electrons of a metal under the influence of an electric field E_e and is observed with most metals. Gold, silver and copper are the most commonly used due to their absorption ranges from near UV to IR.

There are several ways to use plasmon resonance for analytical purposes. The first and oldest method involves using a glass prism, metallized on its backside with a thin layer of gold or silver, to measure the variation in the light incidence angle leading to maximum absorption of scattered light. This variation in the intensity of the reflected light due to an SPR is extremely sensitive to even a slight variation in the refractive index of the dielectric in contact with the metallized surface. This variation in the refractive index is used to detect biological molecules via the Kretschmann device. Based on LSPR, the second method uses gold or silver NPs. On the one hand, it was used in colorimetric assays based on the NP agglutination phenomenon and, on the other hand, in the production of meta-surfaces adapted to SERS spectroscopy.

4.4.5.1.1. Plasmon resonance on a metal film

The interaction of light with a metal film produces an oscillation of the electrons. With the Kretschmann device (glass prism metallized on its base), the incident light polarized parallel to the plane of incidence passes through the prism. It is reflected on the underside at an angle of θ_0 corresponding to a total internal reflection (Figure 4.34(a)).

With the underside being covered with a thin metal film (gold or silver) of a few tens of nanometer, the interaction of light with the metal gives rise to an oscillation of conduction electrons corresponding to a surface plasmon. This oscillation of electrons produces an evanescent electromagnetic wave that propagates along the prism-metal interface and attenuates with the propagation distance. It is accompanied by an electric field perpendicular to the metal surface, whose intensity disappears beyond 300 nm after interacting with the sensitive layer adsorbed on the metal.

The creation of the plasmon produces an absorption of light, which results in a significant reduction in the reflected intensity, which is minimal for the angle of incidence θ_1. The variation in the refractive index, caused by a modification to the composition of the medium in contact with the sensitive layer, modifies the angle of incidence for which the reflection is minimal and corresponds to a variation in angle incidence of θ_1 to θ_2 (Figure 4.34(b)). Numerous immunological assays have been carried out using this technique by measuring the variation in the angle of incidence, expressed in resonance unit RU (1 RU = 0.0001°) (Figure 4.34(c)).

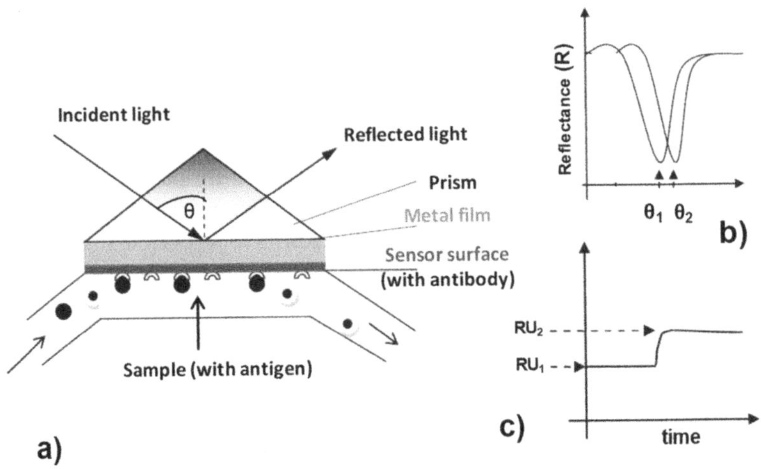

Figure 4.34. *Kretschmann SPR device for immunoassays. (a) The base of the prism is covered with a gold film (thickness ≈ 50 nm) onto which a sensitive layer consisting of antibodies has been grafted. (b) Reflected light curves characterized by their minimum angles of reflection before (θ_1) and after interacting with the antigen to be assayed (θ_2). (c) Time-dependent variation in biosensor response measured in resonance units. Adapted from Narsaiah et al. (2012). For a color version of this figure, see www.iste.co.uk/lacaze/nanotechnologies.zip*

4.4.5.1.2. Plasmon resonance on metal NPs

In the case of metal NPs, the electrons of conduction are confined. They are characterized by their oscillation frequency and can be brought into resonance by irradiating the NPs with a laser whose frequency is close to that of the conduction electron oscillations. The result is then a plasmon resonance localized on the surface of the NP, referred to as an LSPR (Figure 4.35(a)).

This electron excitation leads to a dipolar oscillation and the emission of an evanescent surface electric field, whose intensity decreases quickly and disappears beyond 10 nm from the surface. Measuring the extinction of scattered laser light (sum of plasmonic absorption and scattering) as a function of the wavelength determines the range of plasmonic absorption. It varies considerably depending on NP geometry, particularly when going from a spherical to a rod shape (Figures 4.35(b) and (c)). Any variation in chemical composition near the NP surface causes a variation in refractive index and results in the frequency shift of the absorption peak due to plasmon resonance. This property is exploited for the assay of various compounds, in the case of nanosensors with metal NPs (gold and silver).

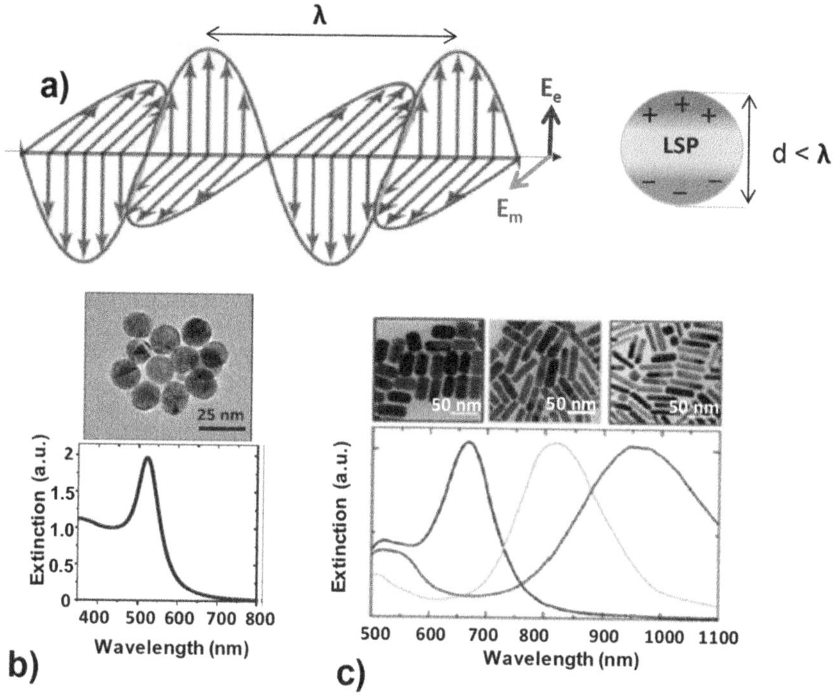

Figure 4.35. *Plasmon resonance of metal NPs. (a) Diagram of the resonance mechanism of electrons of conduction in a spherical NP subjected to the electromagnetic field of light (E_e and E_m are the electric and magnetic fields, respectively). The displacement of the electrons relative to the fixed positive metal ions generates a dipole whose oscillations are in resonance with the light radiation when the laser frequency is close to that of the electrons. (b and c) Plasmon absorption spectra of different types of NPs. (b) Spherical gold NPs with sizes of approximately 10 nm. (c) Gold nano-rods characterized by different length/width aspect ratios (AR). The blue, yellow and green curves correspond to nanorods with length/width of 37/19, 50/12 and 50/8 nm, respectively, and approximate ARs of 2, 4 and 6. Drawings (a) and (b) adapted from Amendola (2017); drawing (c) adapted from Zijlstra (2009). For a color version of this figure, see www.iste.co.uk/lacaze/nanotechnologies.zip*

Furthermore, very high electric field enhancement coefficients (hot spots) can be obtained between two NPs when separated by 1–2 nm. The enhancement coefficients vary between 10^8 and 10^9, which makes SERS detection of a single molecule possible, and compared to conventional SERS, it corresponds to an increase in the signal from 10^2 to 10^3 (Radziuk and Moehwald 2015; Amendola et al. 2017) (Figure 4.36).

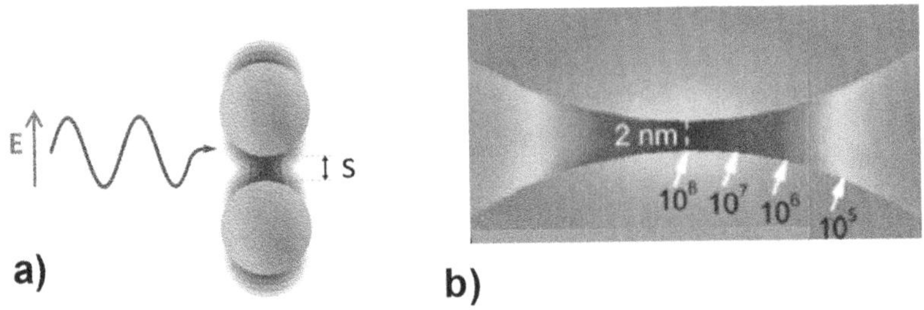

Figure 4.36. *Plasmon resonance "hot spot" between two spherical metal NPs. (a) Location of the maximum intensity of the electric field (dark green between the two NPs) induced by the plasmon resonance. Adapted from Shipp et al. (2017). (b) Evaluation of the enhancement factor in the separation space (s = 2 nm) as a function of the lateral distance of two gold NPs (size 30 nm) for irradiation with polarized light with a wavelength of 559 nm. Adapted from Etchegoin and Le Ru (2008). For a color version of this figure, see www.iste.co.uk/lacaze/nanotechnologies.zip*

This plasmon resonance effect also results in resonant Rayleigh light scattering (RRLS). The effect is particularly important in the case of metal NPs (Au and Ag), and its application to the analysis of biomolecules has made it possible to achieve LODs well below the femtomolar range.

Truong et al. (2013) have thus developed an immunological sensor known as aptasensor used for the detection and assay of an enzyme such as lysozyme[28]. The detection sensitivity is at its maximum when the device is a dual system of two gold NPs in plasmonic interaction (Figure 4.37).

The first NP (40 nm in diameter) is deposited onto a glass microscope slide and functionalized with a particular aptamer capable of hybridizing lysozyme. The other, smaller in size (12 nm), also modified by an antibody specific to lysozyme and in the form of a colloidal solution, is brought into contact with the primary NP, resulting in the association of the two NPs, which then interact plasmonically. The result is a significant reduction in the frequency of the plasmon resonance of both gold NPs, resulting in a red shift of the scattering peaks (Figure 4.37(b)).

28 Lysozyme is an enzyme that is normally found in the human body, in animals, but also in plants and microorganisms. It is a globular protein that, in humans, is made up of 129 amino acids with a very high isoelectric point of 11 and has the function of fighting bacterial infections. It can destroy certain bacteria, which gives it antibiotic properties. In high concentrations, it is a sign of a disturbance due to certain diseases (leukemia, renal insufficiency, or even meningitis).

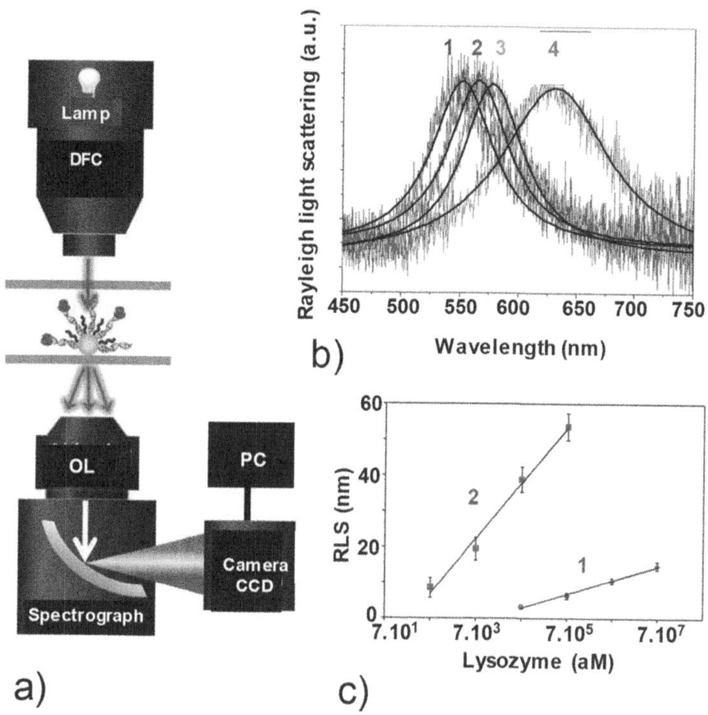

Figure 4.37. *Micro-spectrophotometric device with dark field condenser for measuring Rayleigh light scattering (RLS) using gold NPs (primary). (a) Diagram of the measuring device showing a gold NP (50 nm) deposited onto a glass slide and irradiated with white light in dark field condenser (DFC) mode. (b) Shift toward the IR of the peaks of scattering curves of functionalized gold NPs: (1) unmodified gold NP (50 nm); (2) gold NP after grafting of the specific lysozyme aptamer; (3) hybridization of lysozyme (concentration 7 × 10^7 aM) with the aptamer; (4) addition after (3) of a colloidal solution of gold NPs (12 nm) also modified by an aptamer complementary to lysozyme. (c) Shift (in nm) of the scattering peak as a function of the lysozyme concentration: (1) case where the primary gold NP is only functionalized by the aptamer and (2) after the addition of the colloidal gold solution. (a) Adapted from Song et al. (2013); (b and c) adapted from Truong et al. (2013). For a color version of this figure, see www.iste.co.uk/lacaze/nanotechnologies.zip*

COMMENT ON FIGURE 4.37.– *The scattered light is collected and concentrated by optical lenses (OL) and analyzed in a spectrograph; it is then converted into an electric current by the CCD (charge coupled device) camera, then analyzed on a PC. The substantial shift in the scattering peak observed after adding the colloidal gold solution is due to the interaction between the two gold NPs of different sizes, causing a significant shift in the plasmon resonance.*

The comparison between the device comprising the single gold NP of 50 nm in diameter and the binary system of two NPs coupled together shows, for a lysozyme concentration of 70 fM, a considerable shift in peaks from 14.3 to 53.2 nm, a relative increase by almost 300% which, according to the authors, makes it possible to reach a lysozyme detection limit of 7 aM (7×10^{-18} M). In both cases, a linear variation is obtained between the peak shift and the logarithm of the lysozyme concentration for different concentration intervals, but with considerably improved sensitivity when the two gold NPs are coupled via lysozyme (Figure 4.37(c)).

This shift in plasmon resonance is also observed with networks of NPs. It results in a color change due to the aggregation or spacing of NPs (Figure 4.38) (Haynes and Van Duyne 2001).

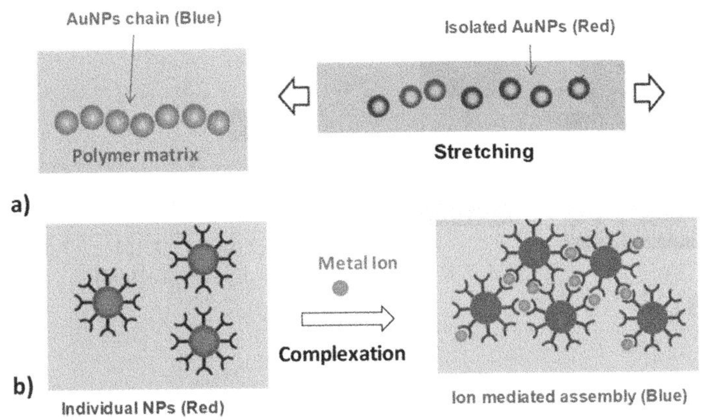

Figure 4.38. *Change in color of a set of gold NPs due to physical stress or chemical complexing. (a) Change in color from blue to red due to stretching a polymer film containing a chain of aligned gold NPs. Adapted from Han et al. (2014). (b) Colorimetric detection of a metal ion. The change in color from red to blue is induced by the agglutination of the gold NPs resulting from the complexation of the metal ions by the gold NPs onto which complexing groups of the ion to be detected have been grafted. Adapted from Saha et al. (2012). For a color version of this figure, see www.iste.co.uk/lacaze/nanotechnologies.zip*

This color change can be caused either by a physical action (e.g. variation in the stress exerted on a polymer containing aligned NPs; Figure 4.38(a)) or by a chemical process (complexation and agglutination of NPs in solution; Figure 4.38(b)). This agglutination phenomenon has been widely exploited in the case of gold NPs to rapidly detect various analytes (metal ions and biological compounds). Many examples can be found in the review article by Saha et al. (2012).

4.4.5.2. *SERS spectroscopy*

SERS spectroscopy differs from ordinary Raman spectroscopy only in its use of nanostructured surfaces, designed to cause a plasmon resonance with an increase in the surface electromagnetic field. The SERS effect is also characterized by an exaltation factor (EF) measured for ordinary Raman scattering, per the relationship: $EF = (I_{SERS}/N_{SERS})/(I_{NRS}/N_{NRS})$, where I_{SERS} and I_{NRS} represent the respective intensities of Raman SERS and normal Raman scattering; N_{SERS} and N_{NRS} are the numbers of corresponding active centers. Compared to conventional Raman scattering, this EF can vary between 10^4 and 10^8, the highest values being obtained for plasmon resonance hot spots, which, in this case, makes it possible to achieve the necessary sensitivity for single-molecule detection (Asiala and Schultz 2011).

From a practical point of view, the surface enhanced Raman spectra, and particularly the exaltation factor EF, depend on the surface preparation method, which implies that the number of active centers as well as the regularity with which they are arranged are well defined. Different methods have been recommended to produce such reproducible SERS surfaces. Nanosphere lithography (NSL), using a polystyrene nanosphere monolayer deposited regularly on a flat surface, has been widely used and involves evaporating a metal such as gold or silver under a vacuum (Nguyen et al. 2017). After dissolving the nanospheres in a solvent, a regular NP lattice of controllable thicknesses is obtained (Figure 4.39).

Numerous variations in the method make it possible to obtain various shaped and sized NP lattices, generating many "hot spots" within the material volume, thus increasing detection sensitivity (Su et al. 2022; Tahir et al. 2020; Palermo 2021).

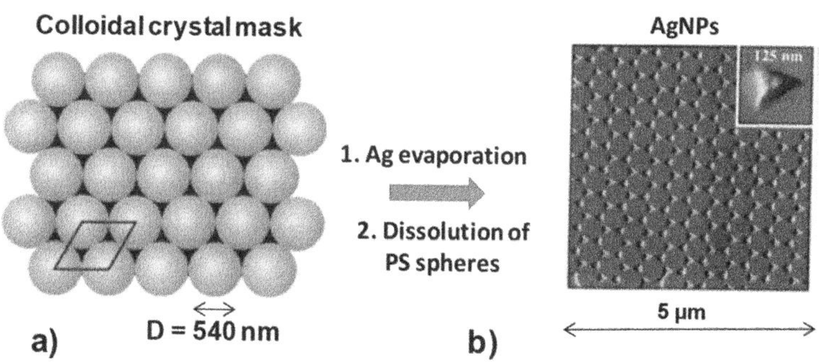

Figure 4.39. *Principle of nanosphere lithography (NSL). (a) Production of a mask consisting of a polystyrene sphere (PS) monolayer deposited on a solid substrate (mica, glass). (b) Obtaining a regular Ag NP lattice (size 100 nm and thickness 40 nm) by vacuum evaporation of the metal followed by dissolution of the PS spheres. Adapted from Haynes et al. (2001). For a color version of this figure, see www.iste.co.uk/lacaze/nanotechnologies.zip*

The EF of a nanostructured "SERS surface" compared to an ordinary surface is very high and varies between 10^5 and 10^8 depending on the metal used (Su et al. 2022). Su et al. have thus developed a three-dimensional microporous substrate denoted 3D ZnO/Ag, obtained by calcination of a particular MOF (ZIF-8 or zeolite imidazole framework) followed by various chemical steps leading to the SERS substrate consisting of silver NPs (size 70 nm) on a micro-porous ZnO film (Figure 4.40(a)).

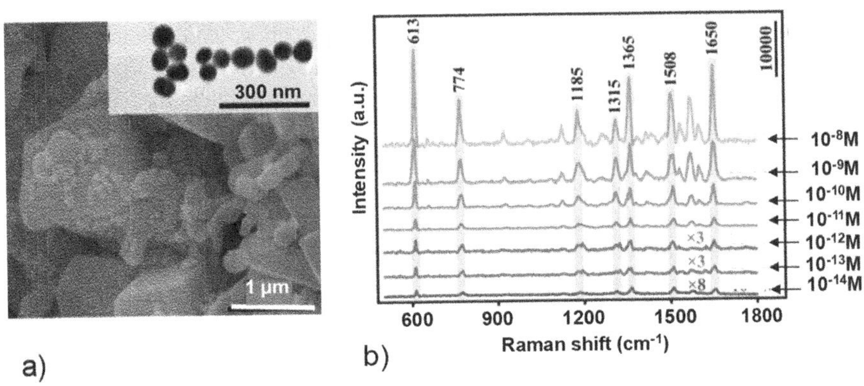

Figure 4.40. *SERS spectrum of rhodamine 6-G (R6G) obtained on a microporous ZnO film covered by Ag NPs. (a) SEM image of the ZnO/Ag film. The micrometer-sized crystals correspond to the ZnO particles on which the Ag NPs are deposited. (b) SERS spectra of rhodamine 6G for concentrations varying between 10^{-14}M and 10^{-8} M. The peak intensity at 613 cm^{-1} (the most intense) varies linearly according to the equation: logI = 6.427 + 0.3131 log C. Adapted from Su et al. (2022). For a color version of this figure, see www.iste.co.uk/lacaze/nanotechnologies.zip*

LODs of between 10^{-8} and 10^{-14} M are obtained depending on the molecule type. As an example of trace detection, Rhodamine 6G is thus characterized by a SERS spectrum with a peak at 613 cm^{-1} whose logarithm of intensity varies linearly with the concentration between 10^{-8} and 10^{-14} M (Figure 4.40(b)).

4.5. Conclusions

The various techniques described in this chapter differ not only by their greater or lesser complexity, but also in the limits of detection and specificity that each can achieve. These characteristics will guide the choice of the method, which will differ depending on whether the measurements can be carried out in a center equipped with heavy equipment or whether, on the contrary, the POC-type measurement must be resolved in situ and in real-time, far from a specialized center. The detection and analysis of an analyte contained in a mixture can be done using separation and identification methods such as chromatography and electrophoresis (see Chapter 2) or by using a specific sensor. This problem of specificity which arises for mineral salts and biological molecules is based on recognition systems, in which the signal – optical or electrical – provided by a specific sensor is characteristic of a single analyte. This technique currently prevails, and its advantage lies in its rapid execution. In the future, it could be in competition with the use of sensor networks, devoid of specificity with respect to a compound, but whose signals can be processed by artificial intelligence methods and are thus capable of providing the desired information.

PART 2

Environmental and Biological Sensors

5

Ion and Gas Sensors

For a very long time, the determination of acidity and ionized species in solution, as well as that of gaseous compounds has been a main line of research, directly related to environmental, agri-food and biomedical issues. The measurement of toxic ions, most often as traces, but also Na^+, K^+ and Cl^- (the "big three") ions considered vital in the human metabolism have been the focus of significant innovations and have led to numerous commercialized devices, constantly improved and increasingly oriented toward portable sensor systems connected to the Internet. Among the many transduction modes considered for ion assays, electrochemical systems based on the membrane's selective permeability to ions have been the most widely used. The first membranes, made with glass, were adopted for measuring pH and specific metal ions. Subsequently, membrane materials diversified considerably with the appearance of polymer membranes modified by complexing agents, specific for ions, which later gave rise to solid–state sensors. The nanostructuring of materials was introduced more recently with the design of portable miniaturized devices, most often produced using printing techniques such as inkjet or screen printing (screen printed electrodes [SPEs]). In gas analysis, chromatography coupled with mass spectrometry remains the gold standard, but at the expense of heavy equipment. This has led to the development of innovative, portable, stand-alone systems, which are increasingly in demand for monitoring atmospheric quality and the presence of toxic gases, as well as for monitoring the quality of food products. In the case of the latter, their deterioration over time (food spoilage), due to the presence of bacteria, is most often accompanied by the appearance of low concentrations of H_2S, NH_3 and other volatile amine compounds (named biogenic amines). Considered as indicators of food freshness, their detection at the point of storage is a public health imperative that requires the use of specific, portable and widely distributed sensors.

5.1. Membrane electrodes for potentiometric pH measurement

It is a long story, dating back to the very early 20th century, marked by the experiments of Cremer (1906) and particularly by those of Haber and Klemensiewicz[1], who in 1909 were the first to carry out a potentiometric acid–base titration curve, by measuring the potential difference between the two sides of a thin glass membrane separating two acid or alkaline solutions of different concentrations. These glass membranes, selective for H^+ ions, correspond to glass based on sodium silicate, resulting from the combination of sodium oxide and silica (Na_2O and SiO_2). Very quickly, the device became the analytical tool for measuring pH[2]. In the case where the glass electrode (GE) is strictly selective for protons, the electrode potential is then linked to the pH of the solution by the Nernst relation:

$$E = \text{Constant} + (RT/F) \ln a_{H+} \text{ that is, } E = \text{constant} - 2.3(RT/F)\, \text{pH}^3 \qquad [5.1]$$

where R represents the thermodynamic ideal gas constant (8.31 J mol^{-1} K^{-1}), T is the absolute temperature and F is the Faraday (96,485 coulombs).

Subsequently, various types of GEs were proposed, but the commercial launch of GEs only happened after a complete pH meter was marketed by Beckman in 1932, who proposed at the same time as the GE, a vacuum tube voltmeter with a high input impedance, essential for accurately measuring the potential of a glass membrane with a very high resistivity.

One of the important points of these electrode materials is that their selectivity to protons depends on the glass composition. As previously indicated, glass based on silica and alkali metal oxides are best suited to optimize pH measurement (Eisenman 1962). Introducing boron-based oxides (B_2O_3) modifies this selectivity and allows measurement of alkaline cations in solution such as Na^+ and K^+. Various theories

1 The advent of the discovery of glass electrodes and the universal nature of their use have been well described by Marczewska and Marczewski (2010).

2 pH was defined by the Danish chemist Sörensen (1909); it is a measure of the activity of H^+ ions in solution, that is, $pH = -\log_{10} a_{H+}$. The activity a_I of an ion I in solution is defined by the relation: $RT \log_{10} a_I = \mu_I - \mu_I°$, where $\mu_I°$ and μ_I represent the chemical potential of species I in the pure state and solution, respectively.

3 The potential of the glass electrode (GE) is measured relative to a reference electrode (Calomel type, $Hg\text{-}Hg_2Cl_2$), which is immersed in the test solution, involving a sequence of several solid/liquid and liquid/liquid phases such as [$Hg/Hg_2Cl_2/KCl$ saturated | Test solution | GE | HCl 0.1 M/AgCl/Ag]. The constant of relation [5.1] integrates the reference potentials of the Ag/AgCl and Hg/Hg_2Cl_2 electrodes as well as the different junction potentials. The range of measurable acidity depends on the composition of the glass, and generally, it is between 0 and 10 pH units.

were developed in the 1930s by Nikolsky et al., then 20 years later by Eisenmann et al. to interpret the functioning of GEs, particularly in the presence of interfering alkaline ions. This slow progress was described from a chronological point of view by Belyustin on the GE's 100th anniversary (Belyustin 2011).

It is now accepted that the functioning of glass membranes is based on the presence on its surface of thin hydrated layers (denoted as m' and m'' in Figure 5.1), with thicknesses <100 nm.

Proton exchanges with the solution occur in these layers, while the core of the membrane, several microns thick, is dry and impermeable to protons. The latter contains mobile sodium ions that diffuse into the membrane and are exchanged with protons at the interfaces (α | m') and (β | m''), thus maintaining the continuity of the ionic conductivity throughout the thickness of the glass.

In the presence of interfering ions in the solution (for example, Na^+), the Nernst relation that links the pH of a solution α to the electrode potential then includes a corrective term corresponding to the product of the activity of the interfering ion (a^{α}_{Na+}) and a dimensionless constant ($K^{pot}_{H+,Na+}$), called the potentiometric selectivity coefficient. The equation relating the GE potential to the pH of the solution α is then:

$$E = \text{Constant} + (RT/F)\ln\left(a^{\alpha}_{H+} + K^{pot}_{H+,\,Na+}\, a^{\alpha}_{Na+}\right) \qquad [5.2]$$

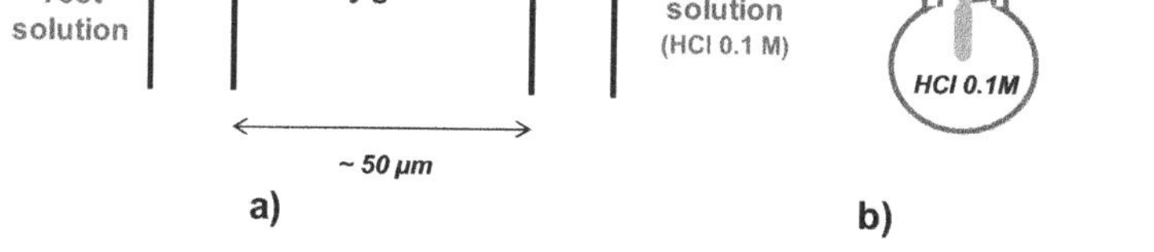

Figure 5.1. *Structural diagram of a glass electrode (GE). (a) Structure of a glass membrane sensitive to H^+ ions. (b) Simplified representation of a GE consisting of the bulb-shaped glass membrane and an Ag/AgCl reference electrode in contact with the 0.1 M HCl internal solution. For a color version of this figure, see www.iste.co.uk/lacaze/nanotechnologies.zip*

COMMENT ON FIGURE 5.1.– *The outer layers of the GE (m' and m''), several nanometers thick, are hydrated and permeable to H^+ ions unlike the core of the* *membrane (m), which is impermeable to protons. The pH is deduced from the voltage measured between a reference electrode (Calomel type, Hg/Hg_2Cl_2) immersed in the test solution and the Ag/AgCl electrode of the GE, immersed in the internal 0.1 M HCl solution. Adapted from Bard and Faulkner (2001).*

Its selectivity to protons will be greater the smaller the constant $K^{pot}_{H+,Na+}$ [4]. The value of this constant depends on the nature of the GE and the interfering ions and, in the case of electrodes dedicated to pH measurement, it is generally less than 10^{-4}, which then makes the interference with other ions negligible.

5.2. Ion selective electrodes (ISE)

Different glass membranes were subsequently developed for a more general analysis of cations (heavy metals) and anions. Significant progress was made in the 1960s with the introduction of pseudo-liquid and solid membrane electrodes, specific for metal ions, but particularly after the introduction of complexing or ion exchange compounds into a membrane (generally of polymer type). Known as ionophores or ion carriers, they have made it possible to considerably extend the field of application of these membranes, both for measuring pH in the most diverse environments and for all cations and anions. These ionophores are specific complexants for particular ions I^+, and their lipophilic nature means that they remain confined in the membrane. They can be neutral or electrically charged and keep the I^+ ions in the membrane at an almost constant and very low concentration, resulting in an interfacial potential between the membrane and the solution depending only on the concentration of the I^+ ions in the test solution (Figure 5.2).

Valinomycin[5] was one of the first ionophores used for potassium assays, followed by a series of other ionophores (calixarenes, crown ethers, amine complexing agents, etc.), a description of which we can find in several review articles, notably in Bakker et al. (1999).

---

4 a^{α}_{H+} and a^{α}_{Na+} correspond to the activities of the H^+ and Na^+ ions in the α phase (test solution), and the constants R, T and F have their usual meaning. The selectivity constant $K^{pot}_{H+,Na+}$ is deduced from thermodynamic considerations and is equal to $(a^{\alpha}_{H+}\ a^{m'}_{Na+})/(a^{m'}_{H+}\ a^{\alpha}_{Na+})$ (Bard 2001).

5 Valinomycin (Val) is a powerful antibiotic comparable to a crown-shaped molecule, comprising on its internal face many oxygen atoms, responsible for the very strong complexation of alkaline cations, in particular potassium ions K^+. This last property, combined with the hydrophobic nature of the molecule, allows the Val-K^+ complex to cross the bacterial cell wall, releasing K^+ ions inside the cell and destroying it by inverting the cellular ratio of K^+/Na^+.

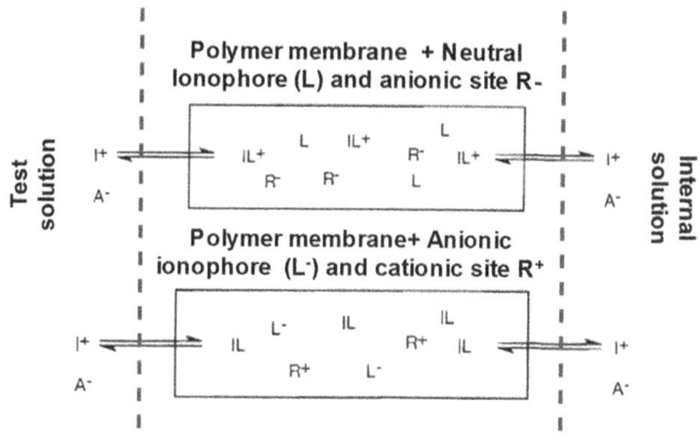

Figure 5.2. *Representation of the ionic equilibria between a membrane selective for I^+ ions and the test and internal solutions. The membrane is doped by a neutral (L) or charged (L^-) ionophore and carries R^- or R^+ ionic sites attached to the membrane. Adapted from Bakker et al. (1997). For a color version of this figure, see www.iste.co.uk/lacaze/nanotechnologies.zip*

As for GEs, the concentration of the primary ion is determined using thermodynamic considerations in the phase boundary model based on the equality of the chemical potentials of an ion in thermodynamic equilibrium between two distinct phases in contact (solution/membrane) (Bakker 2004). In the case where a single metal cation I^{z+} is in solution, the activity (a_I) of I^{z+} is linked to the potential E_I of the electrode by a Nernstian-type relation[6]:

$$E_I = \text{Constant} + (RT/zF)\ \ln a_I^{z+} \quad [5.3]$$

As with pH measurement, a corrective factor must also be introduced when other interfering cations are present (Nägele et al. 1999).

The mechanisms of action of these membranes have been the subject of numerous studies, applicable to ions of different charges, and are described in several review articles by Bakker and Pretsch (2005) and Bakker et al. (1997). The activity of the ion can then be related to the electrode potential by the semi-empirical and simplest equation [5.4], like that of Nikolsky and Eisenman established in the case of GEs where K_{IJ}^{pot} represents the potentiometric selectivity coefficient and J is the interfering ion.

$$E = E_I^0 + (RT/z_IF)\ \ln(a_I + \Sigma K_{IJ}^{pot}\ a_J^{zI/zJ}) \quad [5.4]$$

6 As for glass electrodes, E_I^0 in relation [5.4] integrates the different parameters in the measurement chain, which connects a reference electrode to the selective membrane electrode via several electrolytic junctions, both immersed in the test solution.

The electrochemical cell used to measure the potential and determine the activity of an ion in solution is similar to measuring pH and involves several junctions between different phases corresponding to different potential values. An example of such a cell, consisting of a reference electrode, an indicator electrode and an ion-selective membrane is represented in Box 5.1.

Many electrodes have been designed using different types of membranes and ionophores. At one point in the 1960s, the company Orion released a new electrode each month, selective for a particular ion (Radu et al. 2013).

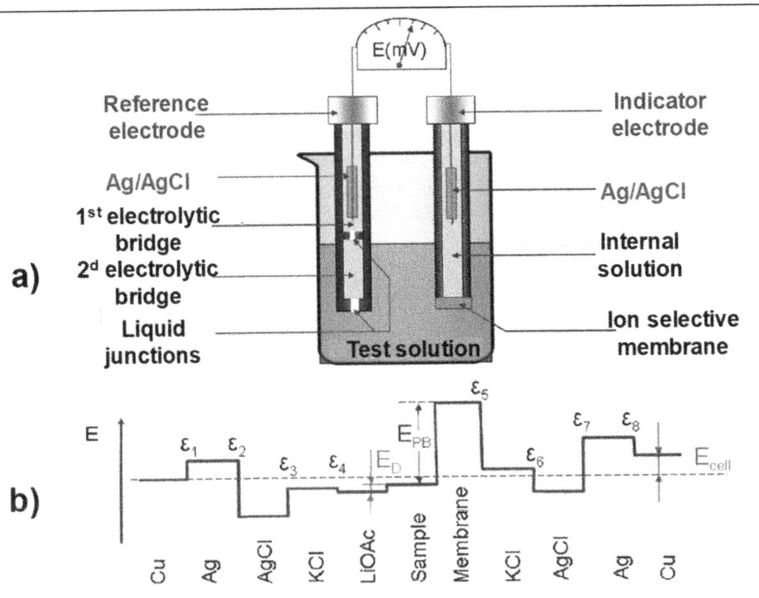

a) Typical representation of a cell potential measurement E_{Cell} with respect to a reference electrode Ag/AgCl with a double electrolyte bridge

b) Relationship between the potential E_{Cell} and the different interphase potentials. E_D (junction potential between solutions KCL and LiOAc) and E_{PB} (interfacial potential between the membrane and the solution or Phase Boundary Potential) are the only potentials that depend on the test solution, E_D being small with respect to E_{PB}. KCL and LiOAc are equitransferring electrolytes used to minimmize junction potentials. Adapted from Lindner et al.(Lindner 2013)

Box 5.1. *Schematic of a potentiometric measuring cell comprising an indicator electrode with an ion-selective membrane. For a color version of this figure, see www.iste.co.uk/lacaze/nanotechnologies.zip*

However, it should be noted that until the end of the 1990s, ISE electrodes could not reach detection limits beyond 10^{-6} M. This is the main reason why they were not used for environmental applications, where the analyzed concentrations of toxic products can be very low, often below ppb[7]. New ideas then emerged and led to a considerable improvement in these limits of detection (LOD). Given their ease of use and possibility of miniaturization, ISEs established themselves in the environmental field to such an extent that they caused a "new wave" of ion-selective electrodes, which generalized and considerably extended their field of application (Pretsch 2007).

5.2.1. *Toward ISEs with very low detection limits*

Detecting traces of pesticides, herbicides or heavy metals (Pb, Cu, Cd, etc.) in soil or in water is one of the main undertakings in environmental protection and management. This requires the use of analysis techniques that detect concentrations less than ppb, easily achieved with gas or liquid chromatography techniques coupled with mass spectrometry analysis or, using ultra-sensitive techniques such as surface enhanced Raman spectroscopy (SERS), flame atomic absorption spectrometry (FAAS), perfectly proven techniques but sometimes cumbersome to implement.

With increasing pollution inspection locations, the need to communicate the information collected at different control centers over the Internet in real-time involves significant deployment logistics. ISEs offer undeniable advantages in terms of cost and manageability and can reach beyond nano-molar detection limits.

As mentioned above, in the 1990s such values were far beyond the capabilities of the ISE, one of whose preferred fields was clinical chemistry applied to the determination of Ca^{2+}, Na^{+}, K^{+} and Cl^{-} ions in blood, serum or plasma, where ionic concentrations are relatively high and much higher than micro-molar concentrations.

For ion (I) concentrations below 10^{-6} M in the presence of an interfering ion J with a potentiometric selectivity coefficient $K_{I,J}^{pot}$ between 10^{-2} and 10^{-5}, the potential curve $E = f(a_I)$ no longer has a linear relationship between the potential and the logarithm of the activity a_I. The greater the selectivity coefficient K_{IJ}^{pot} and the lower the I^{zi} ion concentrations, the greater the potential differences, counteracting the achievement of very low LODs (Figure 5.3).

7 For a product with a molecular mass of 500, a concentration of 1 ppb corresponds to 1 µg of product per liter of water or per kg of material, that is, a concentration of 2 nM.

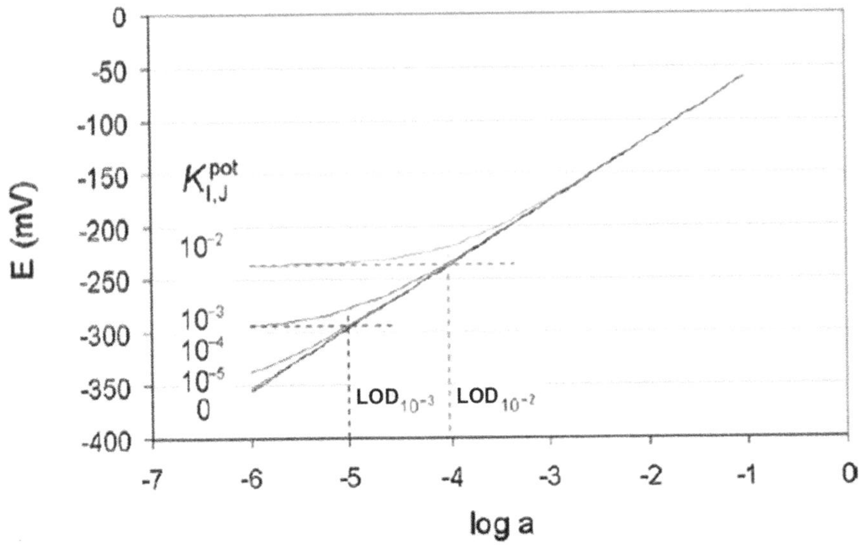

Figure 5.3. *Potentiometric curves of an ion I^{zI} in the presence of an interfering ion J^{zj} and for different potentiometric selectivity constant values $K^{Pot}_{I,J}$. According to Lindner et al. (2013). For a color version of this figure, see www.iste.co.uk/lacaze/nanotechnologies.zip*

COMMENT ON FIGURE 5.3.– *The activity a_I of ion I^{zi} varies between 10^{-1} and 10^{-6} M. Different potentiometric curves were obtained for four potentiometric selectivity constant values (10^{-2},10^{-3}, 10^{-4} and 10^{-5}) and the interfering ion J^{zJ} with an activity of 10^{-2}M. The higher detection limits of I^{zi} for the $K^{Pot}_{I,J}$ constant equal to 10^{-2} and 10^{-3} are 10^{-4} M ($LOD_{10^{-2}}$) and 10^{-5} M ($LOD_{10^{-3}}$), respectively.*

In the case of an ISE comprising a classic liquid membrane and a solution of an ion I in the presence of an interfering ion with a concentration of 10^{-2} M, DLs are relatively high and increase from 10^{-5} to 10^{-4} when the selectivity constant increases from 10^{-3} to 10^{-2} (Lindner and Pendley 2013).

The need for selective membranes with very low selectivity coefficients $K^{Pot}_{I,J}$ is however insufficient to reach LODs greater than 10^{-6} M. Another phenomenon, specific to traditional ISEs, disrupts the measurement, since the internal reference solution, which contains the primary ion (the ion to be measured) at a high concentration compared to those of the solution and of the membrane, is likely, precisely, to pollute the primary solution. The concentration gradient between the internal solution and the membrane causes ions to flow in this direction toward the membrane, and from the membrane toward the solution containing the primary ion. This pollutes the test solution and makes reaching very low LODs impossible under these conditions (Sokalski et al. 1997, 1999; Szigeti et al. 2006; Pretsch 2007; Hu et al. 2016).

Pretsch and colleagues solved this problem and showed that reversing the concentration gradient of the primary ion between the membrane and the internal solution could suppress the ion flow toward the test solution and thus achieve LODs around the ppb (Sokalski et al. 1997). This principle was applied by introducing a Pb^{2+} complexing agent into the internal solution, considerably lowering its concentration in the internal solution (~10^{-12} M) and thus allowing Pb to be detected in very low concentrations around ppb (Sokalski et al. 1999)[8].

In conclusion, polymer membranes and specific ionophores allowed most cations and anions to be practically measured using the classic two-electrode setup until the late 1990s, as shown in Box 5.1. However, this type of device is difficult to miniaturize or make portable. This is why all-solid systems were considered, whose external and internal reference electrodes are replaced by solid contacts, also favoring nanomaterials such as graphene and carbon nanotubes (CNTs).

5.2.2. *Solid contact ISEs*

Replacing the characteristic electrolyte phases of external and internal references with solid contacts is an undeniable simplification (Figure 5.4), acceptable, however, only if the ion-electron transduction is as efficient as in the case of an electrolytic junction between the selective membrane and the internal reference solution.

The first solid-contact ISEs were produced in the early 1970s with a Ca^{2+} ion sensitive electrode and involved directly depositing a PVC membrane and its ionophore onto a platinum wire, thus eliminating the internal solution (Cattrall 1971). It compares very well with a traditional commercial electrode (Orion, Ca^{2+}), which sparked significant research into other types of selective membranes as well

8 The ISE electrode consists of a polyvinyl chloride (PVC) membrane, thickness between 50 and 200 μm, containing 1% by weight of the Pb ionophore (a commercial calixarene, known as Pb ionophore IV), 0.8% of NaTFPB (buffer solution that maintains a slightly acidic pH, around 4), as well as a significant quantity (64.5%) of a lipophilic ether oxide o-NPOE (2- nitrophenyl octyl ether), which has a plasticizing effect on the membrane. The internal solution contains Pb^{2+} in very low concentrations (around 10^{-12} M) due to the action of the complexing agent Na_2EDTA 10^{-2} M (ethylenedinitrilotetraacetic acid disodium salt dihydrate) added to a solution of $PbCl_2$ 10^{-3} M. This solution also contains a very high concentration (2M) of NaCl, which ensures the connection with an internal Ag/AgCl reference cell immersed in a 3 M KCl solution.

as all-solid-state reference electrodes[9] and reached detection limits close to those of traditional cells with liquid junctions and double salt bridges.

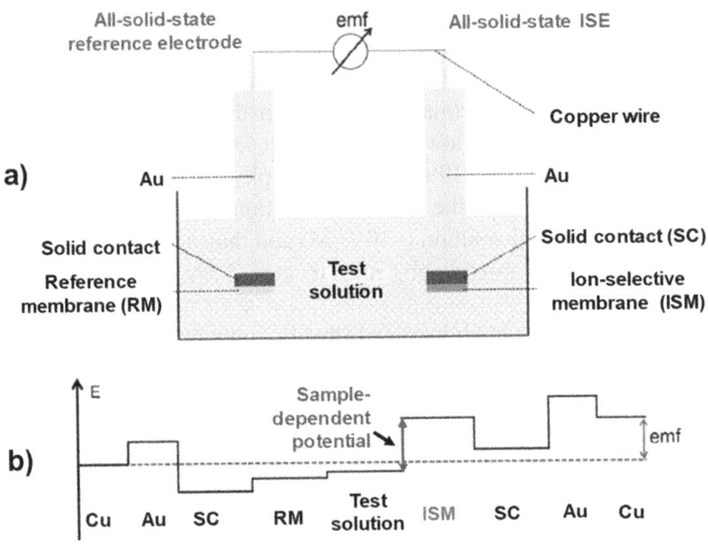

Figure 5.4. *Schematic representation of an electrochemical cell for potentiometric ion measurements comprising all-solid-state electrodes. (a) Device for measuring the voltage (electromotive force [EMF]) between an all-solid-state ISE electrode and an all-solid-state reference membrane (RM). (b) Relationship between the voltage and the different successive interphase potentials, of which only one "solution/ISM" depends on the ionic composition of the solution. Adapted from Hu et al. (2016). For a color version of this figure, see www.iste.co.uk/lacaze/nanotechnologies.zip*

Until the mid-2000s, conductive polymers (see Box 5.2) such as polypyrrole, polyaniline, PEDOT (poly (3,4-ethylenedioxythiophene) and POT (polyoctylthiophene) were used in the design of solid-contact ISEs and are described in several review articles (Bobacka 2006; Michalska 2012; Hu et al. 2016). In direct contact with the selective membrane, they have the advantage of being both ion and

9 Sokalski and colleagues (Mousavi 2013) proposed an all-solid-state reference electrode consisting of a silver wire coated with silver chloride (Ag/AgCl) and an insoluble polymer/salt composite (polyvinylacetate-KCl), in contact with a second composite (polyvinylacetate–LiOAc). According to the authors, such a system (equivalent to an Ag/AgCl reference with a double electrolytic junction) is highly stable over time and does not require frequent calibration. After conditioning in KCl solution, potential drifts do not exceed ± 0.5 mV over several months.

electron conductors, which makes them very good transduction elements for ion-electron conversion.

The redox mechanism at the interface between the conductive polymer in its oxidized state (CP^+) (deposited on a metal constituting the solid contact SC) and the membrane (Mbne) containing the ionophore is thus defined by the electron exchange reaction:

$$CP^+A^-{}_{(SC)} + M^+{}_{(Mbne)} + e^- = CP^0A^-M^+{}_{(SC)}$$

$$\text{or } CP^+R^-{}_{(SC)} + e^- = CP^0{}_{(SC)} + R^-{}_{(Mbne)}$$

in which A^- is a PSS (polystyrene sulfonate) non-mobile counterion, R^- is a small anion and M^+ is the ion to be measured. Similar techniques have also been developed to produce highly stable solid-contact reference electrodes (Kisiel 2005; Rius-Ruiz et al. 2011).

POT proved to be particularly effective in forming a solid-contact ISE electrode, achieving a subnanomolar LOD in the case of K^+ ions (Sutter et al. 2004)[10]. This polymer shows very good stability over time, attributed in part to its pronounced lipophilic character, which considerably reduces the adsorption of water between the polymer film and the solid contact and suppresses the accumulation of ions at the interface, causing erratic responses. Moreover, the device can be stored in ambient air for long periods and retains its selectivity and reproducibility properties.

Subsequently, significant improvements in stability were made with macroporous and conductive carbonaceous layers, obtained using the nanosphere lithography technique. Developed by the Bühlmann group and called 3DOM-carbon (three-dimensionally ordered macroporous carbon), these conductive carbon layers (thicknesses of a few tens of nanometer) are made up of an ordered network of spherical pores interconnected between them and of sub-micrometric size (diameters of a few hundred nanometer) (Lai et al. 2007; Hu et al. 2016). Placed directly in contact with a selective ionophore membrane (PVC + valinomycin), the all-solid-state electrodes formed are selective for K^+ ions in a concentration range between 10^{-1} and 10^{-6} M. Furthermore, unlike ISEs with conductive polymers, they are insensitive to light and redox pollutants such as O_2. They are also characterized

10 The all-solid ISE electrode consists of a gold disk on which the conductive polymer poly(n-octyl-3 thiophene) (POT) is deposited. Obtained by direct electropolymerization of the monomer (n-octyl-3 thiophene) or by deposition of the polymer in solution, it is then covered by a poly(methyl methacrylate)/poly(decyl methacrylate) (MMA/DMA) copolymer membrane containing the ionophore (valinomycin).

by very high stabilities over time[11]; potential drifts are significantly lower than those of previous systems and have been evaluated at approximately 0.3 mV per day.

The design of low-cost, easy to handle all-solid-state electrochemical systems, particularly suitable for customized biological point of care (POC) analyses, is also an increasingly popular area of research, including the creation of microfluidic devices based on the use of commercial filter paper.

Whitesides et al. were the first to produce microfluidic circuits on paper for the rapid detection of various biological compounds (glucose, proteins, etc.), most often adopting a colorimetric detection mode (Martinez et al. 2007)[12]. A few years later, Bühlmann, Whitesides et al. showed that the paper support technique could also be used to produce electrochemical paper-based analytical devices (EPAD) for the selective analysis of chloride ions (Lan et al. 2014) (Figure 5.5).

An EPAD device is composed of a hydrophilic paper support, on which hydrophobic barriers (partitions to ensure watertightness against the aqueous medium) have been printed, for example, a screen-printed hydrophobic wax delimiting the thin electrochemical paper cell (Figure 5.5(a)). The solution to be analyzed (a few microliters) and the reference solution are located respectively in the left and right parts of the cell; they are connected electrochemically by a salt bridge located in the middle of the device, also obtained by impregnation with a concentrated KCl solution (Figure 5.5(e)). An ion-selective membrane, soaked with a specific ionophore, is placed above the sample area, itself capped with an Ag/AgCl inner reference electrode (Figures 5.5(b)–(e)).

The very low diffusion rate of ions in the paper makes it possible to effectively separate the different zones without risk of overlap and guarantees the stability of the ionic concentrations in the different zones of the EPAD (Lan et al. 2014).

Numerous technical improvements were subsequently made to the manufacturing methods of these EPADs and the use of papers as a carrier has evolved considerably, with a view to the widespread use of POC procedures, mainly for the analysis and detection of biological compounds in the blood (Li et al. 2019)[13].

---

11 The potential shift is linked to the fact that the electrode is not completely unpolarizable. Considering the potential measurement conditions close to current intensity 0 (in fact, less than 10^{-12} A), potential shift (ΔE) is determined by the capacitance C of the electrode and the residual current i due to the measurement, that is, $\Delta E/\Delta t = i/C$. It is all the lower as the capacity of the electrode is high, which justifies the use of microporous carbon membrane electrodes with very high capacitances (Lai et al. 2007; Hu et al. 2016).

12 These are generally paper used in thin layer chromatography, first used by Whitesides and colleagues to simultaneously measure glucose and a protein colorimetrically (Martinez et al. 2007).

13 According to Meticulous Research, the POC sensor market is expected to grow considerably, exceeding 30 billion dollars in 2024.

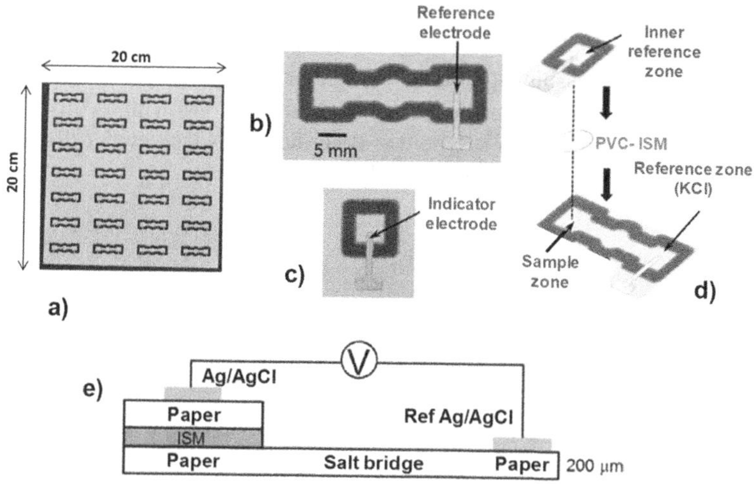

Figure 5.5. *Schematic representation of an EPAD with an ion-selective membrane. (a) Image of 28 EPADs printed on a sheet of paper (gray) and delimited by wax (black) hydrophobic "walls" obtained by printing. (b) Detail of the lower part of an EPAD for the detection of K^+, Na^+ or Ca^{2+} ions including a printed Ag/AgCl reference electrode (right part of the EPAD). (c) Ag/AgCl indicator electrode placed above the selective PVC-ISM membrane (left part of the EPAD). (d) Assembly diagram of the different parts of the EPAD. The ion-selective membrane (ISM) is inserted between the lower part of the EPAD (sample zone) and the indicator electrode. (e) Cross-sectional view of the EPAD indicating the presence of a salt bridge located halfway between the areas in which the sample and the reference electrode are located. Adapted from Lan et al. (2014). For a color version of this figure, see www.iste.co.uk/lacaze/nanotechnologies.zip*

In the case of paper-based ion analysis and detection, the use of graphene and CNTs as solid contacts have significantly improved the transduction system. Rius-Ruiz et al. produced an ISE type sensor[14], designed in the form of a plastic strip on which the different sensor constituents are printed in successive thin layers (ISM and ionophore, solid paper contact made of an agglomerate of SWCNTs, indicator electrode and reference electrode produced by screen printing) (Crespo et al. 2008; Rius-Ruiz et al. 2011).

14 The device is produced by screen printed electrodes (indicator and reference) on a hydrophilic polyacrylic support. The solid contact is composed of an assembly of single-walled CNTs (SWCNTs), functionalized with octadecyl amine (ODA). The polyacrylic support helps to significantly reduce ion diffusion and makes it possible to achieve very low LODs.

Hydrophobic and conductive CNTs, in contact with the K^+ ion-selective membrane, have an amplifying effect on the sensor response. The system makes it possible to analyze potassium solutions in various media (blood, drinks, saliva, etc.) and for a range of concentrations between 10^{-5} M and 10^{-2} M with a detection limit of $10^{-6.5}$ M. A system of the same type was described by Novell et al. (2014) for the detection of lithium in blood[15] for POC analyses (Figure 5.6).

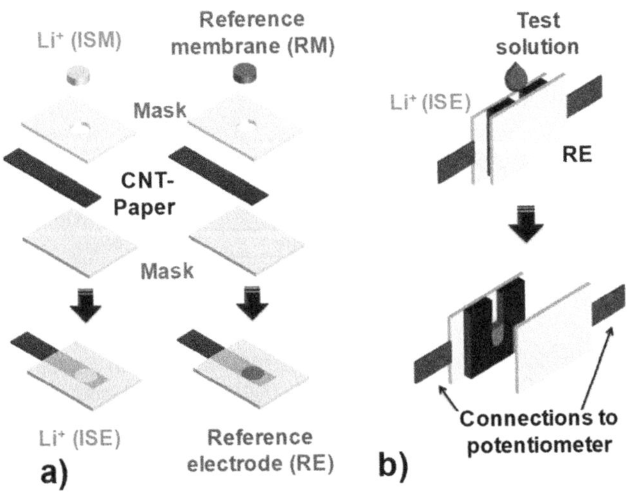

Figure 5.6. *Manufacturing diagram of an ISE type potentiometric paper sensor for the detection of Li^+ ions in blood. (a) Successive manufacturing stages of indicator and reference electrodes. The indicator electrode (left) is made by adhesion of a strip of CNT paper (black) between two plastic masks (gray), one of them comprising a circular cavity into which Li^+-ion-selective membrane is inserted (yellow, polyacrylate with added lithium ionophore Li^+(ISE). The reference electrode (RE, right) is constructed according to the same diagram using a reference membrane electrode. (b) Assembly diagram of both electrodes indicating the location of the solution to be analyzed in a U-shaped receptacle (CNT paper) sandwiched between the two electrodes. Adapted from Novell et al. (2014). For a color version of this figure, see www.iste.co.uk/lacaze/nanotechnologies.zip*

More recently, Citterio and colleagues produced all-solid-state thin layer ISE cells for the determination of Na^+ and K^+ ions using the inkjet printing technique on filter paper and using a poly(3,4-dioxythiophene)/graphene composite as the solid

15 Lithium ion is one of the drugs used for the treatment of depressive illnesses. It acts at blood concentrations between 1.5 and 5 mM. Below these concentrations it is ineffective and above 5 mM it can be dangerous, hence the need to be able to control its concentration frequently.

contact (Ruecha et al. 2017). The geometry of the EPAD is the same as that defined by Lan et al. The ionophore electrode is thus directly printed onto a solid support; the all-solid-state reference electrode is a pseudo-Ag/AgCl reference electrode. The advantage of such a system lies in its ease of use and its low manufacturing cost, with the added possibility of large-scale production. Such devices allow the analysis of various ions with only very small quantities of solutions (around a few microliters) (Nilghaz et al. 2016).

The use of ISE electrodes for the potentiometric determination of non-ionic biological compounds is excluded a priori, but may be considered indirectly, as demonstrated by Bakker and Pretsch and their teams, who showed that antibodies (proteins) could be determined by combining an immunological step with indirect potentiometric detection using an ISE microelectrode sensitive to Ag^+ ions (Chumbimuni-Torres et al. 2006).

The example chosen by the authors is that of the determination of the mouse antigen (immunoglobulin G, IgG) by anti-mouse antibodies[16], also of the IgG type, based on the ELISA immunological technique (see Chapter 6), followed by the determination of Ag^+ metal ions by means of a microelectrode of the ISE type. The correlation between the Ag^+ ions and the IgG antigen is obtained by using, on the one hand, a primary IgG antibody adsorbed on a gold substrate, and, on the other hand, a secondary IgG antibody bound to gold NPs coated with metallic silver, whose dissolution by chemical oxidation (H_2O_2) releases Ag^+ ions whose concentration is related to that of the antigen (Figure 5.7).

A good linear relationship is observed between the potential and the logarithm of the mouse IgG concentration, with a slope of 26 mV per log (μg/mL). The very high detection sensitivity of ISE, which in the case of Ag^+ ions is around 10^{-9} M, allowed them to detect the mouse IgG antigen over a concentration range from 0.25 to 7.5 μg per mL of solution, which corresponds to an LOD of around 12.5 picomoles for 50 μL of solution.

The same authors showed that the use of aptamers, simpler than the antibody method, could also be very effective for the detection of proteins using the potentiometric method and ion selective electrodes (Numnuam et al. 2008). As mentioned before, the technique is inspired by a sandwich ELISA assay, but the antibodies are replaced by two aptamers[17], also capable of complexing a protein

16 The mouse IgG antibody is widely used in biology because it is seen by the human body as an antigen, and provokes an immune response, known as the HAMA (human anti-mouse antibody) response, corresponding to production of human IgG anti-mouse IgG antibody.

17 Aptamers are peptide molecules or oligonucleotides used for their ease of binding with biological target molecules. They can be synthesized but also exist naturally. They are

target. One of the aptamers is "tagged" by a nanocrystal, subsequently decomposed into metal ions, which can be measured using a very sensitive ISE (Figure 5.8).

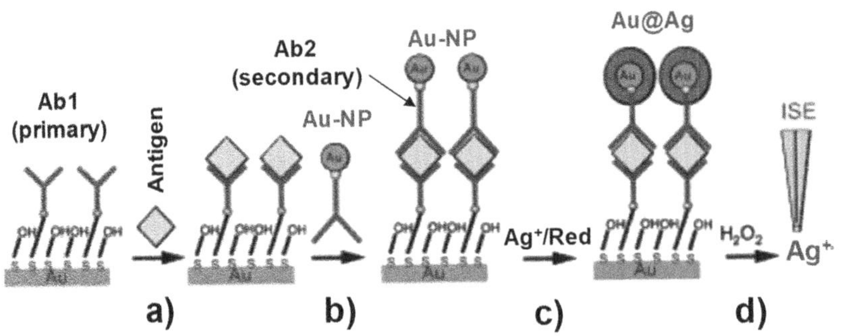

Figure 5.7. *Principle of an immunopotentiometric assay. (a) Addition of the antigen IgG to the anti-mouse primary antibody (Ab1) fixed on a gold substrate. (b) Capture of anti-mouse antibodies (secondary Ab2) tagged with Au-NPs; (c) catalytic deposition of metallic silver on Au-NPs by Ag^+ ion reduction. (d) Dissolution of silver by hydrogen peroxide and release of Ag^+ ions detected by potentiometry using a micro-ISE. Adapted from Chumbimuni et al. (2006). For a color version of this figure, see www.iste.co.uk/lacaze/nanotechnologies.zip*

COMMENT ON FIGURE 5.7.– *The surface on which the thiolated (primary) antibody is adsorbed has vacant spaces, which must be made inaccessible to the (secondary) antibodies tagged by the gold NPs. This is achieved by adding a thiolated alcohol (6-mercapto-1-hexanol), which blocks the free sites on the gold surface and prevents the antigen from being adsorbed.*

The example chosen by the authors is that of thrombin[18] complexed by two aptamers, one of which is tagged with the CdS nanocrystal. Its oxidation by H_2O_2 releases Cd^{2+} ions, which are measured by an ISE comprising a Cd^{2+}ion-specific membrane. According to the authors, an LOD of 28 fmole (28×10^{-15} mole) of thrombin in 200 μL of solution is reached, therefore a concentration of 0.14 nM (1.4×10^{-10} M).

---

classified into aptamers comparable to fragments of RNA or DNA, and into peptide aptamers (short peptide chain attached to a protein). They are widely used in research but also industrially for their clinical applications, notably their ability to vectorize macromolecular drugs.

18 Thrombin is a protease enzyme found in the blood and is recognized as one of the factors involved in a chain of reactions, leading to the formation of fibrinogen, then fibrin, responsible for the appearance of clots, which can lead to thrombosis.

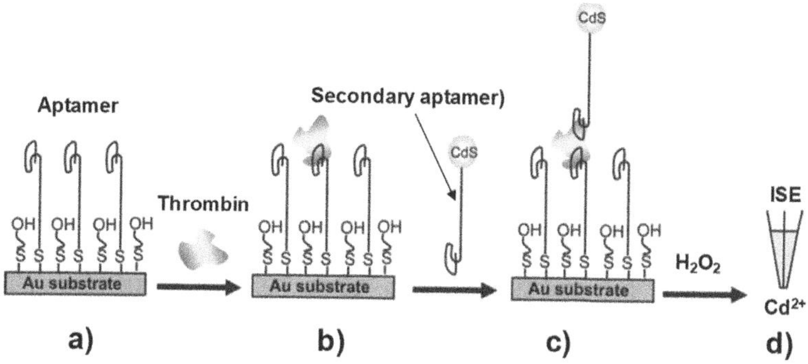

Figure 5.8. *The use of aptamers for the potentiometric assay of thrombin. (a) Formation of a mixed monolayer of thiolated aptamers on a gold substrate. (b) Addition of thrombin to the aptamer. (c) Addition of a CdS NP-tagged aptamer to thrombin. (d) Oxidation of CdS by H_2O_2 and determination of Cd^{2+} ions by a Cd^{2+}-ion-specific ISE. Adapted from Numnuam et al. (2008). For a color version of this figure, see www.iste.co.uk/lacaze/nanotechnologies.zip*

Compared to other physical analysis techniques (optical, piezoelectric, etc.), the use of electrochemical devices based on micro-ISEs make it possible to achieve very low detection limits, with advantages in terms of ease of handling, response rate and the low cost of the equipment. This type of analysis is essential for many applications, details of which can be found in the review article by Zdrachek and Bakker (2019).

5.2.3. *Conclusions on ISE electrodes*

ISE devices, developed shortly after the GEs to measure pH in the early 1930s, subsequently underwent a series of improvements. The development of ion-selective membranes and the advent of ionophores in the 1960s led to the development of a whole range of ISE electrodes, each specific to a particular ion. Next, the design of all-solid-state systems led to an acceleration of miniaturization and, at the same time, made it possible to produce low cost thin-film devices on a large scale using inkjet screen printing techniques on various paper supports, aiming to generalize POC analysis devices. Using these electrodes for assays of non-ionizable biological compounds became possible by following an indirect procedure involving a concentration relationship between the biological compound and an ionizable metal associated with the target. Very low concentration thresholds can thus be achieved, but to the detriment of experiment simplicity.

5.3. Gas sensors

Gas sensors are and have always been an important topic due to the need to protect the population against toxic gases in the atmosphere and in confined spaces. This involves simple, rapid and reliable detection methods, necessary to define effective prevention measures, also applicable to the detection of explosives based on their vapor pressures. They have also established themselves in the agri-food sector, where they are used for the detection of volatile compounds resulting from food spoilage, as well as in the biomedical sector for the analysis of certain gaseous biomarkers secreted by living organisms during respiration.

Many methods to detect and measure gases have been recommended, each with advantages and disadvantages and differing by their detection selectivity for a particular gas and the transduction technique used in the development of the sensor. Milone et al. give a good description of this in a recent review article (Milone 2023). Metal oxide semiconductors such as ZnO, SnO_2, NiO, CdO, TiO_2, WO_3, Fe_2O_3, In_2O_3, etc., which can be easily prepared at low cost in the form of thin layers using physicochemical methods, have been widely used to produce early gas sensors. In this regard, SnO_2 was one of the first metal oxides considered for a prototype. Subsequently, the nanostructuring of the sensitive layer carried out with nanowires (NWs) of metal oxides, graphene (G), graphene oxide (GO), CNTs, conductive polymers, as well as carbon composites of these oxides was considered with the aim of improving the intensity of the transduction signal. In this context, many assay examples have been described and concern the most common gases (O_2, H_2, CO_2), toxic gases (H_2S, NH_3, NO_2, CO), air humidity and volatile organic compounds (VOCs) (hydrocarbons and organic solvent vapors). Their low selectivity, the main limitation of these metal oxides, can be considerably improved using materials with controlled porosity as receptors such as metal organic frameworks (MOFs) (see Chapter 1) and molecularly imprinted polymers (MIPs) (see Chapter 3). Another approach, focused on the detection selectivity of a particular gas, particularly when the gas is contained in a mixture, and no sensor is sufficiently selective, involves using sensor networks, each generating a distinct transduction signal. Statistical and computer processing of the different signals makes it possible to deduce the concentration of the desired gas, a procedure equivalent to that of the electronic nose (odor sensors).

5.3.1. *Gas sensors based on nanostructured metal oxides*

Early sensors for detecting gases were based on the variation in conductivity of a metal oxide semiconductor in contact with a gas. Barsan, Weimar, and colleagues described them in detail in the case of tin oxide (SnO_2), and considered the prototype of gas analyzers due to large variations in conduction caused by both reducing and oxidizing gases (Barsan et al. 1999, 2007; Barsan and Weimar 2001).

The device marketed in the 1970s[19] is relatively simple. Schematically, the sensor is composed of two metal electrodes deposited in the vapor phase onto the sensitive layer (SnO_2), itself attached to an insulating substrate. In the case where gas analysis involves ionization of the sensitive layer (e.g. O_2 transformed into O_2- at high temperature), an additional heating device is provided to bring the sensitive layer above 150°C, necessary for ionization of O_2 to O_2^-, O^- and O_2^- and promote redox reactions with other gases. The variations in conductivity are directly linked to the charge transfer between the adsorbed species and the sensitive semiconductor layer. With a reducing gas, such as CO mixed with ambient air, the reaction with O^- ions adsorbed on the surface of a SnO_2 NW (n-type SC) leads to the formation of CO_2 and an electron transfer in the SnO_2 conduction band ($CO + O^- \rightarrow CO_2 + e^-$), which induces a significant increase in the SC conductivity (Kolmakov et al. 2003). For an oxidizing gas such as NO_2, the opposite reaction occurs with the transfer of an electron from the SC to NO_2 ($NO_2 + e^- \rightarrow NO_2^-$), causing a depletion of the electron density in the SC (Barsan et al. 2007; Shen et al. 2009).

These results led to classifying gaseous chemical species into two groups: oxidizing gases such as NO_2, and reducing gases such as NH_3, H_2S, CO, H_2 and water vapor. In the first case considered as electron acceptors, they deplete the SC electron density. When the oxide is an n-type semiconductor (ZnO, SnO_2 and In_2O_3), there is a reduction in conductivity; when the oxide is a p-type semiconductor (CuO or a conducting polymer), an increase in conductivity is observed, which corresponds to an increase in hole density. In the second case, the reducing gases are electron donors, and the conductivity variations in the n and p oxides are then reversed compared to with oxidizing gases (Huang and Wan 2009).

The operating conditions (polarization voltage, nature of the electrodes, operating temperature) and the conditioning of the sensitive layer (porous or compact structure, doping of the oxide with a metal) are all parameters likely to influence the phenomenon of transduction.

Yamazoe (1991) was among the first to show the influence of the microstructure of the oxide layers on the transduction function. In the case of SnO_2 particles, the grain size (D) relative to the thickness (L) of the space charge layer plays a crucial role in the sensor sensitivity. Furthermore, adsorption onto SnO_2 of PdO or Ag_2O promoters, formed in air from the metals Pd and Ag, contributes to the formation of

19 The first commercial gas detectors based on the use of metal oxide semiconductors date to the 1970s. The first filed patent stated that the operation of a gas detector was due to N. Taguchi (U.S. Patent, No. 3,631,436, 1971).

an electron-depleted space charge layer, which increases the sensitivity of SnO_2 to gases (see Chapter 3, section 3.1.6.2).

The use of metal oxide NWs is another way to significantly increase conductivity variations, measured using an FET or a resistive device with interdigitated electrodes (Ra et al. 2008) (Figure 5.9).

The sensors developed by Ra et al. show a sensitivity to CO five times better than observed on a ZnO film, the same improvement being observed with H_2. The presence of networks of parallel NWs, requiring the implementation of a sophisticated lithography technique, is however not essential and similar results are obtained with a disordered arrangement of NWs or even with sensors composed of a single NW with a resistive or FET assembly.

Despite good sensitivity, these metal oxide-based sensors unfortunately are not very selective and, also, require high operating temperatures. This not only complicates the practical production of the sensor but requires significant energy consumption, a challenge when manufacturing a portable electronic system (Bag and Lee 2019).

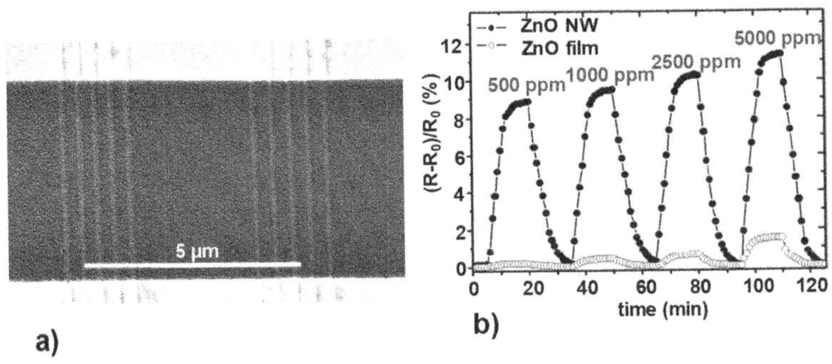

Figure 5.9. *ZnO nanowire networks used for the conductometric determination of carbon monoxide (CO) at 200°C. (a) Scanning electron microscopy (SEM) image of ZnO nanowires deposited on an insulating amorphous carbon film; the width and thickness of each nanowire is 70 and 100 nm, respectively, with an interwire distance of 700 nm. (b) Comparison of the relative variations in conductivity of a ZnO film (100-nm thick) and ZnO nanowires obtained for successive additions of CO (between 500 and 5000 ppm). R_0 represents the resistance of a ZnO nanowire in contact with dry air, and R is the resistance in the presence of CO. Adapted from Ra et al. (2008). For a color version of this figure, see www.iste.co.uk/lacaze/nanotechnologies.zip*

For all these reasons, new materials have been sought to improve selectivity and, in particular, to enable gas analysis at room temperature. A number of two-dimensional materials, such as graphene and transition metal dichalcogenides (TMD), as well as conductive polymers, have been the subject of numerous studies, in some cases leading to promising results for a new generation of sensors. As we will see below, their operation is based measuring the variation in conductivity, observed with resistive or FET type assemblies. For each device, the sensitive layer comprises nano-structured composites, most often resulting from an assembly between inorganic NPs and organic or mineral compounds arranged in thin layers or in the form of microwires or NWs.

5.3.2. *Gas sensors based on conductive polymers*

Conducting polymers are a family of macromolecules with unique electrical properties, described for the first time in the 1970s for polyacetylene $(C_2H_2)_n$. Heeger, MacDiarmid and Shiragawa showed that this compound can evolve from the insulating state ($1.7 \times 10^{-9}\ \Omega^{-1}\ cm^{-1}$ in its *cis* conformation) toward a conductive state ($38\ \Omega^{-1}\ cm^{-1}$) by simply doping the material through the chemical oxidation with iodine[20].

Subsequently, new conducting polymers were discovered, the most used being polypyrrole (PPy), polyaniline (PANI), polythiophene (PT) and poly-3,4-ethylenedioxythiophene (PEDOT). An additional advantage over poly-acetylene was that these polymers can be prepared electrochemically and obtained in the form of films of controllable thickness by oxidation of the monomer on the metal electrode. Monomers can also be functionalized, increasing the possible chemical reactions of the corresponding polymers, which can be exploited in the design of a new sensor (Box 5.2).

Combining nano-composites with NPs or inorganic conductive or semiconductor NWs is the most used technique for making gas sensors. The principle is like that of a p-n heterojunction formed between the p-type semiconductor polymer and n-type semiconductor NPs. Several sensors of this type have been produced according to this concept mainly using PANI, but also PPy, and PT, generally arranged so they coat various semiconductor NPs (ZnO, SnO_2, TiO_2, etc.). They have been exhaustively listed in recent review articles (Yan et al. 2020; Preethichandra et al. 2023). Many sensors have been designed for detecting NH_3 and H_2S[21], with application in food spoiling control. Due to the nature of the inorganic nanoparticles, very different detection thresholds are obtained, varying between ppt and ppb, as has

20 Heeger, MacDiarmid and Shiragawa received the Nobel Prize in 2000 for this discovery.

21 NH_3 and H_2S are two toxic gases dangerous to human health. Exposure to NH_3 vapors can lead to blindness, pulmonary edema and respiratory irritation (Liu et al. 2014a). Beyond 10 ppm, it causes eye irritation; exposure beyond 2000 ppm for a few minutes carries danger of death (Yamazoe 2005) (http://www.epa.gov/Air/Nitrogenoxides/). H_2S has similar detrimental effects as carbon monoxide (CO) and according to INERIS (National Institute of Industrial Environment and Risks), exposure to 700 ppm of CO in air for 10 min can cause death.

been shown with the composites PANI/TiO_2 NW (Gong 2010) and PANI/SnO_2 NP, (Li et al. 2019).

Conductive polymers form a particular class of polymers, the first discovered being polyacetylene (PA) in 1977 by Heeger, MacDiarmid and Shirakawa. Their conjugated polyene structure permits chemical or electrochemical oxidation, which transforms them from an electrically insulating state to a conducting state. The most common-conducting polymers are polypyrrole (PPy), polythiophene (PTh), PEDOT (poly-3,4 ethylene dioxythiophene) and polyaniline (PANI) for their applications and ease of production.

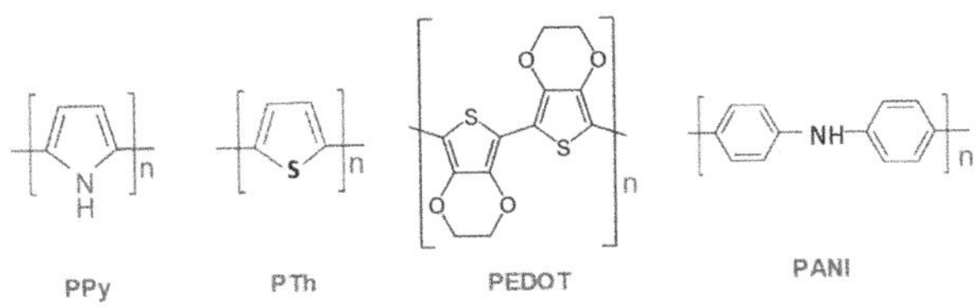

Their conductive state corresponds to a doping process carried out by the chemical or electrochemical oxidation of the neutral polymer. The first oxidation stage (by electron exchange) leads to a positive polaron structure (radical cation), then to that of a bipolaron (dication), in accordance with the diagram below, in the case of PPy:

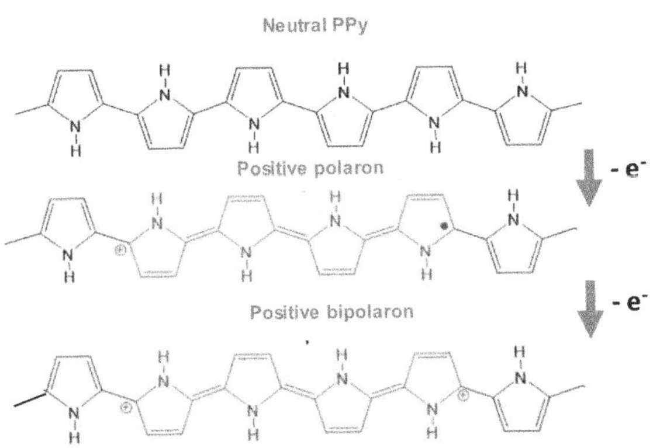

An advantage of these polymers is that they are easy to obtain in the form of thin films, with controllable thickness (few tens to several hundred nm), by electrochemical oxidation of the monomer on a noble metal electrode, or in the case of PPy, on oxidizable metals, after adaptation of the aqueous electrolytic medium (Petitjean et al. 2005). In their oxidized state, and for a random chain organization, the average conductivity value is around 400 $S.cm^{-1}$ (Le et al. 2017).

Box 5.2. *Structural characteristics of some conductive polymers. Adapted from Le et al. (2017). For a color version of this figure, see www.iste.co.uk/lacaze/nanotechnologies.zip*

The NH_3 sensor built by Gong et al. comprises TiO_2 microfibers obtained by electrospinning, on which PANI nanograins are deposited[22] (Figure 5.10(a)), and is characterized by an electrical operation that can be modeled on that of a p-n heterojunction (see Box 5.3).

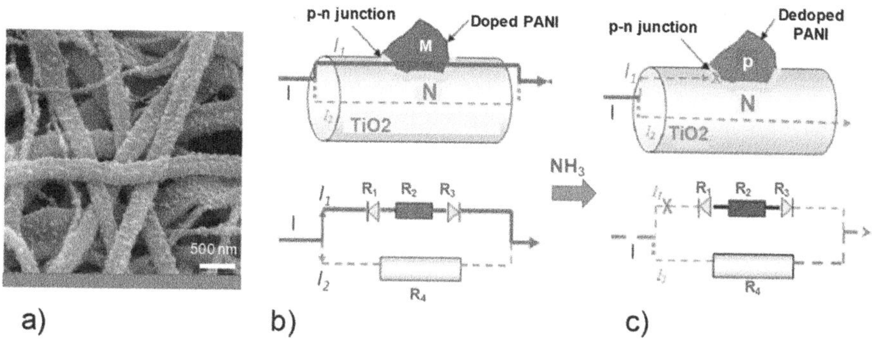

Figure 5.10. *Operating principle of TiO_2/PANI for NH_3 sensors. (a) SEM (scanning electron microscopy) image of the TiO_2 fiber/PANI grain composite. The PANI grains appear as slight protrusions (light gray) distributed on the surface of the fibers. (b) Electrical diagram of the device before contact with NH_3. The PANI grains are in the metallic state (M) and the surface current flows successively through the TiO_2/PANI (R_1) and PANI/TiO_2 (R_3) junctions and the low resistance-PANI (R_2). (c) After contact with NH_3, the PANI grains become almost insulating, which, compared to (b), induces a higher overall resistance and a significant decrease in current. Adapted from Gong et al. (2010). For a color version of this figure, see www.iste.co.uk/lacaze/nanotechnologies.zip*

22 Electrospinning is a technique involving stretching under an electric field, which allows the continuous production of polymer threads with diameters between a few μm and a few tens of nm. TiO_2 microfibers are obtained by calcination at 600°C of a fibrous precursor obtained by drawing a viscous mixture of Ti(IV) isopropoxide, polyvinylpyrrolidone (PVP), acetic acid and ethanol. The TiO_2 microfibers are then immersed into an aniline solution in the presence of ammonium persulfate and hydrochloric acid, which produces nanometric polyaniline (PANI) grains (40 nm) distributed on the surface of the TiO_2 microfibers (diameter 600 nm). These PANI grains correspond to the oxidized form and are therefore conductive.

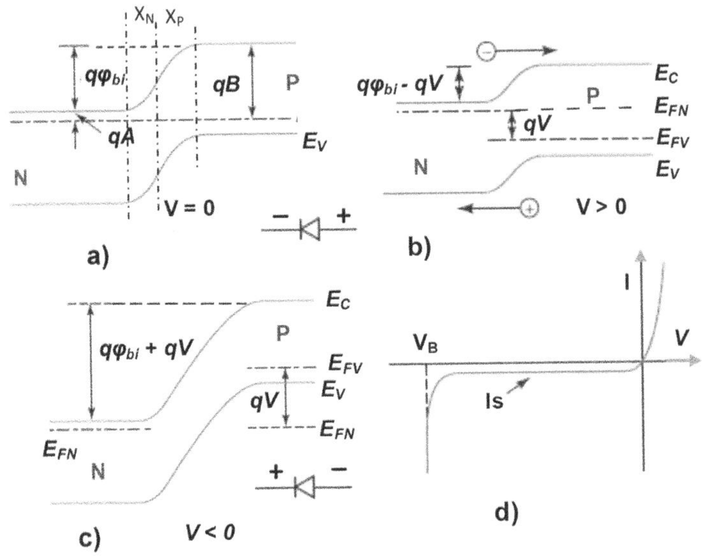

a) After aligning with Fermi levels, the junction between the n- and p-semiconductors (SC) forms a band structure comprising two depletion zones, with widths X_N and X_P, due to the diffusion of electrons and holes toward the respective P and N SCs. They constitute space-charge zones resulting from the respective presence of fixed donor ions D^+ and acceptor ions A^-, which create an electric field at the n-p interface, directed from n to p, which opposes the diffusion of holes and electrons, and stabilizes the junction with an energy barrier $q\phi_b$ for electrons and holes (q, electron charge).

b) With direct polarization V > 0, the band structure is modified, which separates the two Fermi levels E_{FN} and E_{FV} and causes a simultaneous decrease (qV) in barrier energy, increasing the current I, which varies according to the law $I = I_0\,(e^{qV/kT} - 1)$, I_0 being the saturation current Is value for a very negative value of V.

c) The opposite effect is observed when the diode polarization is reversed (V < 0), which induces an increase in barrier energy and a saturation current Is with a low intensity.

d) General shape of the I/V curve of a p-n diode with direct and reversed polarization. For a very negative V_B value, corresponding to the breakdown voltage, there is a rapid increase in current intensity, typical for a breakdown diode.

Box 5.3. *Operating principle of a p-n diode. Adapted from Hu (2010). For a color version of this figure, see www.iste.co.uk/lacaze/nanotechnologies.zip*

The TiO_2 nanofibers are n-type semiconductors and the PANI grains have conductivities varying with the level of doping, which transforms the polymer from a quasi-metallic conductive state (M) to a p-type semiconductor state corresponding to the dedoped state. Under the effect of NH_3 vapors, an electron donor compound, the energy structure of the TiO_2/PANI junction is modified: the injection of electrons reduces the PANI hole density (positive charges), and therefore increases the resistance of the junction. Given the high volume resistivity of TiO_2 microfibers, the main current, high before reaction with NH_3, is a surface current favored by PANI conductive grains (Figure 5.10(b)). After contact with NH_3, PANI is dedoped and acquires a SC-like resistance (Figure 5.10(c)), resulting in an increase in the energy barriers for electrons and holes, and therefore an overall increase in resistance and a decrease in current intensity.

According to the authors, the resistive PANI/TiO_2 composite sandwiched between two indium tin oxide (ITO) electrodes makes it possible to achieve very low detection thresholds of 50 ppt of NH_3 in air, with good reproducibility, rapid response times (<1 min), with sensitivities S around 2% for NH_3 concentrations of 10 ppb[23] (Figure 5.11). These results differ from those obtained with PANI/SnO_2 heterojunctions (Li et al. 2019a) for which the authors state higher NH_3 detection limits of around 10 ppb and recovery times of several minutes at room temperature.

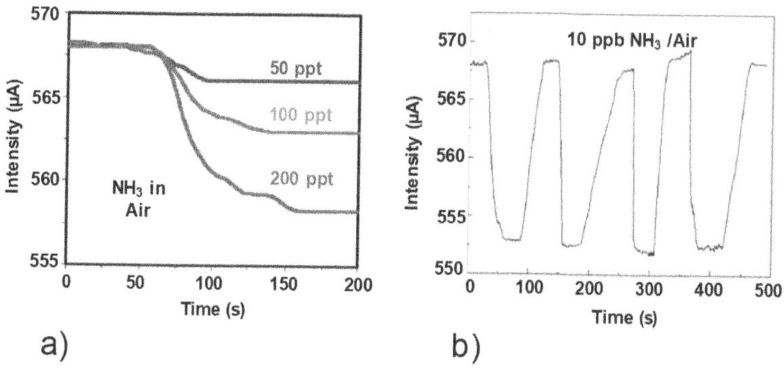

Figure 5.11. *NH_3 titration curves with the TiO_2/PANI sensor. (a) Current intensity/time curves for increasing quantities of NH_3 in air. (b) Reproducibility of the sensor for repeated injections of 10 ppb NH_3 into the air. Adapted from Gong et al. (2010). For a color version of this figure, see www.iste.co.uk/lacaze/nanotechnologies.zip*

23 $S = \Delta I/I_0$, where I_0 is the current intensity measured in air, ΔI is the difference between I_0, and the intensity I after contact with NH_3 mixed with air.

In all cases, the formation of a heterojunction between the PANI and an n-type semiconductor metal oxide is an undeniable improvement compared to a sensor with only the conductive polymer. Good specificity of the PANI/SnO_2 sensor to NH_3 compared to many other gaseous products such as formaldehyde, toluene, acetone, acetylene, ethylene, methane and carbon monoxide is also observed (Li et al. 2019).

5.3.3. *Gas sensors based on two-dimensional thin-film materials*

After the discovery of graphene, its use in the production of sensors has been a very active research topic for many years, developing toward two-dimensional mineral materials, considered as graphene analogues (Xu et al. 2013). Particular attention was devoted to TMD, such as MoS_2, WS_2, $MoSe_2$ and WSe_2 (Chhowalla et al. 2013, 2015; Choi et al. 2017a), and more recently to new two-dimensional materials such as phosphorene, silicene and MXenes (Choi et al. 2017a), the latter becoming the subject of promising studies.

5.3.3.1. *Gas sensors based on graphene and its derivatives*

Shortly after Geim and Novoselov discovered graphene in 2004 (see Chapter 1) and demonstrated its extraordinary mechanical and electronic properties, they described in 2007 (Schedin et al. 2007) the first analytical application corresponding to the titration of NO_2 and NH_3 using an FET-type device. They observed relative variations in graphene resistivity of –4% and +4% for NO_2 and NH_3, respectively, for a concentration of 1 ppm in air.

Paradoxically, a single sheet of pure graphene is insensitive to most gases, as shown by Dan et al. (2009), which would seem to contradict previous results. The authors thus show that by using an FET-type sensor and a non-contaminated graphene, the variation in the drain current is very low, and only of the order of 1%, while the NH_3 concentration is very high (1,000 ppm).

The explanation probably lies in the fact that Schedin et al. used single-layer graphene sheets obtained by mechanical exfoliation of graphite, followed by ion etching in an oxygen plasma. This produces a particular type of graphene, characterized by a charge mobility of only 5,000 $cm^2\ V^{-1}\ s^{-1}$, indicative of partially contaminated graphene (G). This results in improved reactivity with gaseous analytes such as NO_2 and NH_3 which, presumably, is due to an opening of the band gap, which evolves from a “zero gap” to a gap of a few tenths of 1 eV and contributes to increase the sensor sensitivity. This is also probably the reason why most sensors proposed that are based on graphene are composed of either graphene functionalized by reactive groups, or graphene derivatives such as graphene oxide (GO) or reduced graphene oxide (rGO), the latter with numerous hydroxyl and carboxylic groups on its surface, inclined to adsorb various analytes (see Chapter 1).

From these considerations, it follows that graphene-based resistive sensors are either functionalized or combined with other organic or inorganic materials. This solution has been envisaged for the detection of gases such as NH_3, NO_2, H_2, CO, SO_2 and H_2S and has led to research into various composites based on graphene or one of its derivatives, listed and described in several review articles (Basu and Bhattacharyya 2012; Yuan and Shi 2013; Wang et al. 2016a; Ghosh et al. 2022).

However, depending on how G is prepared, resulting in materials that are doped with different impurities, more or less significant variations in sensitivity are observed between sensors, which can be a challenge for future exploitation. The technique of synthesizing graphene on a copper film by CVD and transferring the film onto a Si/SiO_2 support partially overcomes these drawbacks (Choi et al. 2020).

As such, many graphene-based sensors have been made using the graphene derivatives, GO oxide and its reduced form rGO, both easily accessible in large quantities. They are used as composites comprising metal or metal oxide NPs, or even conductive polymers in various structural forms (films, microwires and NPs). The list of these sensors based on graphene composites and its derivatives is very long and a comparison of their characteristics (LOD, response and recovery time, selectivity and sensitivity) has been well described and analyzed in a recent review article (Ghosh et al. 2022).

As an example, we show the behavior at room temperature of sensors made with two different nano-structured hybrid composites rGO/Co_3O_4 used to detect NH_3 (Feng et al. 2016b) and NO_2 (Zhang et al. 2018b)[24]. Cobalt oxide Co_3O_4 (p-type semiconductor) is a material widely used for its catalytic properties, particularly when it is incorporated into a conductive matrix such as rGO. The composite used to form the sensor consists of hybrid rGO/Co_3O_4 microwires, 1–5 μm long and 200–300 nm in diameter where Co_3O_4 NPs (100–200 nm in size) are encapsulated by porous rGO sheets[25], themselves coated with amorphous and porous C (Figures 5.12(a) and (c)). This composite is then transformed into a paste by adding a few drops of water, then applied to an array of interdigitated electrodes and dried

24 NO_2 is a toxic gas with primarily emissions originating from thermal vehicles (with combustion engines) and industrial sites. NO_2 is responsible for the formation of acid rain and can cause bronchitis, emphysema and respiratory irritation; its tolerance limit in the air has been set at 53 ppb in the United States.

25 The Co_3O_4 composite is in the form of microwires obtained by electrospinning a viscous product containing GO, Co_3O_4 suspended in a mixture of solvents and polyvinyl pyrrolidone (PVP) to which a small amount of $Co(NO_3)_2$ is added. These filiform composites are heat-treated at 550°C then 800°C to reduce GO to rGO. The proportion of rGO in the composite is approximately 1% by weight, while the C content in the nanowire is 40%, implying that the microwires also contain a significant amount of porous C from the carbonization of PVP.

at room temperature. The sensor is thus characterized by a very good detection sensitivity of more than 50%[26] for an air NH_3 concentration of 50 ppm, which is better than most sensors featuring rGO and oxides other than Co_3O_4.

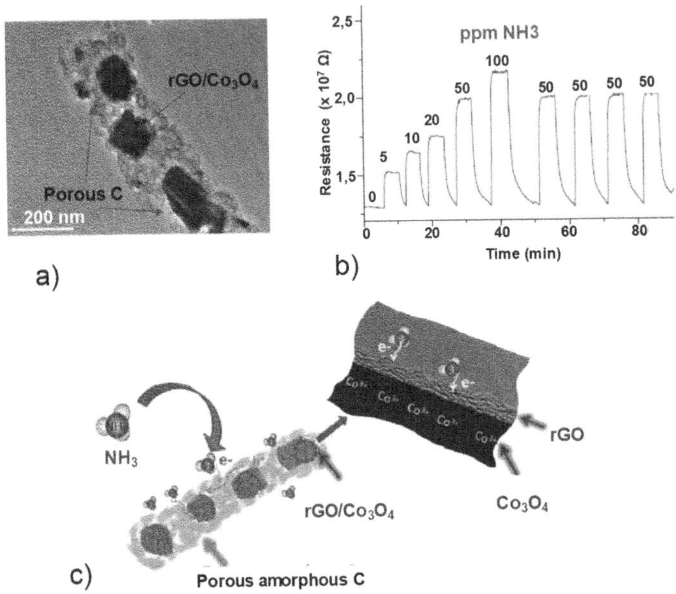

Figure 5.12. *Resistive sensor based on the composite rGO/Co_3O_4 for detecting NH_3. (a) TEM (transmission electron microscopy) image of the hybrid composite rGO/Co_3O_4. (b) Resistive sensor response to successive injections of air-NH_3 mixtures with, after each injection, a return to the initial state corresponding to contact with air. (c) Illustration of the electron exchanges between NH_3 (electron donor) and rGO (electron conductor), then Co_3O_4, type P semiconductor. Adapted from Feng et al. (2016b). For a color version of this figure, see www.iste.co.uk/lacaze/nanotechnologies.zip*

COMMENT ON FIGURE 5.12.– *Note the very particular structure of the rGO/Co_3O_4 microwires where it appears that the Co_3O_4 NPs are encapsulated by rGO sheets, themselves coated with amorphous porous C, thus facilitating the accessibility of NH_3 to rGO and Co_3O_4. The resistance/time curves after NH_3 injection show a response time of only 4 seconds to reach 90% saturation, but a much longer recovery time of about 5 minutes is required to return to the initial signal in air.*

---

26 The sensitivity of the sensor is defined by the relative change in resistance $S = (R - R_0)/R_0$, R_0 being the resistance of the sensor in air and R that of the sensor in contact with the gas–air mixture.

The presence of the conductive rGO sheets coating the oxide is crucial. In the absence of rGO, the sensor sensitivity is only 20% for the same NH_3 concentration. This confirms a very strong interaction between NH_3 (electron donor) and rGO (electron acceptor) reducing the concentration of positive charges (holes) in Co_3O_4 and increasing the resistance of the microwire.

Furthermore, for a mixture of different gaseous compounds with a concentration of 50 ppm in air, including VOCs such as methanol, ethanol, formaldehyde, toluene, benzene and acetone, the sensor turns out to be 10 times more sensitive in detecting NH_3 than VOCs. According to the authors, this could be due to the non-diffusion of VOCs toward the rGO sheets and to their adsorption at the periphery of the microwires, or to a lesser donor effect than that of NH_3.

The response and recovery curves also showed good reproducibility for a 50 ppm concentration (Figure 5.12(b)). However, sensitivity to NH_3 appears lower than that obtained with sensors with conductive polymer sensitive layers. On the other hand, its good signal reproducibility as well as its insensitivity to air humidity are definite advantages.

The same hybrid composite rGO/Co_3O_4 prepared using a method slightly different from the previous was also used by Zhang et al. (2018) to detect NO_2 at room temperature, which is an oxidizing gas and therefore an electron acceptor. The composite structure is also different from the composite previously used to detect NH_3. It is no longer filiform but instead a structure composed of Co_3O_4 lamellar fragments to which rGO nanosheets are anchored (Figure 5.13(a)).

The sensor sensitivity measured using the relative variation in conductivity (Figure 5.13(b)) is around 28% for 5 ppm of NO_2 in air at ambient temperature (25°C), much higher than for NH_3, which is only 2.5% for 5 ppm of NH_3. Note that the response and signal recovery times are slow. Thus, for an injection of 5 ppm of NO_2, 90% of the response is obtained after 1.5 min, and the signal recovery time in the presence of dry air is around 30 min at 25°C.

Regarding its selectivity to NO_2 in the presence of interferents such as benzene, toluene, acetone, ethanol, methanol and formaldehyde, the presence of rGO also proves critical. The variation in resistance for each VOC is indeed very strong in the case of Co_3O_4 without rGO, while it becomes negligible with the rGO/ Co_3O_4 composite.

According to the authors, the interaction mechanism between NO_2 and the (p/p) isotype rGO/Co_3O_4 heterojunction derives from the electron transfer from Co_3O_4 to rGO and NO_2, which, overall, increases the density of positive charges (holes) in the

Co_3O_4 SC and therefore reduces the sensor resistance. The absence of a signal from the VOCs could result from a screening effect due to rGO, causing a blockage and their non-diffusion toward Co_3O_4.

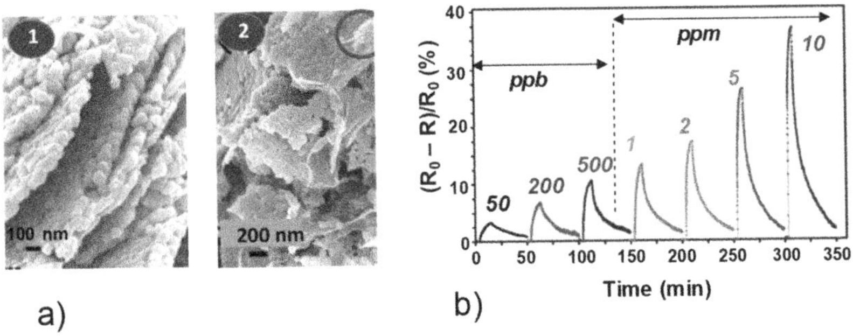

Figure 5.13. *Resistive rGO/Co_3O_4 sensor for detecting NO_2. (a) FESEM (field emission scanning electron microscopy)[27] images of undoped cobalt oxide (1) and after doping with rGO (2). The red circle indicates the deposition of an rGO fragment onto Co_3O_4. (b) Relative variations in resistance for NO_2 injections into air in increasing from 50 ppb to 10 ppm. Adapted from Zhang et al. (2018b). For a color version of this figure, see www.iste.co.uk/lacaze/nanotechnologies.zip*

In conclusion, graphene is rarely used in its pure state, but always either functionalized by chemical groups that increase the sensor selectivity, or in the form of a composite, which can be a hybrid between graphene and metal or metal oxide NPs, or even combined with conductive polymers, for which controlled porosity can be an additional element of selectivity.

5.3.3.2. *Dichalcogenide gas sensors*

In this family of compounds, we mainly find sulfur, selenium and tellurium dichalcogenide (MoS_2, $MoSe_2$, TeS_2, WS_2, WSe_2, etc.), but it is molybdenum disulfide MoS_2 that has been most frequently used, due to its well-established semiconductor properties and the ease to obtain it in the form of nanosheets[28].

---

27 Field emission scanning electron microscopy (FESEM) differs from the classic scanning electron microscopy (SEM) technique by the nature of the electron beam. It is generated by applying an electric field instead of being produced by a simple thermionic emission. This device makes it possible to reduce the width of the electron beam and obtain better image resolution.

28 MoS_2 is widely used in industry, particularly in the aerospace field for its dry lubricant properties (Krishnan et al. 2019).

Unlike resistive devices based on oxide semiconductors that require the oxide to be brought to an elevated temperature of at least 150–200°C, most sensors based on MoS_2 are designed to operate at ambient temperature, which offers greater simplicity, particularly in the design of FET-type sensors for the analysis of gaseous nitrogen compounds (NO, NO_2, and NH_3) (Li et al. 2012; He et al. 2012; Liu et al. 2014a; Late et al. 2013) (Figure 5.14).

The FET channel is thus composed of a film of MoS_2 nanosheets, with a thickness between 0.8 and around 10 nm, the equivalent of one (1L) or several nanosheet layers (nL). It is in contact with a rigid (Si/SiO_2) or flexible (polyethylene terephthalate-PET) insulating substrate; the drain and source electrodes, deposited on MoS_2, are metallic (Au or Ti) or carbon (graphene or reduced graphene oxide (rGO)). The entire device can be made flexible when the insulating support is PET.

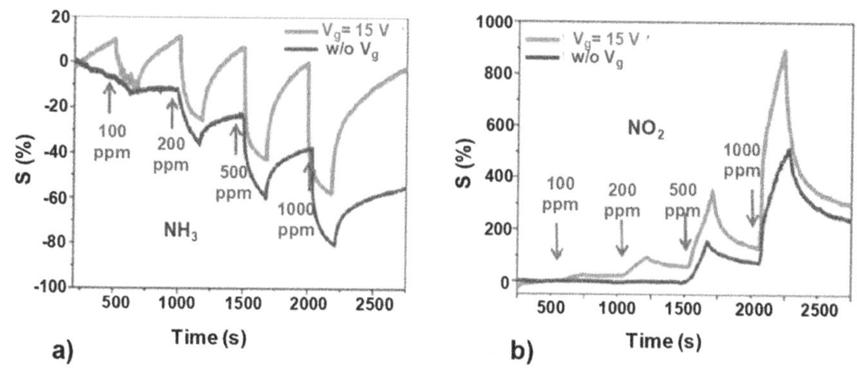

Figure 5.14. *Time–sensitivity curves of an MoS_2 (2L)-based FET-type sensor subjected to injections of air-NH_3 and air-NO_2 mixtures. (a) Relative variations in sensitivity as a function of the amounts of NH_3 injected in the case of a gate voltage V_g of 15 V (curve in green) or without grid voltage (curve in red). (b) Same experience with NO_2. The sensitivity S (ΔR/R°) corresponds to the relative variation in channel resistance (ΔR = R – R°, R° and R being the respective resistances measured in air and in the presence of NH_3 or NO_2). The blue arrows indicate the moment when the gas is introduced and begins to decrease (NH_3) or increase (NO_2) the channel resistance; on the contrary, the increase (NH_3) or decrease (NO_2) in S corresponds to the introduction of nitrogen or dry air intended to purge the sensor. Adapted from Late et al. (2013). For a color version of this figure, see www.iste.co.uk/lacaze/nanotechnologies.zip*

COMMENT ON FIGURE 5.14.– *As with the previous sensors, the resistance variations are slow, around 300 s, both during stabilization of the measurement in contact with*

NH_3 or NO_2 and during the recovery of the initial state, caused by purging FET by air or nitrogen. We also notice a greater sensitivity of the sensor with respect to NO_2, which could be due, compared to NH_3, to a stronger interaction of NO_2 with MoS_2. We also note a greater sensitivity of the sensor operating in FET (gate polarized at 15 V) than when the device operates as a resistive system (no gate voltage).

When analyzing NO_2 and NH_3 with a FET with a "back gate" configuration and with a channel comprising two layers (2L) of MoS_2 sheets, Late et al. (2013) show that the sensor sensitivity S is a function of the applied gate voltage and is very different for NH_3 and NO_2. With NH_3 being an electron donor and MoS_2 an n-type semiconductor, the injection of electrons by NH_3 into the channel increases its electron density and, therefore, reduces its resistance, which is observed when NH_3 comes into contact with MoS_2. The opposite effect is induced with NO_2, an electron acceptor and the resistance increases.

The sensitivity varies depending on how the FET is made and the MoS_2 sheets are functionalized. He et al. (2012) thus observed an evident improvement in the sensitivity of NO_2 detection when the MoS_2 nanosheets are coated with metallic or semiconductor NPs. For a concentration of 1 ppm NO_2, the relative variation in conductivity obtained by depositing platinum NPs onto MoS_2 nanosheets is 16%, whereas with uncoated nanosheets, it is only 6%, almost three times less.

Much better results were recently obtained by Han et al. (2018) with a resistive sensor, by coating the MoS_2 nanosheets with ZnO semiconductor NPs. The device sensitivity at room temperature is significantly improved, which is mainly due to the formation of hetero-nanostructures between the two semiconductors n-ZnO and p-MoS_2, characterized by respective work functions of 4.5 and 5.2–5.4 eV and respective band gaps of 3.4 and 1.2–1.9 eV, implying a transfer of electrons from ZnO to MoS_2 (Figure 5.15).

The contact zone between the two SCs forms a local energy barrier and a space charge layer. Depending on the nature of the electronic exchanges, the adsorption of a gas can lead to a reduction or increase in the energy barrier height, as well as a reduction or increase in the thickness of the ZCE. In the case of nitrogen dioxide (NO_2), which is highly electrophilic, there is electron transfer from the ZnO conduction band to NO_2, which reduces the thickness of the space charge layer as well as the energy barrier size between the two SCs. The variation in conductivity of the hetero-nanostructure is much greater than for MoS_2 alone. For 5 ppm of NO_2, the relative variation in conductivity $\Delta I/I_0$ is exceptionally high, around 30 (i.e. 3000%), whereas it is 11 times lower for a sensor without ZnO NPs.

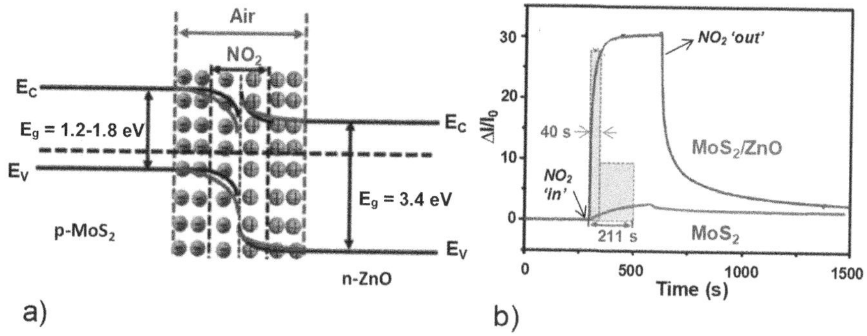

Figure 5.15. *Working principle and response curve of a resistive sensor to NO_2, whose sensitive layer consists of a MoS_2/ZnO hetero-nanostructure. (a) Energy diagram of the p-MoS_2/n-ZnO junction showing the lowering of the energy barrier and the narrowing of the space charge layer when NO_2 is present in air. (b) Comparison between the conductivity variation curves of two sensors interacting with 5 ppm of NO_2 with different sensitive layer structures (MoS_2 nanosheets and MoS_2/ZnO hetero-nanostructure). Adapted from Han et al. (2018). For a color version of this figure, see www.iste.co.uk/lacaze/nanotechnologies.zip*

COMMENT ON FIGURE 5.15.– *The resistive sensor comprises two interdigitated metal electrodes (see Figure 4.5), spaced 20 µm from each other and on which the MoS_2 nano-sheets are first deposited, then the ZnO NPs. A voltage of 500 mV is applied between the electrodes. A linear relationship is observed between the logarithm of the NO_2 concentration (between 50 ppb and 5 ppm) and the relative variation in intensity. The curve in Figure 5.15(b) is obtained for an optimal concentration of ZnO NPs. Also note the rapid rise in the sensor signal, which, 40 s after the introduction of NO_2 (NO_2 "in"), reaches more than 90% of its maximum value, while the same sensor without ZnO NPs has a weaker and much slower response (211 s).*

The selectivity toward NO_2 is also increased compared to other gases such as NH_3, H_2S, and SO_2, for which the relative variations in conductivity $\Delta I/I_0$ (in %) are only –43%, –46% and 81% for 1,000 ppm in air, while that of NO_2 is around 3000% for a concentration of 5 ppm. Similarly, saturated vapors of methanol and acetone in air react weakly with the sensor and only induce slight variations in conductivity (respectively, –53% and –66%).

The detection limit for NO_2 is estimated at around 50 ppb for $\Delta I/I_0$ = 5% and, according to the authors, this value outperforms numerous results obtained by other groups using resistive sensors, FET devices and those based on MoS_2 sheets associated with various metal particles comparable to hetero-nanostructures.

5.3.3.3. *New 2D materials for gas sensors*

The discovery and success of graphene for its exceptional electrical and mechanical properties has led many teams to explore new avenues of 2D materials, that can also be easily processed in the form of atomic monolayers, with the hope to discover new applications, particularly in the field of sensors for gases and biomolecules.

The list of these new materials is particularly long; however, some of them are very popular thanks to their specific electrical properties, making them ideal candidates for the fabrication of sensor devices. In addition to TMD, which have been used for a long time to make sensors, new 2D materials have emerged and are the focus of current research. Much work has been done on sensors using boron nitride and carbon derivatives (hBN and g-C_3N_4), silicon derivatives (silicene), phosphorus derivatives (phosphorene) and metal hybrids (MXenes). As in previous years, much of the research applied to gas detection was devoted to toxic gases, with a particular focus on H_2S, NO_2 and NH_3.

5.3.3.3.1. Phosphorene-based sensors

Phosphorene, described in Chapter 1, is a novel 2D material, with a structure comparable to that of graphene, but characterized by a non-zero band gap that can increase from 0.3 to 2.2 eV when moving from a bulk state to a monolayer (see Chapter 1). In practical terms, this makes it possible to consider resistive or FET devices well suited to produce phosphorene-based sensors. According to Cho et al. (2016), resistive sensors comprising a phosphorene sensitive layer perform significantly better than those with graphene or MoS_2 layers, both in terms of sensitivity and response time (Figure 5.16).

The authors thus observe that for 1 ppm of NO_2, the sensitivity ($\Delta R/R_0$) with phosphorene is –80%, while with MoS_2 and G, it is only –20% and –12%, respectively. Also, the response time with phosphorene is significantly shorter than that observed with MoS_2 and graphene, with, however, still very long recovery times of several minutes for all three sensors. This can be attributed to higher adsorption energies of the analyte with the sensitive layer.

It should, however, be noted that compared to a sensor comprising a p-n hetero-nanostructure sensitive layer, such as MoS_2/SnO_2, sensors made of bare layers of graphene, MoS_2 and phosphorene have significantly worse responses to a gas like NO_2. Han et al. showed that with the MoS_2/SnO_2 sensor, its sensitivity $\Delta R/R$ to 1 ppm of NO_2 is 2240%, whereas with phosphorene it is –80%.

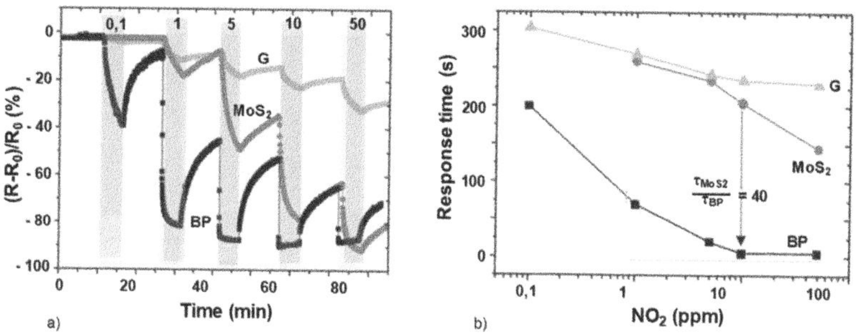

Figure 5.16. *Comparison of response curves for NO_2 with resistive sensors based on phosphorene (BP), MoS_2 and graphene (G) sensitive layers. (a) Response curves as a function of time for successive additions of 0.1, 1, 5, 10 and 50 ppm NO_2. (b) Response times of the three sensors for injections of NO_2 into air between 0.1 and 100 ppm. We note that the response with phosphorene is much faster than with MoS_2 and G (reduction by a factor of 40 for 10 ppm NO_2), but the recovery times are still very slow (several minutes). Adapted from Cho et al. (2016). For a color version of this figure, see www.iste.co.uk/lacaze/nanotechnologies.zip*

As with MoS_2, new properties can also be observed with phosphorene when its surface is decorated with metallic gold or platinum NPs (Cho et al. 2017) (Figure 5.17).

This structure can be produced very simply by directly immersing the BP sheets in aqueous solutions of $NaAuCl_4$ and Na_2PtCl_4 salts, produced from the spontaneous reduction reaction $AuCl_4^- + 3e^- \leftrightarrow Au + 4Cl^-$ and $PtCl_4^{2-} + 2e^- \leftrightarrow Pt + 4Cl^-$[29]. BP nanosheets modified in this way with gold and platinum NPs, with sizes between 5 and 10 nm (referred to as Au/BP and Pt/BP), acquire completely different electronic properties than pure BP and can be compared to n-type SCs, resulting in responses to VOCs (acetone, ethanol, acetaldehyde, hexane, toluene), H_2 and NO_2 gases that are opposite to those observed with unmodified BP. The properties of the sensors made with Au/BP and Pt/BP composites are also clearly different in terms of NO_2 and H_2 detection.

29 The work function of BP is around 4.5 eV, that is, equivalent to the potential energy level of the reference redox couple H^+/H_2. The gold and Pt salt solutions are characterized by respective redox potentials of 1 and 0.74 V relative to the same H^+/H_2 couple, making the Fermi level of BP well above the energy levels of the $AuCl_4^-/Au$ and $PtCl_4^{2-}/Pt$ redox potentials. Spontaneous electron transfer can occur between the BP sheets and the gold and Pt salt solutions leading to their reduction and the formation of metallic NPs.

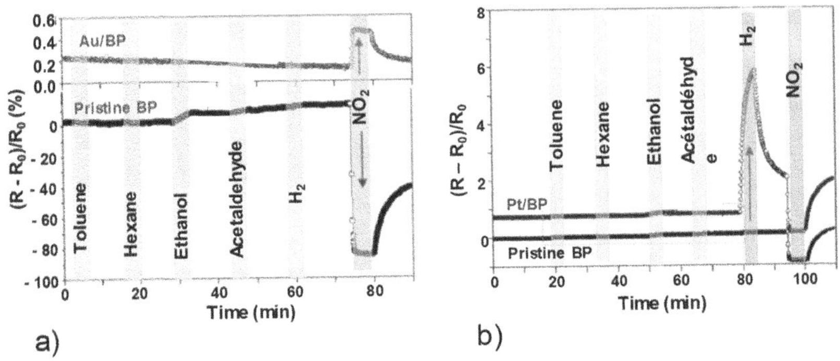

Figure 5.17. *Resistance variation of BP, Au/BP and Pt/BP sensors after injection of N_2/VOC (toluene, hexane, ethanol and acetaldehyde), N_2/NO_2 and Air/H_2 mixtures. (a) Comparison of responses (in % of $(R-R_0)/R_0$) of resistive sensors based on BP and Au/BP. (b) Comparative sensitivities (in absolute values) of BP and Pt/BP sensors to the same analytes. In both cases, the concentration of VOCs in nitrogen is 1000 ppm, H_2 is 1% and NO_2 is 100 ppm. Adapted from Cho et al. (2017). For a color version of this figure, see www.iste.co.uk/lacaze/nanotechnologies.zip*

In the case of Au/BP, since the p-type SC BP evolves toward an n-type SC when the covering the BP sheets with gold NPs is high, the presence of NO_2 is accompanied by an increase in resistance. This is due to the transfer of electrons from the SC to NO_2 and, therefore, the depletion of electrons from the n-type SC. We note that the two BP and Au/BP sensors have a high selectivity for NO_2 but very different sensitivities: very high with BP (relative variation of R around 80%) and very low for Au/BP (around 0.2%). For H_2 and VOCs, the resistance variation is almost non-detectable, which increases the selectivity of these sensors to NO_2 and other analytes.

Pt/BP shows very different behavior from Au/BP toward the analytes. The H_2 sensitivity is extremely high (more than 500% for 1% H_2 in air)); sensitivity of NO_2 is also high, around 200% for 100 ppm. The very large variation in resistance with Pt/BP is due to the catalytic effect of Pt which, according to the authors, dissociates H_2 into atomic $H^\bullet$; this lowers the work function of BP and induces a flow of electrons toward Pt, resulting in the depletion of electrons from the SC, inducing an overall increase in resistance.

Another significant effect of metallic NPs is that they increase the stability of these sensors. Pt/BP remains stable during use in ordinary environmental conditions (humid air) over several months. This is a significant progress for this type of sensor, probably due to the blocking of the phosphorus electron pair by the metals Au and Pt.

5.3.3.3.2. MXene-based sensors

Described in Chapter 1, MXenes are a new family of 2D materials with the general formula $M_{n+1}X_nT_X$. Due to their good conductivity properties, MXenes also considered as transition metal carbides or nitrides are often compared to graphene, with properties that depend on the nature of their surface Tx groups (see Chapter 1). As with graphene, their use in gas sensors is only envisaged in the form of varied composites giving rise to heterojunctions, which increase chemical interactions with the analyte and induce strong variations in conductivity.

The detection of nitrogen oxides NOx was thus carried out by Sun et al. (2021) with a sensor comprising a $Ti_3C_2T_X$ sensitive layer, modified by cobalt oxide Co_3O_4 NPs, corresponding to the formation of a p-n heterojunction[30]. The prior modification of the MXenes by depositing branched polyethylene imine (PEI) over their surface ensures the complexing of the cobalt oxide Co_3O_4 NPs and homogenize their dispersion.

In this case, $Ti_3C_2T_X$ (surface modified by F atoms and by PEI) and Co_3O_4 complexed by PEI are semiconductors, respectively, n and p types, characterized by a work function of 4.79 eV and 6.48 eV with respective energy band gaps of 1.14 eV and 2.08 eV. The operation of this sensor is comparable to that described by Han et al. (2018), which also has a p-n MoS_2/ZnO heterojunction and was used for NO_2 detection (see section 5.3.3.2).

Mechanistically, based on surface XPS analyses, the authors consider a series of electron exchanges between O_2^- (adsorbed oxygen reduced to the sensor surface by electron transfer from Ti_3C_2Tx to O_2), Co_3O_4 and the nitrogen species NO and NO_2, which induces significant variations in sensor resistance. While Ti_3C_2Tx alone is characterized by a weakly resistive behavior (~6 kΩ), the addition of Co_3O_4 strongly increases the sensor resistivity by more than a factor of 1,000 (7 MΩ), which is due to the electron transfer from Ti_3C_2Tx to Co_3O_4. It is even more pronounced after the adsorption of O_2 onto Ti_3C_2Tx due to another transfer of electrons. This results in the formation of a space charge layer formed between Ti_3C_2Tx and Co_3O_4, corresponding to a layer of accumulated negative charges on Co_3O_4 and a depletion layer of negative charges in Ti_3C_2Tx, resulting in a higher resistance of the sensor when in contact with air (Figure 5.18).

---

30 The sensitive layer of the NOx gas sensor is produced from a series of chemical reactions carried out using Ti_3C_2Tx sheets (thicknesses between 2.5 and 3.1 nm), functionalized on the surface by branched polyethyleneimine (branched PEI, characterized by a linear main $H(NHCH_2CH_2)_nNH_2$ chain with lateral branches ($CHCH_2NH_2$ and $-CHCH_2N(CH_2CH_2NH_2)_2$), linked to the NHs of the main chain. This surface modification homogeneously disperses the Co_3O_4 NPs, which are deposited directly onto these composites using cobalt nitrate ($Co(NO_3)_2$ in a basic medium. Prolonged heating leads to Co_3O_4@PEI/Ti_3C_2Tx sheets.

The presence of nitrogen species that adsorb onto Co_3O_4 causes the opposite effect by absorbing the electrons accumulated in Co_3O_4. This reduces the thickness of the space charge layer and, therefore, the sensor resistance.

For optimal preparation of the Co_3O_4@PEI/Ti_3C_2Tx sensitive layer corresponding to a specific concentration of Co_3O_4 NPs (2.4 mg per composite, CoPM-24), there is a significant reduction in sensor resistance between an atmosphere of humid air (26% relative humidity at ambient temperature) and air containing 100 ppm of NOx. The ratio R_{Air}/R_{NOx} is around 30 (R_{Air} = 0.8 MΩ and R_{NOx} = 0.026 MΩ), which approximately corresponds to a $\Delta R/R_0$ sensitivity of 97%. The detection limit indicated by the authors is 30 ppb of NOx, with only a 3% detection sensitivity for this concentration. The sensor response time is fast: less than 2 s for 100 ppm of NOx with, however, a slower resistance recovery time in air around 70 s. The selectivity of NOx to gases such as H_2, CH_4, NH_3 and CO is also excellent due to their low intensity response (R_{Air}/R_{Gas} less than 1) compared to that with NOx.

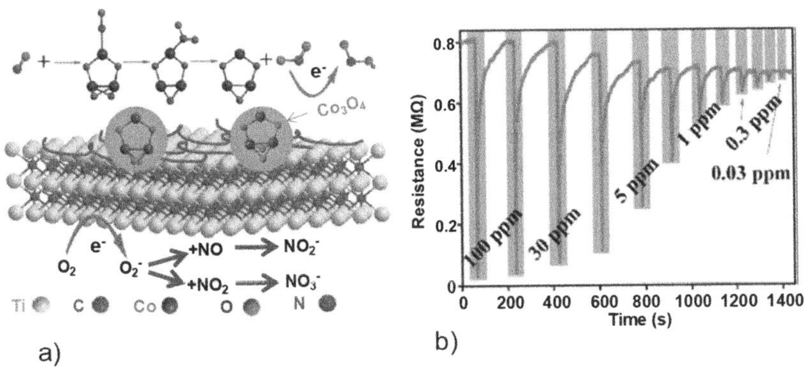

Figure 5.18. *Operating principle of the Co_3O_4@PEI/Ti_3C_2Tx sensor. (a) Schematic structure of the sensitive layer indicating the adsorption of Co_3O_4 and O_2 onto the Ti_3C_2Tx layer. (b) Evolution of the sensor resistance for successive additions of NOx (decreasing ppm) preceded by injections of humid air. Adapted from Sun et al. (2021). For a color version of this figure, see www.iste.co.uk/lacaze/nanotechnologies.zip*

Other examples of gas sensing using MXenes have been described and are listed in the review article by Xia et al. (2023). Most of which involve the development of a sensitive layer made of binary MXene/NPs mixtures (metallic or semiconducting). Recently, NH_3 sensors operating at room temperature have been developed using MXene Ti_3C_2Tx combined with metal oxides such as TiO_2, SnO_2 or $Ni(OH)_2$.

The use of transition metals other than titanium is rarer, although it has been reported for the detection of methane with vanadium carbide V_2CTx.

5.3.3.4. *Conclusions on 2D materials*

Obviously, thin layered 2D materials, conductors or semiconductors are materials of choice to produce gas sensors at room temperature. Chalcogenides and graphene, generally associated with metallic nanomaterials or metal oxides, are probably the oldest 2D materials used in the manufacture of sensors, with which very good results have been obtained for the determination and detection of gases and VOCs. The quality of the results obtained with these compounds is probably at the origin of new research into new 2D materials, likely to compete with graphene and chalcogenides. New 2D materials have thus emerged, such as silicene, phosphorene and MXenes, the most successful being phosphorene and MXenes. Silicene, similar in structure to graphene, is composed of Si atoms in a low-buckled hexagonal honeycomb arrangement, is still in a developmental stage and has mainly been used for theoretical studies (Nguyen et al. 2022). MXenes with varied structures can be easily mass-produced, increasing their field of application. They can also be conductors or semiconductors and prove well-suited for gas sensors. In all cases, it turns out that the operation of these materials as a gas sensor is only optimal when they are combined with other materials in such a way that they create p and n SC/SC heterojunctions (p-n diode) or SC/conductor junctions (Schottky junction). The signal induced by the gas is generally a result of electron transfer between the analyte and the sensitive layer, resulting in a change in the resistance, whose amplitude and sign will depend on the donor or electron acceptor character of the gas molecule. Desirable sensor characteristics, often called the "three S" (stability, sensitivity and selectivity) (Milone 2021), are difficult to achieve simultaneously. This can explain the wide range of systems explored and the wide dispersion of results for the same analyte.

5.3.4. *Sensor networks and their use in artificial noses*

Analyzing an analyte mixture to identify and determine the relative composition of each element is one of the major challenges in analytical chemistry. The ideal solution would be to have a network of sensors, each with a specific response for a different element in the mixture. This situation is difficult to achieve in practice; most sensors rarely have specific responses, and many compounds are likely to interfere.

The problem can, however, be solved by adopting a signal processing approach consisting of identifying the constituents of a gas mixture and their concentrations using a shape recognition system and databases (see Chapter 3). This can be achieved by using different transduction devices, some being colorimetric systems based on the use of various dyes, which change color during an interaction with a particular gas, others functioning like an electronic nose (e-nose), like the olfactory system of mammals and humans, a particularly effective model for odor recognition (Axel 1995) which, however, does have its drawbacks due to its insensitivity to certain odorless gases such as carbon monoxide and water vapor.

5.3.4.1. *Electronic sensor networks*

Creating a system of sensors that functioned like an artificial nose was first imagined in the 1980s by Persaud and Dood (1982). Subsequently, enormous progress was made in the design of such sensors and their transduction systems (resistive, gravimetric, electrochemical, colorimetric, etc.) and signal analysis techniques, integrating artificial intelligence (AI) methods, coupled with those of statistical analyses. An e-nose is thus defined by Gardner and Bartlett (1994) as a network of partially selective chemical and electronic sensors with pattern recognition software capable of recognizing the odor of a gas or a complex mixture from pre-recorded signals in a database.

This new approach to analyzing gas mixtures is increasingly used in different fields. Among these, the healthcare sector has seen considerable developments in the early detection of diseases, for example, through analyzing gases exhaled by a patient. Electronic noses are now used daily in the food industry to monitor the freshness of food, in security applications for detecting explosives and nerve gases, and in the environmental sector to control atmospheric pollution. As the list of electronic nose devices identified in the literature is too long to mention, we recommend reading recent review articles by Hu et al. (2019b) and Wilson (2018). Given their importance, we described a few examples of e-nose applications in the biomedical, agri-food and security sectors.

5.3.4.1.1. Gas sensor networks in the medical field

One of the avenues explored is research into the correlation between the gases exhaled by a patient and warning signs of an illness. This non-invasive technique allows an early medical diagnosis. This approach was recently carried out by interdisciplinary medical teams (17) spread across five countries, bringing together doctors from various specialties, including chemists and biologists (Nakhleh et al. 2017). Using e-noses built according to a specific protocol, their objective was to establish a correlation between the nature of the gases exhaled and the disease specific to each patient. For this, a total of 1704 individuals were tested; 813 patients (men and women) suffering from 17 distinct diseases were compared to 591 healthy individuals who showed no signs of illness.

The sensor networks (20) used were resistive and made in a conventional manner by depositing around ten pairs of interdigitated electrodes by evaporation onto a Si substrate, oxidized on the surface by a thin layer of SiO_2 (300 nm thick). The electrodes were then covered by a composite sensitive layer of single-walled carbon nanotubes (SWCNTs) with various functional groups, onto which gold NPs (diameter 3–4 nm) were grafted.

This study, carried out over several years, revealed that the 13 main VOCs form a mixture whose composition differs from one disease to another. For each sensor, two to four sensing features (SFs) appeared for each gas mixture exhaled by a patient suffering from a particular disease, which ultimately led to 59 SFs, represented as a statistical data heat map, labeled from SF1 to SF59 (Figure 5.19).

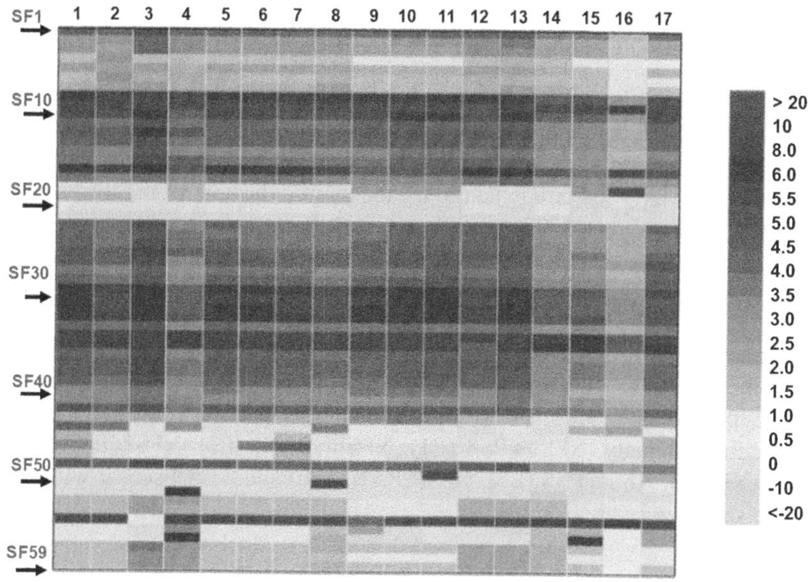

Figure 5.19. *Heat map of the 59 main characteristics (SF1 to SF59) recorded for each mixture of gases exhaled by patients suffering from different diseases. The selected diseases (columns 1–17) were treated in different hospital departments. They corresponded to different diseases, with 1–8 being eight different cancers (lung, colon, neck and head, ovarian, bladder, prostate, kidney and stomach). Note that 150 gases were detected by GC-MS (gas chromatography coupled with mass spectrometry analysis). Thirteen gases (2-ethylhexanol, 3-methylhexane, 5-ethyl-3-methyloctane, acetone, ethanol, ethyl acetate, ethylbenzene, isononane, isoprene, nonanal, styrene, toluene and undecane), considered most characteristic of the diseases studied, were selected for the DFA statistical analysis. Adapted from Nakhleh et al. (2017). For a color version of this figure, see www.iste.co.uk/lacaze/ nanotechnologies.zip*

In this heat map, each square represents the mean characteristic for a particular sensor, calculated over an entire population of patients and for a specific disease. The color of each square indicates the relative and standardized intensity of each signal. Note that an isolated characteristic SFi,n attributed to a specific disease (column n) cannot be used to establish a clear correlation between SFi,n and the disease. On the other hand, all the SFi (59) for a specific disease, corresponding to a particular column n of the heat map, present numerous criteria that distinguish them from other columns. Discriminant factor analysis (DFA) then makes it possible to attribute, with a high probability, a correspondence between these signals and the type of disease and, ultimately, identify a specific signature for a particular disease with a high prediction rate[31].

5.3.4.1.2. Gas sensor networks in the agri-food sector

This area concerns the food quality and its conservation. Food spoilage is caused by various factors (temperature, humidity, oxidation, etc.), which leads to the formation of toxic VOCs, among which we find amines (methylamine [MA], trimethylamine [TMA]) and ammonia (NH_3), the main indicators of the freshness of fish and meat. Andre et al. (2022) recently showed that it was possible to create an "intelligent" e-nose to detect these VOCs with seven different resistive sensors arranged in a network (Figure 5.20).

The sensors have different sensitive layers, comprising SiO_2 nanofibers, doped with indium (In_2O_3) or zinc oxides (ZnO), then functionalized with polyaniline (PANI) or polystyrene sulfonate (PSS). Ultimately, the network includes six different sensors, namely, NFSI@PSS (E1), NFSI@PANI (E2), NFSZ@PSS (E3), NFSZ@PANI (E4), NFS@PSS (E5), NFS@PANI (E6) and a final sensor with only PSS for the sensitive layer (E7)[32].

---

31 The diseases researched in this study were cancerous, inflammatory and neurological, as well as a few other specific diseases. For each sensor, the sensing features discovered had no obvious physical meaning and were simply particular points on a conductivity variation curve (not interpretable) that were observed with a sensor as a function of time and for a gas mixture exhaled by a patient (see Chapter 3, section 3.2). The values obtained correspond to the peak intensity of the resistance, the median and the final values, as well as the area of the curve $R = f(t)$ for a determined duration. It is this set of raw data that is processed to identify a correlation between the disease and the signal fingerprint. An overview of discriminant analysis can be found in Ramayah et al. (2010).

32 The notation NFSI@PSS means that a silica nanofiber (NFS) is doped with indium oxide (NFSI) and is then functionalized by PSS (NFSI@PSS). The same notation convention is applied to zinc oxide (NFSZ).

With these sensors being resistive, a resistance spectrum was obtained by impedancemetry for these three VOCs for three concentrations (5, 15 and 25 ppm) in the air; the resistance measured at 1,000 Hz was found to be the most discriminative.

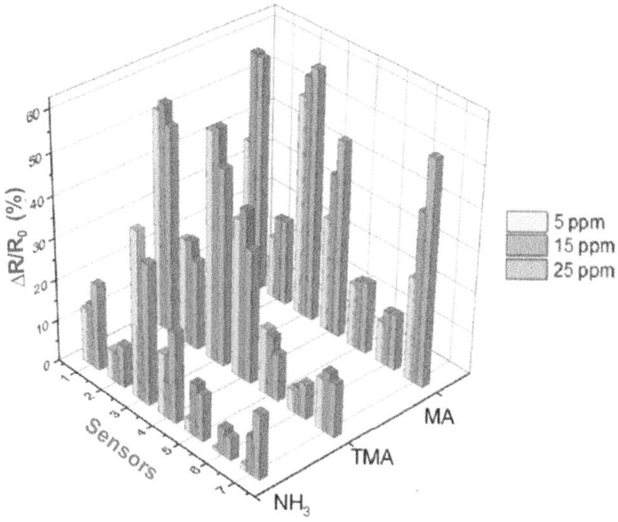

Figure 5.20. *Spectrum of relative resistance variations ($\Delta R/R_0$) between air and air/VOC mixtures, measured at 1,000 Hz for each sensor previously described. Adapted from Andre et al. (2022). For a color version of this figure, see www.iste.co.uk/lacaze/nanotechnologies.zip*

By applying multidimensional projection techniques to this dataset, the authors showed that the device satisfactorily discriminated between standard analytes as well as VOC actual samples from spoiled food.

5.3.4.1.3. Sensor networks for chemical warfare agent detection

This is another example of an e-nose built based on four surface acoustic wave (SAW) sensors, each comprising a different sensitive layer of metal oxides ZnO, TeO_2, SnO_2 or TiO_2[33] (Figure 5.21).

33 The sensitive oxide layers are produced by projecting at room temperature and under reduced pressure a gaseous mixture of oxygen and argon ions onto Zn, Te, Sn and Ti targets using the radiofrequency sputtering technique. The layers formed are nanostructured and are composed of an assembly of nanocrystals, unlike TiO_2 layers which are devoid of them.

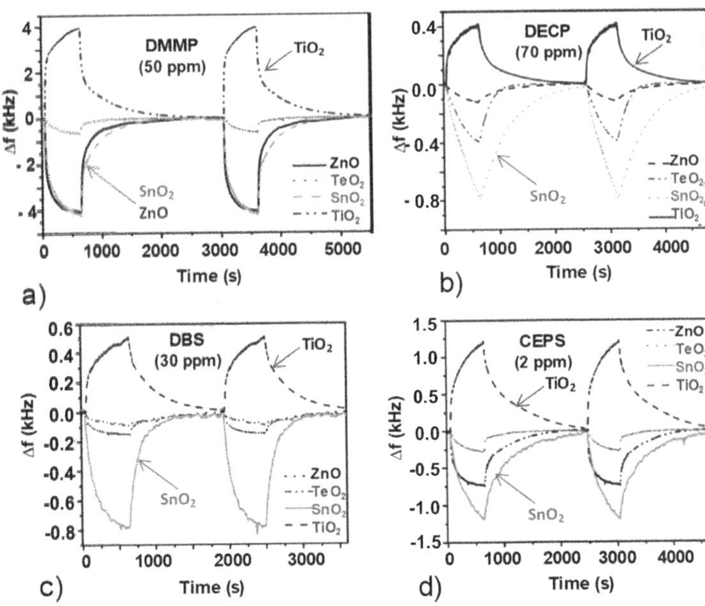

Figure 5.21. *Variations Δf of resonance frequencies obtained on SAW sensors comprising the sensitive layers ZnO, TeO_2, SnO_2 and TiO_2 for simulation gases DMMP, DECP, DBS and CEPS. Adapted from Raj et al. (2013). For a color version of this figure, see www.iste.co.uk/lacaze/nanotechnologies.zip*

COMMENT ON FIGURE 5.21.– *Good reproducibility of the sensors was found over two consecutive cycles but with slow response and recovery times of several minutes. Furthermore, unlike the quartz balance, where the variation in frequency as a function of the adsorbed mass is always negative, in the case of the TiO_2 sensor, there is a positive variation in frequency due to an elasticity effect, which overrides the mass effect, unlike the other three sensors. This is an important discriminatory effect that significantly improves the discriminating power of the e-nose and differentiates SAW sensors from QCM.*

The objective is to have an e-nose that detects simulated versions of the neurotoxic organophosphorus gases: sarin (2-[fluoro (methyl)phosphoryl] oxypropane), soman (pinacolyl methylfluorophosphonate) and tabun (N-dimethyl amino ethyl cyanophosphite), as well as sulfur mustard gas simulation products[34].

34 Organophosphate products are extremely powerful neurotoxic agents that inhibit the action of the enzyme acetylcholinesterase by blocking the hydrolysis of acetylcholine (nervous system neurotransmitter), which causes its accumulation in the body and paralysis of nerve

Dimethylmethyl phosphonate (DMMP), diethyl chlorophosphonate (DECP), dibutyl sulfide (DBS) and 2-chloro ethyl phenyl sulfide (CEPS) are the four compounds simulating these combat gases that Raj et al. (2013) chose to evaluate the detection performance of their device, by measuring the frequency shift of an acoustic wave interacting with a mixture of air and simulant gases.

From these experimental data (raw frequency variations) obtained for a given population, the idea is to transpose them from a multidimensional space into a two- or three-dimensional representation system while retaining the maximum amount of information, distinctly localizing the position of the different analytes in the plane or space. From this new data representation, it becomes possible to identify a new product using a recognition software (see Chapter 3, section 3.2.1).

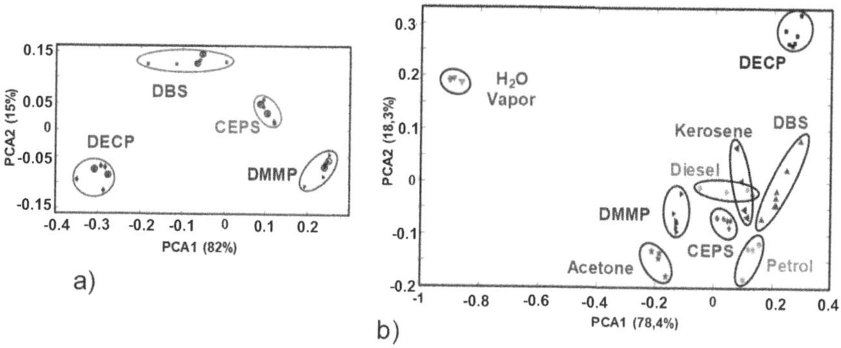

Figure 5.22. *PCA diagrams for mixtures of combat gas "simulants" and interferents. (a) PCA diagram of the four simulant gases DECP, DBS, CEPS and DMMP. The concentrations of the simulant gases vary between 20 and 150 ppm in the air and are located within separate ellipses in the plane PC1-PC2; moreover, gases of the same family, initially unidentified, are found located inside the corresponding ellipses (circled colored points). (b) PCA diagram in the presence of the same simulant gases and the interferents H_2O, acetone, diesel, kerosene and petrol. The "simulants" in various concentrations are located within separate "clusters" without overlapping with the interferents. The variance rate is high at 97% in (a) and 96.7% in (b). This indicates that the statistical transformation retains more than 95% of the information. Adapted from Raj et al. (2013). For a color version of this figure, see www.iste.co.uk/lacaze/nanotechnologies.zip*

Among the different transformation systems, the two-dimensional principal component analysis (PCA) is one of the most frequently used and involves

centers. Mustard gas $(ClCH_2CH_2)_2S$, a combat gas used by the German army during the First World War, acts on DNA and although it does not cause immediate death, it causes serious damage to the body.

representing all the data in one plane to statistically maximize the data variance along the axes PC1 and PC2, which discards non-significant data. Excellent discrimination between the different compounds is thus obtained; points for the same simulant are closely grouped and located inside ellipses, separated from the others (Figure 5.22(a)). The presence of interferent gases (omnipresent on battlefields) such as petrol, diesel, kerosene, water vapor and acetone have also been tested and show distinct results compared to combat gas "simulants". However, the PCA diagram is very different from the previous one, which shows that introducing new components means new statistical calculations and different PCA diagrams (Figure 5.22(b)).

5.3.4.2. *Colorimetric sensor networks*

Using vision to detect odors is undoubtedly the simplest and most original way to identify gaseous compounds. This visual identification technique was first demonstrated by Rakow and Suslick (2000) and was subsequently rapidly modernized with the adoption of digital color identification, replacing ocular vision (Suslick 2004). Current colorimetric networks are thus made up of many colorimetric sensors, each corresponding to a colored compound that changes color by interacting with an analyte.

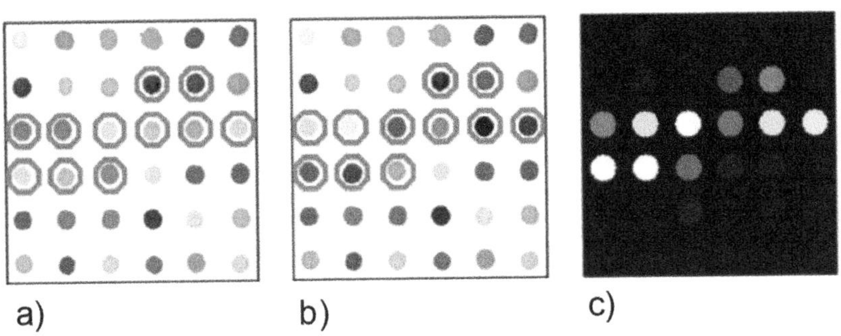

Figure 5.23. *Operating principle of a network of colorimetric sensors interacting with ammonia. (a) A network of 36 sensors was obtained by depositing a micro drop of a specific colored reagent onto a 9 × 9 mm square plate containing a commercial hydrophobic gel. A gray circle identified the sensors with the most significant color variations. (b) Changes in sensor coloring upon contact with 300 ppm of NH_3 (concentration chosen for its dangerousness, referred to as IDLH [immediately dangerous for life and health]). c) Color difference map obtained by subtraction of digitalized images from (b) and (a). According to Askim et al. (2013). For a color version of this figure, see www.iste.co.uk/lacaze/nanotechnologies.zip*

These receptors are chosen to cover an extensive range of physical and chemical molecular interactions, each time giving rise to changes in the optical properties of the sensors, which can be measured before and after interaction with the analyte. The result is a color difference map produced using a digitized color representation of the sensitive dye layers (Figure 5.23).

The Suslick group (University of Illinois), a pioneer in the field of colorimetric sensors, proposed a wide variety of dyes from different chemical families, including tetraphenyl metalloporphyrins, solvatochromic dipolar compounds, azo derivatives, dyes corresponding to pH indicators, as well as chromogenic redox compounds (Figure 5.24).

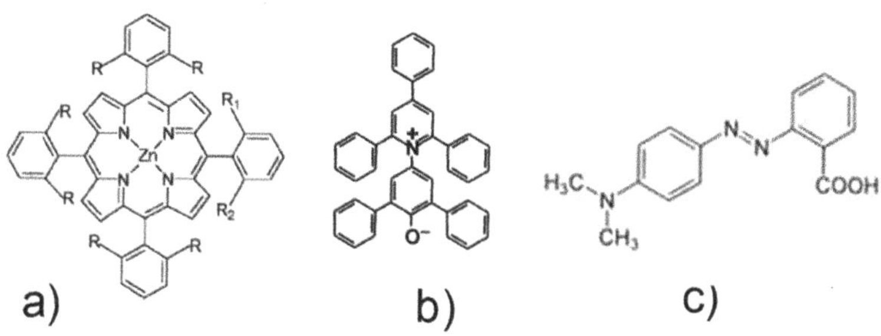

Figure 5.24. *Examples of chromogenic molecules belonging to different families of compounds. (a) Variously functionalized Zn tetraphenyl metalloporphyrins. (b) Solvatochromic dipolar molecules (zwitterions). (c) pH-sensitive azo dyes (methyl red). According to Suslick et al. (2004)*

COMMENT ON FIGURE 5.24.– *The tetraphenyl metalloporphyrins represented in (a), functionalized by various groups R, R_1, R_2 and coordination metals Zn, Co and Fe, are comparable to Lewis acids, capable of interacting strongly with basic analytes via the central coordination metal, causing intense color changes. The family of compounds shown in (b) are highly polar zwitterions, characterized by a permanent dipole, which gives them solvatochromic properties. Significant color changes are observed upon interactions with polar analytes. The compounds represented in (c) belong to the family of azo dyes, which change color according to the pH.*

This wide choice of compounds, which gives rise to all kinds of reactions with various analytes, characterized by reaction enthalpies of between 100 kJ and a few kilojoule, contrasts with the e-nose technology where arrays generally include a smaller number of sensors, and where interactions with the sensitive layer are most often low energy adsorption reactions. In the first case, the greater number of colored sensors means a greater detection selectivity and a much better discriminatory power for different analytes than e-noses. On the other hand, it should be noted that when the reaction of the dye with the analyte is irreversible (high ΔH), the sensor cannot be reused, which makes it a disposable device. This drawback is minor since these sensors can be produced quickly, in large quantities, and at low cost, thanks to a system of grouped micropipettes that simultaneously inscribe the corresponding number of new sensors on a gel support.

This type of sensor is characterized by extremely high discriminatory properties, which are clearly visible when comparing the color images induced by different VOCs on a colorimetric sensor array (Figure 5.25).

This discriminatory power results from the digitalized color detection mode in the RGB (red, green, blue) system with 256 shades per channel (primary color). With 24 colorimetric sensors and three primary color channels comprising 72 dimensions, the possible number of characteristic images is 256^{72}, which is considerable even after limiting the various differentiating factors. According to Suslick, by limiting the experiment to a dozen essential dimensions and a dozen shades, the number of characteristic images would still be 10^{12}.

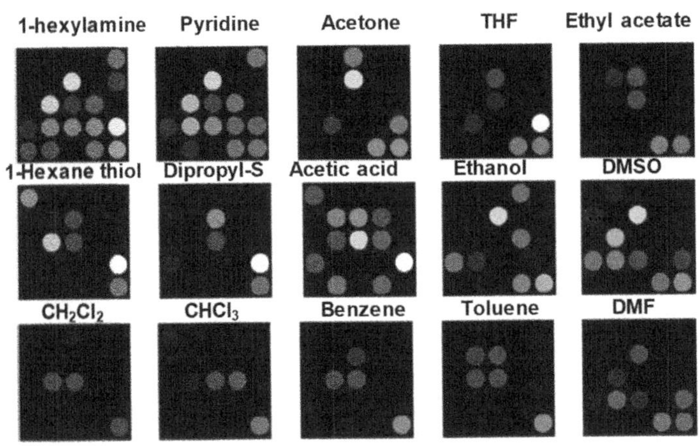

Figure 5.25. *Characteristic images of different volatile compounds interacting with a network of 24 colorimetric sensors. The images in the figure are digitalized and correspond to the difference in sensor coloring before and after interaction with VOCs in the saturated vapor state at room temperature (300°K). Each VOC is characterized by a specific image that allows it to be identified, like a fingerprint (THF, DMSO and DMF correspond, respectively, to tetrahydrofuran, dimethylsulfoxide and dimethylformamide). Adapted from Suslick et al. (2004). For a color version of this figure, see www.iste.co.uk/lacaze/nanotechnologies.zip*

Regarding sensitivity, very different images are also obtained for slight variations in concentrations. Lin and Suslick thus showed that a powerful explosive such as TATP (triacetone triperoxide)[35], characterized by a high vapor pressure, relatively easy to prepare, can be detected and measured over a range of concentrations from 50 ppb to 10 ppm using an array of colorimetric sensors (Lin 2010) (Figure 5.26). TATP detection is accomplished using an acidic solid reagent that rapidly and quantitatively decomposes TATP vapor into H_2O_2 and acetone. The reaction of H_2O_2 with a dye produces a color change characteristic of the interaction between H_2O_2 and the sensitive surface of the sensor.

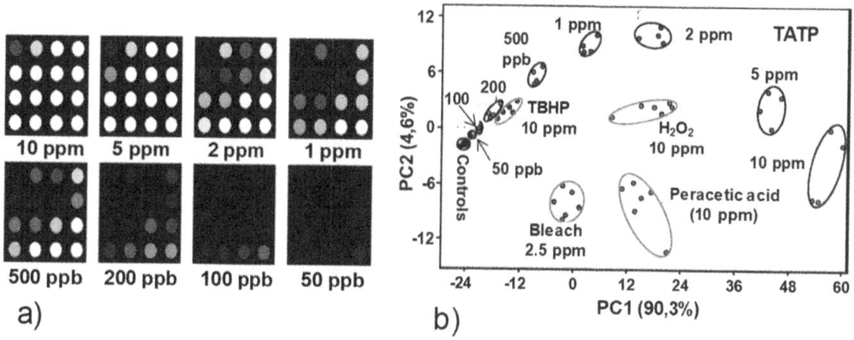

Figure 5.26. *Colorimetric detection of TATP vapors for concentrations between 50 ppb and 10 ppm. (a) Differential colorimetric images were observed for the reaction of H_2O_2 on a network of 16 colorimetric sensors whose sensitive layers were made of different dyes. (b) Statistical processing of the different digital images by the PCA method clearly showed the separation between the different concentrations of TATP and other compounds interfering with TATP (peracetic acid, ter butyl hydroperoxide-TBHP, H_2O_2, bleach HClO). The "controls" correspond to responses at different humidity levels, showing that the locations of the different TATP PCA clusters remain unchanged. Adapted from Lin and Suslick (2010). For a color version of this figure, see www.iste.co.uk/lacaze/nanotechnologies.zip*

In conclusion, this type of colorimetric detector has many advantages over previous electronic noses. The fact that they are disposable devices, reproducible and easy to manufacture (quickly and cheaply) avoids the problems related to transduction signal drift that can occur over time, inherent to e-noses. In the same way, the hygrometric factor of the air, which sometimes intervenes randomly in the

35 TATP is a very dangerous explosive, used several times by terrorist groups in the 2000s, due to its ease of preparation but also because it is difficult to detect by usual spectroscopy techniques (UV and fluorescence).

response of an electronic sensor, does not in a colorimetric sensor and can be considered. Finally, the great diversity of reactions gives rise to different colorations, which is also an additional advantage that allows better differentiation between analytes, making these colorimetric sensors optimal for detecting and identifying various analytes. Considered an optoelectronic nose, it proves effective in many fields of application, such as biomedical, food control and detection of pollutants and gases; many examples are described in a recent review article by Li and Suslick (2021).

5.4. Conclusions

Environmental monitoring has always been a primary focus of society, regulating the presence of harmful substances and protecting the population. Harmful substances are found in air and water and, at the end of the production chain, can also be found in food. As authorized concentration standards become more and more strict, the search for more efficient detection devices has become essential, with a significant orientation toward portable, connected and easy-to-handle sensor systems. Depending on the detection methodologies, two distinct areas must be considered based on the analytical treatment of solutions or products in the gaseous state.

The first case mainly concerns detecting heavy metals, for which various electrochemical techniques based on potentiometry and voltammetry have been developed. After the discovery of ion-selective membrane electrodes (ISE), specific electrodes became commercially available to detect particular ionic species. The trend toward miniaturized all-solid devices using paper supports has also developed, and detection limits initially around the micromole per liter (10^{-6} M) have improved, now reaching the nanomole per liter (10^{-9} M) after assembly improvements that lowered their detection limits.

As far as gases are concerned, resistive techniques based on the interactions of gas molecules with metal oxides, made possible by operating at high temperatures. Subsequently, the replacement of the oxide layers by various composite nanostructures, including the assembly of conductive polymers and semiconductor metal oxides, enabled resistive-type assays at room temperature, with improved detection sensitivities. Quartz microbalance and surface acoustic wave devices have also been used, in this case producing sensitive layers that sequester gas molecules and induce a variation in mass, resulting in a variation in frequency. Many sensors have been designed to detect toxic gases, mainly NH_3, H_2S and nitrogen oxides

NOx, with variable detection limits (between a few ppb and a few ppm), depending on the support and sensitive layer used. This type of device is oriented toward detection selectivity. Electronic noses differ in that they allow a mixture of gaseous compounds to be analyzed using low-selectivity sensors, whose signals are often not physically interpretable but, after statistical processing, can identify the different gaseous components in a diagram whose coordinates are reduced to two or three dimensions. Based on color recognition, optoelectronic noses are another option; they are much easier to produce from sensitive layers composed of chromogenic reactive dyes and allow more extensive sensor networks to be produced.

6

Biosensors for Health

The health sector's place in society is becoming more assertive, and there is considerable medical research in conjunction with large pharmaceutical companies. The need to constantly, and sometimes instantaneously, monitor the presence of particular biomolecules, indicators of medical diseases, has pushed research toward designing new types of sensors adapted for point of care (POC) at-home analysis, with the added possibility of an internet connection with an analysis center. This type of portable and autonomous device, typically not invasive, for which the criteria of simplicity and cost are decisive, means that several systems can be proposed for the same target molecule.

Most detection methods described in Chapter 3 differ from those practiced in analytical laboratories where immunological techniques are favored, generally associated with high-sensitivity emission spectroscopy techniques such as fluorescence, electrochemiluminescence or mass spectroscopy.

The recent focus on the development of easily manipulated and potentially disposable biosensors (Dincer et al. 2019), based on the use of nanomaterials, which can be produced on a large scale and at low cost, is now a research priority. Associating a bio-recognition element with a physical transduction phenomenon is the key to optimal functioning (Morales and Halpern 2018), for which many possible combinations exist. This research is also justified by the new demand for continuous monitoring of specific disease biomarkers using non-invasive detecting systems based, for example, on the analysis of saliva or sweat. Such systems are envisaged to monitor the real-time evolution of biomarkers of cancer, viruses or infectious bacteria or to analyze the presence of toxic substances present in the body, but also for periodic monitoring of substances revealing an individual's state of health.

The physical phenomena used to obtain high-intensity transduction signals vary and are described in Chapter 4. They mainly concern fluorescence, surface-enhanced Raman spectroscopy (SERS), surface plasmon resonance (SPR), electrochemistry and, increasingly, electronic devices based on field-effect transistors, known to be very sensitive. Gravimetry with the quartz crystal microbalance (QCM) and surface acoustic wave sensors are also avenues that have been proposed for the detection of various molecules. For most of these techniques, nanomaterials have proven decisive in some instances. Using graphene and its derivatives as the sensor sensitive layer, on which biomolecules (DNA, proteins, etc.) are adsorbed, led to numerous devices being described in several review articles (Pumera 2011; Szunerits and Boukherroub 2018; Kumar et al. 2019; Bai et al. 2020; Goldoni et al. 2021).

As the field of biosensors is particularly vast, we will limit our description to the main fields of activity where their use has proven to be important. We will thus discuss biosensors designed to control a selection of essential biomolecules responsible for medical diseases (glucose, urea and cholesterol) or those involved in the functioning of the nervous system (neurotransmitters and stress hormones). The biosensors envisaged for the early detection of biomarkers relating to cancers (Ye et al. 2018) and heart disease (Qureshi et al. 2012) will also be described, as well as those for detecting food pathogens, viruses and bacteria.

6.1. Blood sugar, uremia and cholesterol

The analysis of blood and its constituents has always been necessary to assess the health of an individual. Glucose, urea and cholesterol are among the most important, and they can be very harmful to health when their concentrations in the blood exceed certain limits. Therefore, their periodic and rapid control has become imperative to prevent severe health events. Given the very high number of patients suffering from this type of disorder, an increasingly strong demand for portable and autonomous sensors has emerged, enabling individuals to monitor their health, thanks to major pharmaceutical companies that now market such devices.

6.1.1. *Blood sugar*

Glucose, whose concentration in the blood is called glycemia or blood sugar, is a simple sugar essential to the human organism. It is transported by the blood but is stored as glycogen (a high molecular weight glucose polymer) in the liver and muscles. Three hormones control this blood sugar: insulin controls the formation of glycogen, while glucagon and adrenaline control the return of glycogen to glucose. In a healthy individual, blood sugar remains constant, around 5 mmol·L^{-1}. Glucose can be ingested directly in this form during meals (to be limited), but it is mainly released into the blood circulation, by the liver and kidneys, following the retransformation of glycogen.

Diabetes is diagnosed if fasting blood sugar levels are above 7 mmol/L[1]. There are two types of diabetes: type 1 (10% of cases) and type 2 (90% of cases). Type 1 diabetes is innate and appears in young people or from birth: insulin, necessary for the regulation of blood glucose, is not secreted by the pancreas. Type 2 diabetes is acquired and appears in older people: either too little insulin is produced or the body has developed resistance to insulin, which no longer acts sufficiently to properly regulate blood sugar levels.

Type 1 diabetes can be detected very early; however, this is not the case for type 2 diabetes, which can be latent for years before being diagnosed, already having caused irreversible damage. It is therefore essential to have access to simple and financially accessible diagnostic tools for rapid screening. Furthermore, maintaining acceptable blood sugar levels requires that those diagnosed have simple-to-use low-cost measuring devices that give immediate results, which they can use on their own, several times each day. Different techniques have been proposed to measure glucose; some correspond to chemical reactions often with enzymes that can be carried out at home or using spectrophotometry and electrochemistry in the laboratory.

6.1.1.1. *Enzymatic glucose assays*

Diabetes is most often diagnosed in a centralized analysis laboratory where blood-glucose is measured by spectrophotometry or electrochemistry from a blood sample taken from the patient. These techniques differ from each other by the nature of the enzymes and thus correspond to the so-called glucose oxidase-peroxidase (GOD-POD) and hexokinase-GDH (glucose dehydrogenase) assay methods.

The GOD-POD method is based on the combined use of the enzyme GOD and a POD, typically horseradish peroxidase (HRP). Measuring glucose involves measuring the quantity of H_2O_2 formed during the initial oxidation stage, which is achieved by a series of reactions leading to the formation of quinoneimine. Its

1 Diabetes affects around 20 million people in the United States (7% of the population) with a strong upward trend since the risk of developing diabetes in people born after 2000 is 35%! In Europe, 7% of the European population suffers from diabetes, including 3.5 million in France (5%) and 6 million in Germany (7%). Worldwide, 460 million people are affected (6%); this epidemic does not spare developing countries with few resources allocated to public health, whether for prevention, diagnosis or care. Mainly due to a lack of care, a significant proportion of individuals suffering from diabetes develop nervous, kidney, retinal and cardiac disorders.

concentration, measured by spectrophotometry at 530 nm, allows us to trace the glucose concentration in the blood plasma sample indirectly.

$$\text{Glucose} + O_2 + H_2O \xrightarrow{\text{GOD}} \text{Gluconic acid} + H_2O_2$$

$$H_2O_2 + \text{phenolic compounds} + \text{4-aminoantipyridine} \xrightarrow{\text{POD}} \text{Quinoneimine} + H_2O$$

The other method uses the enzyme hexokinase, which reacts with glucose and adenosine triphosphate (ATP) to form adenosine diphosphate and glucose-6-phosphate (G6P). G6P is then oxidized to 6-phosphogluconolactone via the enzyme G6PD (G6P dehydrogenase) and nicotinamide adenine dinucleotide (NAD), also forming the reaction by-product $NADH/H^+$ (reduced form of NAD), according to the sequence reactions as follows:

$$\text{Glucose} + \text{ATP} \xrightarrow{\text{Hexokinase, } Mg^{2+}} \text{ADP} + H^+ + \text{G6P}$$

$$\text{G6P} + \text{NAD} \xrightarrow{\text{G6PD}} NADH/H^+ + \text{6-phosphogluconolactone}$$

A spectrophotometric reading at 340 nm makes it possible to measure the concentration of $NADH/H^+$ and indirectly deduce the glucose concentration in the sample.

Electrochemical methods and photometric methods have been developed. They also use the enzymes GOD or GDH but focus on recycling the enzyme active site by an artificial redox mediator, different from the natural mediator O_2; this mediator is then recycled on a measuring electrode (Figure 6.1).

Electrochemical devices historically use GOD, associated with a mediator (M), which is usually ferrocene (Fc). Glucose quantification then amounts to measuring the quantity of reduced mediator M_{Red}, which is carried out electrochemically as per the following reaction:

$$\text{Glucose} + M_{Ox} + H_2O \xrightarrow{\text{GOD}} \text{Gluconic acid} + M_{Red}$$

$$M_{Red} \xrightarrow{\text{Electrode}} M_{Ox} + 2e^-$$

According to a similar reaction, some commercial devices now use FAD-GDH (this is the case for the OneTouch commercial device detailed below).

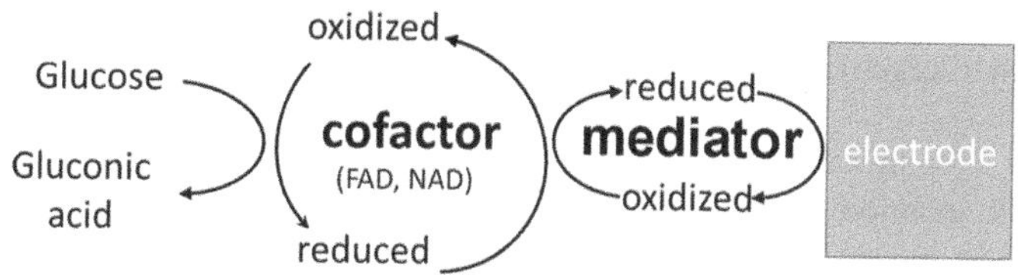

Figure 6.1. *Enzymatic electrochemical determination of glucose. Diagram illustrating the sequence of reactions catalyzed by the enzyme GOD or GDH using an artificial mediator, whose oxidation on an electrode produces a measurable current. Glucose oxidase (GOD) is only known to be associated with the co-factor FAD (flavin adenine dinucleotide). On the contrary, there are several types of glucose dehydrogenases (GDHs), distinct by the nature of the co-factor: NAD-GDH or FAD-GDH*

Further details relating to the use of GOD and GDH enzymes in glucose sensors are given in the review article by Ferri et al. (2011).

6.1.1.2. *Non-enzymatic electrochemical glucose assays*

Commercial devices use conventional enzyme detection techniques and make very little use of nanomaterials; numerous alternative approaches have been studied using non-enzymatic catalysts to oxidize glucose or completely different methods to identify the presence of glucose in a solution and quantify it.

Platinum is the first material that comes to mind for the catalytic oxidation of glucose. Like most non-oxidizable metals, oxidation proceeds by adsorption of the hydroxyl radical to give Pt[OH]ads, which reacts catalytically to oxidize glucose to glucolactone (Toghill and Compton 2010). Unfortunately, platinum has several drawbacks including its high cost and rapid poisoning by other molecules often present in a biological environment (ascorbic acid, amino acids) or its reaction with chloride ions, which quickly reduce the glucose adsorption surface and oxidation kinetics. A drastic increase in this active surface area can nevertheless be obtained using macro-, meso- or nanostructured platinum (which increases the surface/volume ratio). Platinum nanoparticles (PtNPs) have shown excellent results, with reactivities similar to those of enzymatic electrodes (Kim et al. 2013), although they do not compete in terms of selectivity.

Nickel is also widely studied, particularly in the form $Ni(OH)_2$ and NiOOH. This glucose oxidation mechanism also occurs in stages: formation of the hydroxide $Ni(OH)_2$ on the metal surface (in an alkaline medium), then oxidized to NiOOH, which is catalytically active, and transforms glucose into gluconolactone. Nanoporous Ni electrodes gave excellent results (more stable than platinum, despite a lower activity) (Niu et al. 2013). Nickel, unlike platinum, is not poisonous, but it is only active at alkaline pH, excluding any in vivo use (in addition to being a well-known allergen).

Copper, like nickel, is catalytically active in its Cu(II) and Cu(III) forms. There are multiple examples of the use of copper for the catalytic oxidation of glucose, driven in part by its low cost. Here again, using mesostructured or nanostructured copper makes it possible to obtain satisfactory results: we can cite the work of Niu et al. (2014), who demonstrated the catalytic oxidation of glucose on a copper foam. Copper oxide CuO is also widely studied; its catalytic activity is based on the same principles as NiO:

$$\mathrm{CuO} + \mathrm{OH}^- \longrightarrow \mathrm{CuOOH} + e^-$$

$$\mathrm{CuOOH} + e^- + \mathrm{Glucose} \longrightarrow \mathrm{CuO} + \mathrm{OH}^- + \mathrm{Gluconolactone}$$

Finally, more recently, metal oxide frameworks (MOFs) have also been proposed to produce nanostructured catalysts to manufacture glucose biosensors. Li et al. (2019) hydrothermally synthesized a cobalt MOF from aqueous Co^{2+} and terephthalic acid. They obtained a Co-MOF network presenting strong activity for the electrooxidation of glucose in an alkaline environment. The use of MOFs appears to be judicious, given the immense possibilities of nanostructuring this material (see Chapter 1). Their involvement in glucose measurement has given rise to numerous devices, as described by Adeel et al. (2021) in a recent review article.

Given the importance of blood glucose monitoring, a number of portable, stand-alone devices have been introduced to the market, some of which are described below.

6.1.1.3. *Devices usable at home or PoC*

For a device to be effective at home, diabetics must monitor their blood sugar several times a day. This can be done directly by the patient when he is able to do so. However, stand-alone devices are desirable for patients who are not able to monitor their blood sugar levels independently, particularly children with type 1 diabetes.

6.1.1.3.1. Historical method: self-monitoring of blood glucose (SMBG)

The historical method involves pricking the tip of a finger to obtain a droplet of blood, immediately deposited on a strip that is immediately inserted into a reader; a blood sugar value is indicated within a few seconds. There is a wide range of blood glucose meters sold in pharmacies (Figure 6.2).

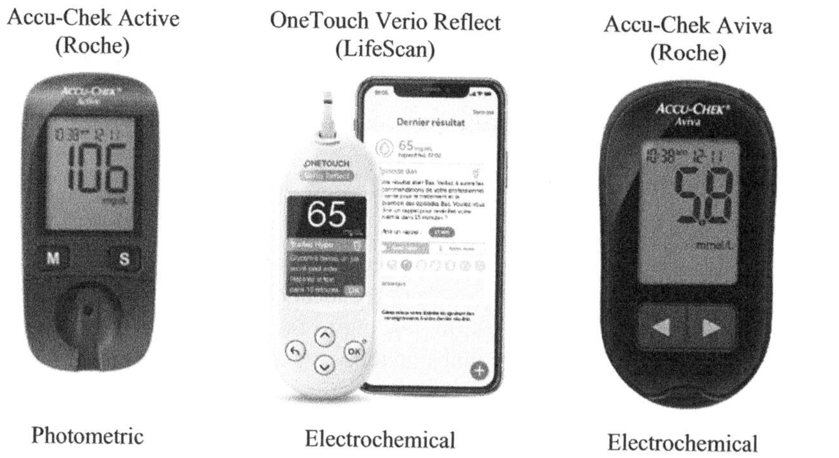

Figure 6.2. *Examples of marketed self-monitoring blood glucose devices. In all cases, the patient must take a droplet of blood (by pricking) and place it on a test strip. The Accu-Chek Active is based on photometric detection, while the OneTouch Verio Reflect and Accu-Chek Aviva are based on electrochemical detection methods. For a color version of this figure, see www.iste.co.uk/lacaze/nanotechnologies.zip*

Their operation is based on a sequence of enzymatic reactions producing either a chromogen quantified by spectrophotometry, or a redox mediator that can be recycled electrochemically on a metal electrode, as detailed in the previous section. Figure 6.3 details the structure of three "electrochemical" strips.

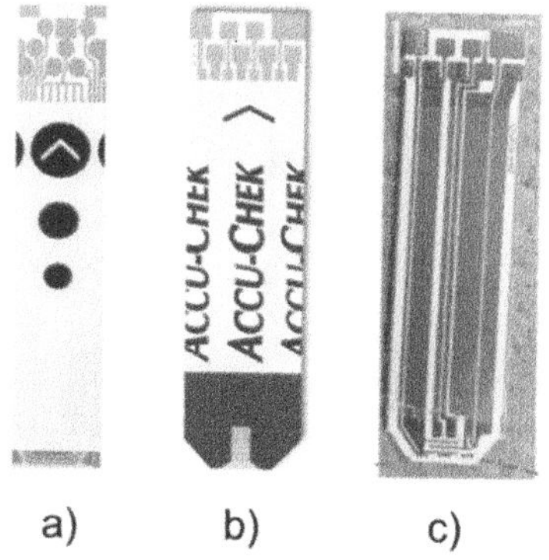

Figure 6.3. *Examples of "electrochemical" strips used in Accu-Chek devices: (a) Accu-Chek Instant strip, (b) Accu-Chek Aviva strip and (c) Accu-Chek Aviva strip with its upper layer removed, revealing the connectors (upper part) and the sensitive area (lower part surrounded by green dotted lines). Adapted from Liu et al. (2020). For a color version of this figure, see www.iste.co.uk/lacaze/nanotechnologies.zip*

It should be noted that these sensors have systematically oversized useful areas, especially the sensitive part. This is due to practical reasons involving easy handling of the strips; the operation of these readers and their associated strips can be found in an application note from Freescale Semiconductor (Garcia Yanez 2013).

6.1.1.3.2. Continuous glucose monitoring (CGM)

Blood sugar self-monitoring based on regular skin punctures is painful and poorly accepted by patients. However, forgetting to perform these tests can be fatal very quickly, and repeated deviations from the ideal concentration can produce very disabling side effects in the long term, leading to the formation of chronic wounds and even amputation of the lower limbs, in those most affected by these side effects.

In recent years, blood sugar monitoring can be done using semi-invasive devices such as sensors placed on the abdomen or on the arm that measure blood sugar continuously (more precisely, occasionally but at close frequencies), day and night, during exercise or during meals. These devices are semi-invasive because they require the insertion of a small needle into the dermis, which remains in place for between 7 and 15 days. The autonomous sensor transmits the data to a receiver equipped with a user interface, which today not only indicates the measured values but, using artificial intelligence, can help the patient take measurement-related decisions (mealtimes and snack times, or when physical activity is necessary or considered dangerous). For example, the free device MCG Freestyle Libre from the company Abbott has been 100% reimbursed by French Social Security since 2017. Figure 6.4 details the three devices on the market today.

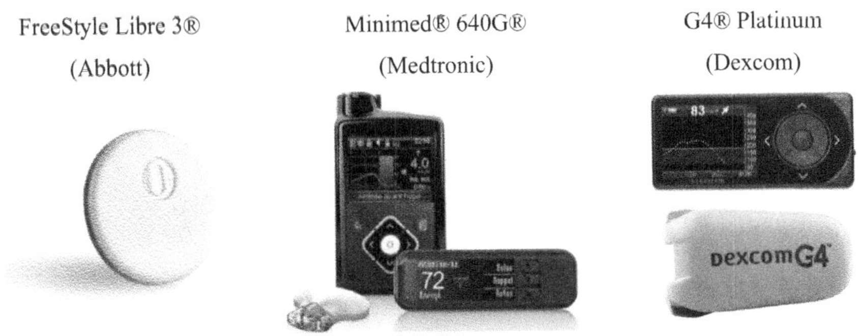

Figure 6.4. *Some continuous glucose measuring devices currently on the French market and reimbursed by their national health service. FreeStyle Libre 3 sensor, released on the market in 2022, is 22 mm in diameter and 3 mm thick, the equivalent of two 50 cent coins stacked on top of each other. For a color version of this figure, see www.iste.co.uk/lacaze/nanotechnologies.zip*

Clarke et al.'s (1987) error grid (Figure 6.5) is one tool developed to determine the accuracy of blood glucose measurements by SMBG and MCG devices, demonstrating manufacturer concern regarding measurement reliability.

The Clarke error grid comprises a scatter plot of a reference blood glucose meter and the blood glucose meter to be assessed, with five regions. Region A corresponds to values located 20% or less from the reference; region B corresponds to a deviation greater than 20% but which would not lead to an incorrect decision by the system or patient; region C corresponds to results that would cause the system or the patient to make an incorrect (but not dangerous) decision; region D corresponds to measurements that could lead to a dangerous decision (e.g. the system does not detect hypoglycemia or hyperglycemia); region E corresponds to measurements that would mistake hypoglycemia for hyperglycemia and vice versa, causing immediately death.

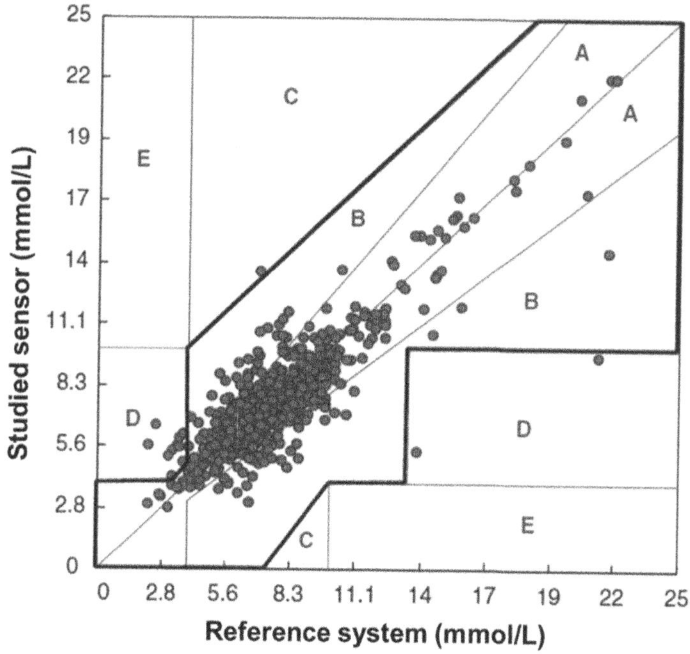

Figure 6.5. *The Clarke error grid compares the blood glucose values given by a Medtronic Minimed continuous glucose monitoring system (Y-axis) and a reference system, the ABL 715 blood glucose analyzer from Radiometer Medical (X-axis). The meaning of the different domains A to E is given in the text. Adapted from Ligtenberg et al. (2010). For a color version of this figure, see www.iste.co.uk/lacaze/nanotechnologies.zip*

6.1.1.4. *Non-invasive methods*

The continuous glucose measuring devices described above are based on quantifying glucose in the blood or interstitial fluid (ISF), typically described as an invasive or semi-invasive method. Research today focuses on non-invasive, strictly external methods. In addition to obvious advantages in terms of comfort for people wearing these devices, this also considerably reduces regulatory constraints. Glucose measurements can be obtained in two main ways: either by quantifying glucose in external body fluids (tears, saliva and sweat) or by non-contact spectroscopy or radiofrequency measurements.

6.1.1.4.1. Quantification in external body fluids

Sweat contains ions and proteins. It also contains metabolites, including glucose. Its concentration is nevertheless significantly lower than in the blood, and its variations in concentration observed in sweat are delayed by around 30 min compared to those measured in blood. This seriously limits its scope because this delay does not allow a fast enough feedback loop, meaning the patient cannot correct hypo- or hyperglycemia emergencies in time. However, sweat is a quickly and continuously accessible fluid, which explains why it is one of the most studied. The device developed by Lee et al. (2017) is one of the most successful and functional, featuring a complete sweat collection device, an enzymatic sensor combined with a temperature, humidity and pH sensor, and a system for delivering an active substance through the skin.

Saliva and tears are also widely studied, although they are fraught with major drawbacks. Thc cycs are extremely sensitive organs; the quantity of glucose in tears is imperfectly correlated with that in the blood, and the phase shift between the concentrations of glucose in these two bodily fluids is approximately 15 min, which is also too long to treat hypo- and hyperglycemia emergencies. For saliva, the correlation is even worse, and glucose cannot be measured during or immediately after a meal due to the strong contamination of the oral environment by food remains, which considerably limits its effectiveness.

Glucose measurements from the ISF[2], without the need for an electrode to penetrate the dermis, can be obtained by reverse iontophoresis. The GlucoWatch (Tierney 2001), a device briefly placed on the market until 2007, used this technique, justifying this approach (Figure 6.6).

---

2 Interstitial fluid is the fluid that occupies the space between the blood capillaries and the cells of the dermis. Cells draw their nutrients and oxygen from it.

The ISF and the blood exchange small molecules by diffusion. The variations in glucose concentration in the ISF follow those of the blood with only a few minutes delay and the correlation between the two concentrations is excellent, much better than for tears or saliva. It is the concentration of glucose in this ISF, and not in the blood, that is currently measured by commercial CGM devices. However, to make the device non-invasive, therefore avoiding penetrating the skin, transdermal reverse iontophoresis is used and allows glucose to be extracted through the skin without using a needle.

The principle of inverse iontophoresis, comparable to an electroosmotic flow process (see Chapter 2) enables glucose extraction through the skin by applying a weak electric current between two electrodes (Figure 6.6(a)). In the interstitial liquid, Na^+ ions higher in concentration than Cl^- ions migrate toward the cathode and carry with them the solvent and neutral molecules such as glucose, which emerges at the skin surface and can then be quantified electrochemically. The process gave rise to a portable device, the GlucoWatch, in the 2000s.

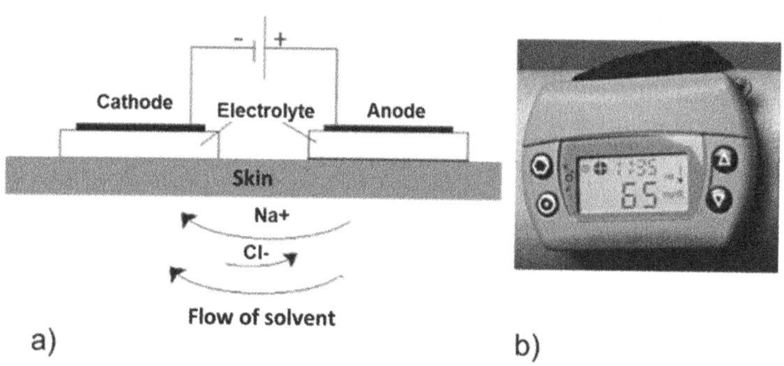

Figure 6.6. *The GlucoWatch blood glucose monitor. (a) Principle of inverse iontophoresis applied to the electrochemical determination of glucose through the skin. (b) Photo of the GlucoWatch device applied to a patient's arm. Adapted from Potts et al. (2002). For a color version of this figure, see www.iste.co.uk/lacaze/nanotechnologies.zip*

Unfortunately, the ionic current passing through the skin can cause significant lesions, requiring the device using this technique to be withdrawn from the market a few years later. Other devices using sweat have also been the subject of portable, flexible and non-invasive equipment (Emaminejad et al. 2017).

6.1.1.4.2. "Contactless" spectrometric or radiofrequency measurements

Contactless measurements have been designed which enable glucose monitoring without a needle or microneedle penetrating a patient's skin; for example, visible light spectrometric methods are well suited for glucose detection. However, they encounter many difficulties due to the strong absorption of light by human tissues[3]. Microwave spectroscopy, corresponding to frequencies between 300 MHz and 300 GHz, is a more promising approach: the absorption spectrum of the dermis varies depending on the glucose concentration and a fine analysis of this spectrum makes it possible to discriminate the influence of other parameters or other species in solution. This detailed analysis uses signal processing methods as well as artificial intelligence strategies that go beyond the scope of this work. However, details of the method can be found in the article by Buonanno et al. (2022).

6.1.2. *Uremia*

Uremia is the concentration of urea in the blood. The expected concentration for a healthy individual is between 2.5 and 7.5 mmol/L (rarely considered in isolation and usually accompanied by the quantification of other blood markers).

Urea is an essential biomarker for medical diagnosis, particularly for detecting and monitoring patients suffering from urea cycle disorders (UCDs), a genetic disease leading to serious functional disorders. Besides UCDs, uremia measurement is also helpful for identifying and monitoring renal dysfunction acquired during life. Urea sensors are, unfortunately, far from technical and commercial maturity like blood glucose sensors.

6.1.2.1. *Enzyme-based detection*

Most urea sensors are based on a catalytic reaction of the hydrolysis of urea to ammonia, most frequently carried out via an enzyme, urease, according to the following reaction:

$$NH_2CONH_2 + 3\ H_2O \xrightarrow{\text{Urease}} 2\ NH_4^+ + OH^- + HCO_3^-$$

This enzymatic activity is often detected by optical or electrochemical methods or by coupling optical and electrochemical methods.

6.1.2.1.1. Optical transductions

SPR is a surface transduction technique widely used in biosensors. It is detailed in Chapter 3 of this work and benefits significantly from nanomaterials. SPR

3 UV wavelengths are absorbed by DNA and proteins, those corresponding to the visible range are absorbed by hemoglobin, and those corresponding to infrared are absorbed by water, which makes spectroscopic analysis of the human body very difficult. However, near infrared (in the range 680–2,500 nm) can penetrate soft tissues to a depth of >0.5 mm.

generally applies to probe-target type systems where the association of the target on the grafted probe significantly modifies the surface reflectivity. However, an accumulating enzymatic reaction product can also generate a change in reflectivity; urea sensors based on this technology have been described, in particular using an optical fiber as the sensitive surface (Bhatia and Gupta 2012).

More original is the use of photonic crystals based on inverse opals composed of polymers with a highly ordered porous structure. Through the phenomenon known as Bragg diffraction, the color of these materials is directly dependent on the size of the cavities they contain. Any stimuli modifying the structure generates a variation in the refractive index and leads to a very significant change in the color of the material, directly visible to the naked eye (Figure 6.7).

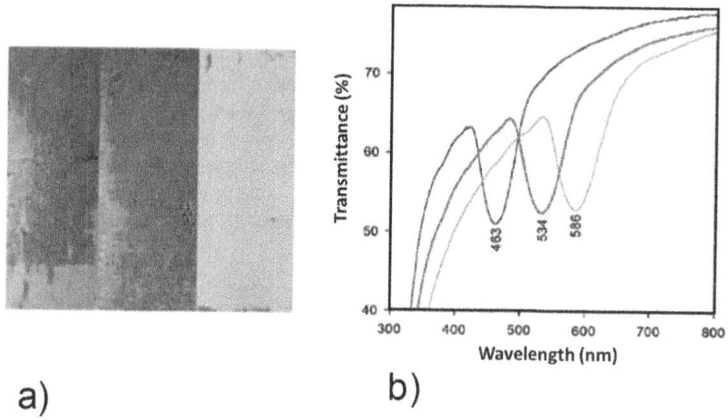

Figure 6.7. *Inverse opal surface colorings. (a) Photograph of thin-film alumina inverse opals obtained from different sizes of PMMA microbeads. (b) Corresponding UV-Vis light transmission spectra. The inverse opal surfaces of blue, green and yellow colors were obtained, respectively, with opals consisting of 313, 352 and 394 nm beads and are characterized by absorption peaks at 463, 534 and 566 nm. Adapted from Waterhouse et al. (2015). For a color version of this figure, see www.iste.co.uk/lacaze/nanotechnologies.zip*

The dimensions of the structures responsible for Bragg diffraction are of the same order of magnitude as the wavelengths of visible light and correspond to cavities several hundred nanometers in size or diameter. Chapter 4 (Figure 4.39) describes the principle of obtaining inverse opals and relates to nanosphere lithography (NSL). This process is widely used to produce surfaces suitable for SERS and various colorimetric assays induced by slight changes in the dimensions of the cavities of inverse opals.

Inverse opals made from polymers lead themselves well to optical transduction applications because they are generally very sensitive to various external stimuli, particularly pH. Li et al. (2016b) showed that an inverse opal made up of a methacrylic acid (MAA) and ethylene glycol dimethylacrylate (EGDMA) copolymer could be used as a urea sensor. Indeed, this copolymer swells due to an increase in pH, caused by urease in the presence of urea, which modifies the diffraction grating and causes a color variation. At a pH of around 5.5, the Bragg diffraction wavelength is 548 nm and shifts toward the red for increasing urea concentrations (Figure 6.8).

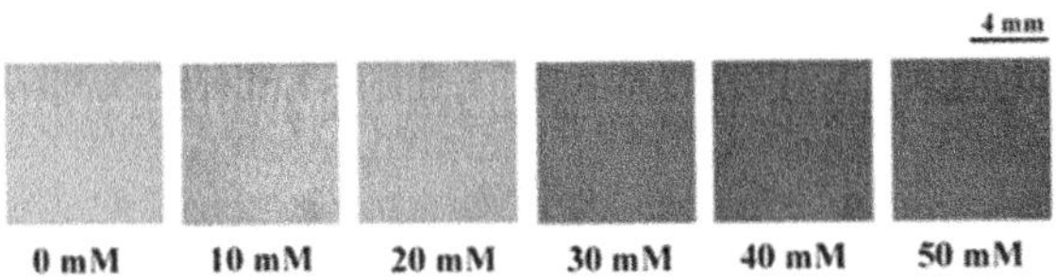

Figure 6.8. *Application of an inverse opal to the colorimetric detection of urea. The different squares represent the color changes of a poly(AA-co-EGDMA) inverse opal film for increasing urea concentrations. Initially centered at 548 nm at pH 5.5, the reflected light shifts to 570, 591, 644 and 668 nm for respective urea concentrations of 10, 20, 30 and 40 mM in the presence of urease. According to Li et al. (2016). For a color version of this figure, see www.iste.co.uk/lacaze/nanotechnologies.zip*

Nanometric quantum dots (QDs) (see Chapter 1), whose confined electronic structure confers particular electrical and optical properties, are also well suited for the photometric detection of biomolecules.

Core-shell CdSe/ZnS QDs, functionalized with D,L-mercaptosuccinic acid (MSA), were used by Huang et al. (2007) to determine the urea concentration using photoluminescence. This is characterized by photoemission proportional to the surrounding pH and has shown to be well suited to measuring urea in the presence of urease for a wide range of concentrations between 0.01 and 100 mM, covering the expected concentrations of urea in the blood and urine. This principle, applied to the enzymatic detection of urea, can be used for many enzymes whose activity produces or consumes protons.

The enzymatic detection of urea in the presence of urease can also be obtained by ratiometry (see Chapter 3) applied to the fluorescence of pH-sensitive QDs compared to an insensitive probe, or the opposite scenario with QDs whose fluorescence is insensitive to pH, while the other probe is sensitive to it. Zhang et al. (2019b) produced a ratiometric device using molybdenum sulfide MoS_2 QDs whose photoemission at 420 nm is insensitive to pH and 2,3-diaminophenazine, used as a probe, whose fluorescence intensity at 568 nm increases gradually with pH. In this specific case, the I_{568}/I_{420} fluorescence intensity varies linearly for urea concentrations between 5 µM and 0.7 mM.[4]

Finally, although there are very few examples, electrochemiluminescence (described in Chapter 3) has also been used in the detection of urea, especially for fundamental in vitro mechanistic studies, such as research by Fen et al. (2023) devoted to the study of the transmembrane transport of urea.

6.1.2.1.2. Electrochemical transduction

Knowing that neither the products of the enzymatic reaction (ammonium, hydroxide, and hydrogen carbonate ions) nor the active site of the enzyme is directly electroactive (urease is a hydrolase and not an oxidoreductase), the detection techniques in amperometric sensors have typically focused on the use of a redox-active molecule, coupled with a molecular probe sensitive to pH. Several polymers have been used as pH probes, the most typical being PAMAM or poly(amidoamine) (Dervisevic et al. 2017). Functionalized by ferrocene, the local increase in pH due to the activity of urease causes a collapse in the dendritic structure of PAMAM (Liu et al. 2009), causing the redox probes to touch the surface of the measuring electrode, increasing the measured redox currents. Overall, however, amperometric techniques remain marginal and not always convincing.

Based on the local modification of pH, resulting from the decomposition of urea into ammonium ions by urease, potentiometric sensors have also been produced using electrodes modified by a layer of polyaniline, polyethylenediamine, or iridium oxide, whose equilibrium potential is sensitive to pH, limited by poor reversibility and long response times. However, these devices remain limited by the intrinsic low sensitivity of the potentiometric technique, their poor reversibility and slow response time.

Sensors based on using ion-sensitive field effect transistors (ISFETs) were extensively studied in the 1990s (see Chapter 3). Operating on the basis of a variation in membrane potential (whether due to NH_4^+ ions or H^+ protons) (Senillou et al. 1999), they are characterized by a significant amplification of the potential variation occurring at the transistor gate level, resulting in a variation in current intensity between the source and drain contacts.

Using the more recent organic electrochemical transistors (OECT) is a promising approach. Their operation is based on the conductivity variation of the conductive

---

4 The authors of this work have not considered the extension of this domain to physiological values but this should not pose any particular technical problem.

polymer constituting the transistor channel. Berto et al. (2018) produced a urea biosensor by immobilizing the urease enzyme in a hydrogel placed on the surface of the PEDOT:PSS channel. The urease activity alkalizes the medium, and the conductivity of the PEDOT:PSS channel is reduced, causing a reduction in the drain current (Figure 6.9).

FETs have also been developed based on CdS, ZnS or graphene semiconductors. Middya et al. developed a FET based on a $CdS\text{-}TiO_2$ nanocomposite. Their device showed an increase in the drain current with increasing ammonia release from urease, explained by an increase in the number of charge carriers due to the adsorption of NH_3 onto the $CdS\text{-}TiO_2$ nanostructures. The detection threshold was 0.85 ppm of urea for an applied gate voltage of 0.5 V (Middya et al. 2019).

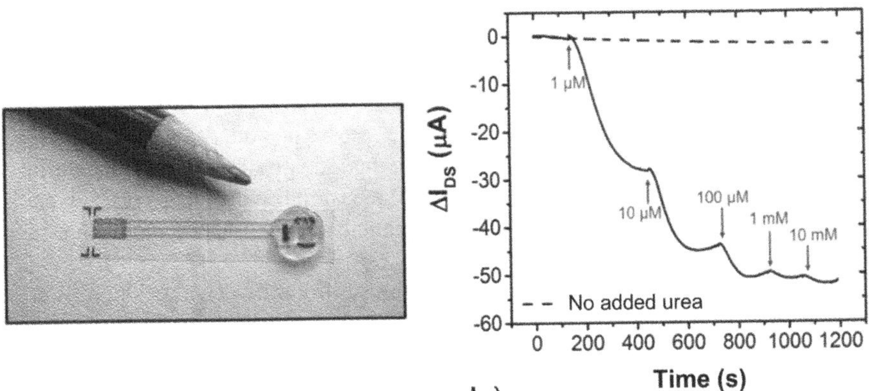

a) b)

Figure 6.9. *OECT biosensor for urea measurement. (a) Photograph of a printed OECT showing the gate (central contact), the source and drain (lateral contacts) and the channel located between the source and drain (greenish color). (b) Variation in the drain current (at constant drain potential) as a function of the quantity of urea in contact with the electrolytic hydrogel. Adapted from Berto et al. (2018). For a color version of this figure, see www.iste.co.uk/lacaze/nanotechnologies.zip*

In conclusion, although numerous devices have been produced, detecting urea in blood or urine is still a current area of research. Botewad et al. (2023) conducted a thorough analysis by reviewing all the publications concerning urea sensors, comparing their analytical characteristics in a tabulated form, and highlighting the contribution of nanomaterials where applicable.

6.1.3. *Cholesterol*

Cholesterol is a lipoprotein systematically quantified during a blood test. Lipoproteins are very large protein and lipid complexes, 5–30 nm in diameter depending on whether they are low-density lipoproteins (LDLs) or high-density lipoproteins (HDLs), and cholesterol is associated with these two LDL and HDL forms. LDLs are considered harmful because they deposit on the arterial walls causing blockages over time, whereas HDLs do not. Cholesterol is found on the outer ring of these lipoproteins and is not dissolved in the blood. HDLs, on the contrary, have a beneficial effect because they "collect" the accumulated cholesterol, all of which is then broken down by the liver.

In addition to cholesterol in the circulating blood, this molecule is also present in our food. It may, therefore, be relevant to measure it in so-called fatty foods, always of animal origin, such as eggs, dairy products and meat.

6.1.3.1. *Electrochemical transduction*

The cholesterol molecule is not easily electrooxidized. However, it has been possible to electrooxidize cholesterol on nanostructured catalytic surfaces like those already described in the previous chapters, but without selectivity. A NiO surface electrooxidized to NiOOH thus makes it possible to oxidize cholesterol to cholestenone (Rengaraj et al. 2015), according to the following catalytic cycle:

$$\text{NiO} \rightarrow \text{NiOOH} + \text{n e}^-$$

$$\text{NiOOH} + \text{cholesterol} \rightarrow \text{NiO} + \text{cholestenone}$$

This reaction mechanism, using nickel oxide and peroxide, requires the electrocatalysis to be carried out in a very alkaline medium (1 M KOH solution), which makes the method inapplicable in situ, and requires the transfer of the sample to an alkaline environment before any measurements can be taken.

Electrochemical methods, therefore, more often combine molecular recognition and a redox probe, which is free to diffuse in solution or remain immobilized on the sensitive surface of the sensor. One solution is to use complexing reagents such as β-cyclodextrin (attached to an electrode), for which hydrophobic cholesterol has a good affinity. The transduction method then involves immobilizing a hydrophobic redox probe in the cavity and then putting the surface in contact with the cholesterol, which moves the redox probe by replacing it (Yang et al. 2015).

The use of enzymes is much more general and provides the desired selectivity. All are based on cholesterol oxidase (ChOx), sometimes coupled with cholesterol esterase (ChEt). The use of ChEt is justified by the fact that part of the cholesterol is present in LDL and HDL in an esterified form; ChEt then initially allows the cholesterol to reform, and then ChOx oxidizes the cholesterol to cholestenone, releasing H_2O_2.

$$H_2O + \text{cholesteryl ester} \xrightarrow{\text{ChEt}} \text{Cholesterol} + \text{fatty acid}$$

$$O_2 + \text{cholesterol} \xrightarrow{\text{ChOx}} H_2O_2 + \text{Cholestenone}$$

Cholesterol is then determined by measuring H_2O_2 in the presence of peroxidase, most often HRP and a redox-active molecule M_{Redox}. A catalyst such as platinum can also be used for the reduction or oxidation of H_2O_2:

$$H_2O_2 + M_{red} \xrightarrow{\text{HRP}} 2\ H_2O + M_{ox}$$

$$M_{ox} + 2e^- \xrightarrow{\text{HRP}} M_{red}$$

or:

$$H_2O_2 \xrightarrow{\text{Pt}} O_2 + 2H^+ + 2e^-$$

These various reactions give rise to amperometric measurements for which, ideally, the measured current is proportional to the quantity of H_2O_2 produced and, therefore, to the quantity of total cholesterol initially present.

Numerous studies have proposed improvements to this fundamental principle, mainly relating to the electrode materials, in order to maximize the hydrogen peroxide oxidation current and increase the sensor sensitivity. In particular, nanozymes, catalysts based on nanomaterials with activities close to enzymes, have been studied to oxidize H_2O_2 (see Chapter 4, section 4.1.5.2). A good overview of the state of the art regarding electrochemical detection of cholesterol can be found in the review article by Amiri and Arshi (2020).

6.1.3.2. *Optical transduction*

Despite the advantages of electrochemical transduction, optical (colorimetric) methods are still preferred, particularly for all standard laboratory analyses of a blood sample.

As explained above, cholesterol does not move freely in the bloodstream: it is linked to LDL and HDL. Blood tests do not quantify cholesterol alone, so they do not provide information on the risk of atheroma (accumulation of cholesterol on arterial walls), but instead the respective quantities of "good" (HDL) and "bad" (LDL) cholesterol.

Thus, for LDLs, one of the methods applied in laboratory analyses involves coupling enzymatic action and colorimetric detection. The first step is to eliminate free HDL cholesterol. The blood sample is added to a buffered mixture of 2-(N-morpholino)ethanosulfonic acid with a pH of approximately 6, containing a surfactant that surrounds the LDL and prevents it from reacting. Cholesterol esterase (ChEt) and cholesterol oxidase (ChOx) then react with unprotected lipoproteins. The hydrogen peroxide produced is eliminated by a catalase (Fe(III)-E), an enzyme comprising an iron(III) complex in its active center, catalyzing the breakdown of H_2O_2 into O_2 and H_2O according to the following sequence of reactions:

$2\ H_2O_2 \rightarrow O_2 + 2\ H_2O$ (complete reaction, decomposed as follow)

$H_2O_2 + Fe(III)\text{-}E \rightarrow H_2O + O{=}Fe(IV)\text{-}E$ (iron oxidation and H_2O production)

$H_2O_2 + O{=}Fe(IV)\text{-}E \rightarrow H_2O + O_2 + Fe(III)\text{-}E$ (iron recycling and H_2O production)

In the second step, another reagent is added (for example, NaN_3), which inactivates the catalase and releases the LDL from the surfactant. Again, ChEt and ChOx produce H_2O_2, which HRP decomposes with a phenolic co-substrate. In its oxidized form, this phenolic co-substrate is colored. The initial LDL concentration can be deduced by measuring the absorbance of the colored solution.

For HDLs, the detection principle is similar, but no protective surfactant is added. As a result, LDL is eliminated by ChEt and ChOx, without this leading to the production of colored markers. Second, a surfactant is added, which solubilizes the HDL and releases the contained cholesterol. The same enzymatic reactions lead to the production of H_2O_2, whose decomposition by HRP produces colored markers. The HDL concentration can then be deduced by measuring the absorbance of the colored solution.

Although the method is not yet used in laboratory analyses, it seemed relevant to describe a less traditional approach, based on the use of metallic gold nanoclusters (AuNC). With selectivity being a particularly important criterion, the possibility of grafting a functional group onto the AuNCs that complex the target molecule while not interacting with many other molecules is a significant advance. Numerous research groups have exploited this by grafting β-cyclodextrin (β-CD) onto AuNCs and exploiting the fluorescence properties in solution of certain organic luminophores, quenched when complexed with β-CD and regenerated when displaced from β-CD by the target molecule with a greater affinity. Such a system was developed to measure cholesterol using rhodamine–B as the luminophore. The rhodamine, initially complexed by the β-CD@AuNCs, does not emit any fluorescence; when displaced from the β-CD@AuNCs by cholesterol, it fluoresces with its intensity being directly proportional to the amount of cholesterol complexed by the β-CDs.

6.1.3.3. *Some practical examples*

On a practical level, new cholesterol sensors continue to be described in the literature, although still reserved for health professionals; some easy-to-use portable instruments already on the market are well described in the review article by Narwal et al. (2019). As an illustration, we have shown in Figure 6.10 a cholesterol biosensor, sold for around €150, designed to provide a result in 45 s.

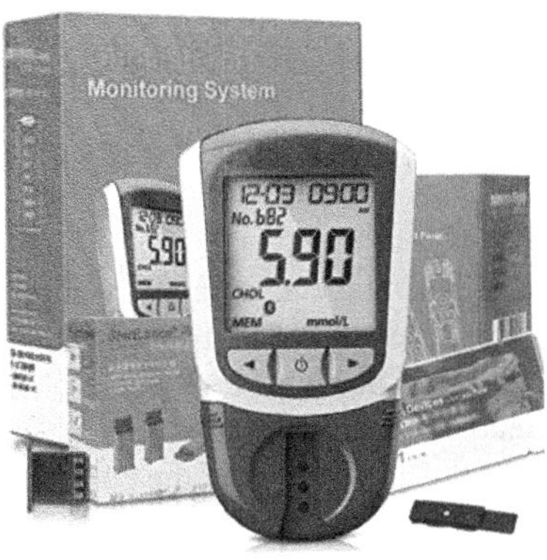

Figure 6.10. *Example of portable blood cholesterol testing devices. For a color version of this figure, see www.iste.co.uk/lacaze/nanotechnologies.zip*

Like blood glucose sensors, no commercial devices use nanomaterials for cholesterol detection; all are based on established principles.

6.2. Biomarkers

The term biomarker appeared in the 1950s and since then, new biomarkers have continued to be discovered, notably due to improvements in analysis techniques. Specific biomarkers can be used to identify a wide variety of conditions: cancers, acute or chronic inflammation, Alzheimer's disease, tuberculosis, heart disease, brain lesions, diabetes, autism, etc.

Biomarker	Biological fluid(s)	Pathology
AFP (α-fetoprotein)	Blood	Liver cancer
B-type natriuretic peptide (BNP)	Blood	Heart condition
CA-125 (Carbohydrate antigen)	Blood	Various cancers, including ovarian cancer
CA-15-3 (MUC1, cancer antigen 15-3)	Blood	Breast cancer
CA-27-29 (Cancer antigen 27-29)	Blood	Breast cancer
Carcinoembryonic antigen (CEA)	Blood	Cancer, liver disease, Crohn's disease
Cardiac Troponin 1 (cTn1)	Blood	Heart attack
Fragment cytokeratin 21-1 (CYFRA 21-1)	Blood, urine	Lung cancer
EGFR (Epidermal Growth Factor)	Blood	Colorectal or lung cancer
ER/PR (estrogen receptor/progesterone receptor)	Blood	Breast cancer
hCG (human chorionic gonadotropin)	Blood	Various cancers
HE4 (Human epididymis protein)	Blood	Ovarian cancer
HER-2/neu (Human epidermal growth factor receptor 2)	Blood	Breast cancer and gastric cancer
High-density lipoprotein (HDL)	Blood	Cardiovascular condition
Homocysteine	Blood	Arterial disease
Human apolipoprotein A1 (hAPOA1)	Blood	Heart condition
Lactate dehydrogenase (LDH)	Blood	Heart attack
Platelet-derived growth factor PGDF/PDGFR	Blood	Leukemia
PML/RAR-alpha (promyelocytic leukemia/ retinoic acid receptor)	Blood	Leukemia
Progesterone	Blood	Breast cancer
Prostate-specific antigen (PSA)	Blood	Prostate cancer
Testosterone	Blood	Prostate cancer
TPMT (thiopurine S-methyltransferase)	Blood	Leukemia, Crohn's disease

Table 6.1. *Different cancer and cardiac biomarkers (excluding nucleic acids), their matrix (body fluid where they are significantly present) and the cause of their (over)expression are classified in alphabetical order*

Biomarkers are chemical substances that quantitatively indicate a medical condition and make it possible to establish a relationship between their detected quantity and disease symptoms. They should not be confused with the symptoms themselves, which are primarily qualitative through their perception by patients or healthcare staff. In this sense, biomarkers are objective and quantifiable medical signs. The challenge today is to determine the reliable relationship between these quantifiable molecules and clinical diagnoses in order to combine analytical sciences and medicine and ultimately connect medicine and Big Data to take a significant step toward Medicine 3.0.

The list of biomarkers known to date and present in blood or body fluids is particularly long, covering many diseases, which can be found in specialized studies such as the one recently published by Ozkan et al. (2021). We will focus on only two families of pathologies and the most significant development examples concerning certain cancers and cardiovascular conditions.

It is striking to note that the overexpression of a molecule, considered in isolation, is rarely a valid marker that signals a condition; on the contrary, it is necessary to consider several molecules whose concomitant overexpression can be associated with a specific dysfunction.

6.2.1. *Cancer biomarkers*

Some biomarkers can be used to determine an individual's risk of developing cancer, for example, when a family history is known. In this case, the specific biomarker is a genetic variation of the genome, such as the BRCA1 gene. Biomarkers can also be used to detect cancer in patients who have not yet developed symptoms. This is, for example, the case of prostate-specific antigen (PSA), even though this biomarker cannot be considered in isolation. In treated patients, biomarkers can guide therapy: for example, in stomach cancer, the overexpression of the HER2 gene indicates resistance to hormonal treatments. In addition to coding DNA sequences, many tumor biomarkers are strands of RNA (miRNA) or circulating proteins, such as miRNA-155, CEA and CA125, characteristic of intestinal, colorectal, prostate or ovarian cancers. They are described in the review article by Jayanthi et al. (2017). As with other types of targets, optical and electrochemical detection methods are often used.

6.2.1.1. *Cancer screening using optical sensors*

The detection of circulating RNA strands (micro-RNAs or miRNAs) is very often carried out using fluorescence methods, particularly those based on fluorescence resonance energy transfer (FRET), described in detail in Chapter 4. Afzalinia et al. (2020) applied this FRET principle to detect strands of the breast cancer-associated biomarker miRNA-155. They used two RNA probe strands, each complementary to part of the miRNA sequence. These two RNA strands are linked to two different types of nanoparticles: one, fluorescent, is a MOF (metal organic framework) containing lanthanum (III), and the other, a FRET acceptor and therefore a fluorescence quencher, is a silver nanoparticle (Figure 6.11). In this specific case, this method makes it possible to detect miRNAs at concentrations between 0.04 ppb (parts per billion) and 50 ppb. The authors were able to assess the method using actual samples from healthy patients and those suffering from kidney or breast cancer.

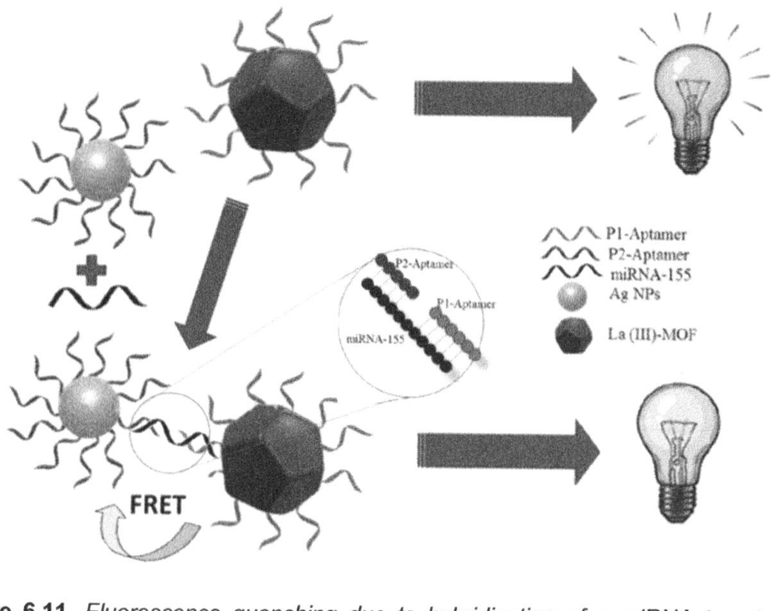

Figure 6.11. *Fluorescence quenching due to hybridization of a miRNA target. Two probe RNA sequences, complementary to part of the miRNA-155 target sequence, are immobilized, respectively, on a fluorescent lanthanum(III) MOF and a silver nanoparticle (AgNP). In the absence of miRNA-155, these nanoparticles are statistically distant from each other so that La(III)-MOFs fluoresce. In the presence of the miRNA-155 target strand, the ensuing hybridization reduces the distance between the MOF and the AgNP. This results in non-radiative energy transfer (FRET), which causes the fluorescence quenching of the MOF. Adapted from Afzalinia et al. (2020). For a color version of this figure, see www.iste.co.uk/lacaze/nanotechnologies.zip*

In addition to circulating microRNAs, cancers are characterized by protein biomarkers in the blood. Among these, PSA is one of the most studied, mainly because prostate cancer is the most common cancer in men. One of the most common methods to detect PSA is based on electrochemiluminescence, which is explained in detail in Chapter 3.

Qi et al. (2014) performed a PSA assay using a specific peptide sequence CHSSKLQK (each letter corresponds to an amino acid) serving as a molecular recognition element, selectively recognized and cleaved by PSA (Figure 6.12). $Ru(bpy)_3^{2+}$ coupled with tri-n-propylamine (TPA) is used as the ECL emitting species, and ferrocene carboxylic acid (Fc-COOH) is used as the ECL quencher. The whole is attached to an electrode modified by a Nafion film containing gold nanoparticles, significantly increasing the electrode surface area and reducing the charge transfer resistances. In the absence of PSA, the ECL is quenched by ferrocene attached to one end of the peptide chain; it reappears after enzymatic cleavage of the peptide sequence by PSA and partial elimination of Fc-COOH.

This approach has demonstrated its relevance, allowing PSA to be quantified in a range of concentrations between 5 pg/mL and 5 ng/mL for a very low detection threshold of 0.8 pg/mL (i.e. molar concentrations of respectively 0.2 pM, 0.2 nM and 0.03 pM).

These optical processes and optical biosensors dedicated to biomarkers can be found in the recent review article by Akgönüllü et al. (2022).

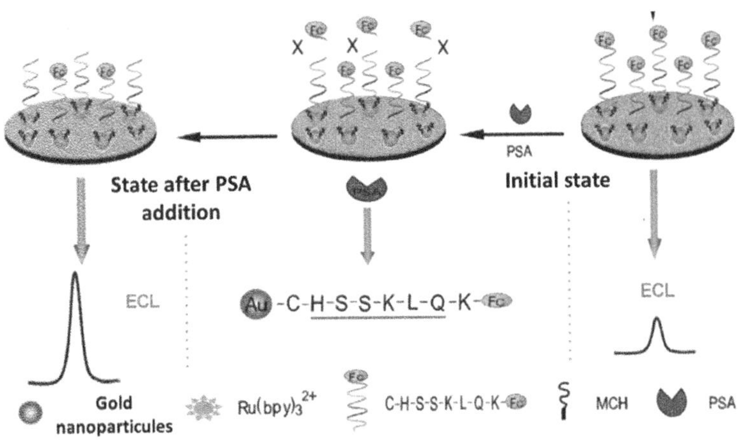

Figure 6.12. *Diagram of an ECL biosensor used to determine PSA. In its initial state, the biosensor comprises a mixture of Nafion and gold nanoparticles on the electrode (glassy carbon) surface as well as the complex $Ru(bpy)_3^{2+}$ and the co-reagent TPA. Mercapto-1-hexanol (MCH) containing SH groups is added to block free sites on the gold NPs. ECL is partially quenched by the proximity of the Fc groups. In the presence of PSA, the peptidase activity cleaves the immobilized peptide, releasing the ECL-quenching Fc group, causing the ECL to increase and signaling the presence of PSA. Adapted from Qi et al. (2014). For a color version of this figure, see www.iste.co.uk/lacaze/nanotechnologies.zip*

6.2.1.2. *Cancer screening using electrical and electrochemical methods*

The development of biosensors based on the use of FET type devices (see Chapter 3) is also a common approach, recently used by Mansouri Majd et al. (2023) to perform an assay of the target gene BRCA1, a biomarker for breast cancer. The system was composed of DNA probe strands complementary to the BRCA1 target gene and bound to nanoparticles. These NPs are themselves deposited onto an electrolytic metal (gold) gate acting on the channel of an electrolyte-gated field effect transistor (EGFET), the channel comprising a molybdenum disulfide MoS_2 film (Figure 6.13). The gate is also modified by a molecularly imprinted polymer calibrated to the BRCA1 target.

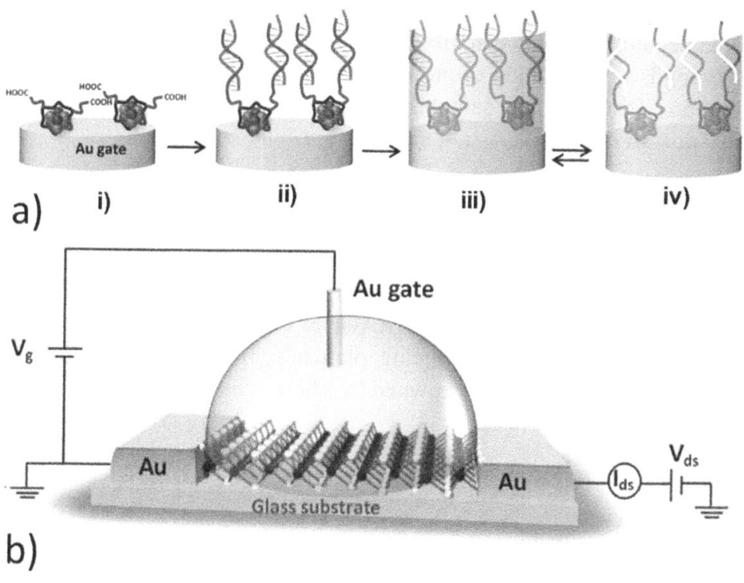

Figure 6.13. *Schematic diagrams of an EGFET-type sensor used to detect the BRCA1 gene (breast cancer biomarker). (a) The (gold) gate of this device is modified by Ag-Au bimetallic nanoclusters (NCs) (i) stabilized by a layer of insulin (Ins) (carrying functional groups (R) COOH and NH_2) corresponding to the Ag–Au@Ins NCs nanostructure; (ii) functionalization of NCs by the probe-target DNA double strand (the probe is an aptamer complementary to BRCA1 and bound to insulin). (iii) Polymerization of o-phenylenediamine on the Ag–Au@Ins/DNA-BRCA1/NCs assembly; (iv) elimination of the target sequence and formation of the molecular fingerprint (MIP). (b) Schematic of the EGFET comprising a semiconductor layer of molybdenum disulfide MoS_2. Recognition of the target strand results in a decrease in the current crossing the channel between source and drain. Adapted from Mansouri Majd et al. (2023). For a color version of this figure, see www.iste.co.uk/lacaze/nanotechnologies.zip*

The originality of this work essentially concerns the principle of molecular recognition, based both on molecular imprinting and on the hybridization of DNA strands. The gold gate of the transistor is modified by bimetallic Ag-Au nanoclusters on which the complementary probe DNA strands to the BRCA1 gene target sequence are immobilized. After hybridization, orthophenylenediamine is electropolymerized to cover the nanoclusters/DNA assembly, then the target strands are eliminated to form a hybrid DNA-MIP receptor. This approach is interesting since EGFET-type transistors are sensitive either to a variation in the interfacial capacitance at the gate or to a polarization of the gate due to the variation in the number of surface charges. The hybridization of DNA strands in the cavities modifies the capacitance and/or polarization at the electrolyte/gate interface, causing a reduction in the drain current. Mansouri et al. observed a linear variation in the drain current as a function of the logarithm of the target DNA concentration for a range of concentrations between 10^{-18} and 10^{-9} M. The detection sensitivity is high and corresponds to a variation of around 0.48 μA per logarithm unit of target DNA concentration. Tests on real samples (serum from patients carrying the BRCA1 gene) also proved conclusive.

This method, and the other methods used for biosensors described in the literature to recognize DNA sequences, must, however, be put into perspective with the technologies used today in analysis laboratories, which almost exclusively use polymerase chain reaction (PCR) described in Chapter 2, such as the CleanPlex BRCA1 detection kit marketed by PARAGON Genomics (paragongenomics.com) and use either the Illumina sequencing platform (fluorescence detection) or that developed by Ion Torrent and now owned by Thermo Fisher Scientific.

6.2.2. *Cardiac disease biomarkers*

Biomarkers of heart disease include proteins such as Troponin 1 (cTn1-cardiac troponin 1), B-type natriuretic peptide (BNP) or even NT-proBNP (n-terminal pro-brain natriuretic peptide type B).

The cTn1 protein is large, comprising around 200 amino acids, with a mass of around 25 kDa. Its presence in the blood at concentrations above 2 nM and its persistence at approximately 20 nM for 1 week indicates a heart attack.

BNP, a smaller protein of 32 amino acids, is approximately 3.5 kDa. Its concentration is relatively stable in the blood, and it is only during acute heart failure that its concentration increases to levels above 500 ng/L. That said, BNP only seems to be linked to left ventricular dysfunction.

NT-proBNP is a little larger with 76 amino acids, approximately 8.5 kDa. The concentration of this protein varies linearly with age. The thresholds for diagnosing heart failure are 450 ng/L for patients under 50 years of age up to over 1,800 ng/L for patients aged over 75; therefore, its analysis cannot be dissociated from the patient age and its concentration may also be affected by renal dysfunction. The combined BNP/NT-proBNP analysis remains the standard for diagnosing heart failure.

Further details relating to these cardiac biomarkers can be found in the review article by Cao et al. (2019). Certain detection principles, optical, electrochemical or based on less common transduction techniques, are discussed below for illustration.

6.2.2.1. *Optical biosensors*

Inspired by the conventional immunogold or nanogold ELISA test, Liu et al. (2022) developed a fiber-optic sensor exploiting surface plasmon absorption of gold NPs for the detection of NT-proBNP. Immunogold detection uses the ELISA principle with gold NPs functionalized by numerous antibodies, forming a crown around the NPs.

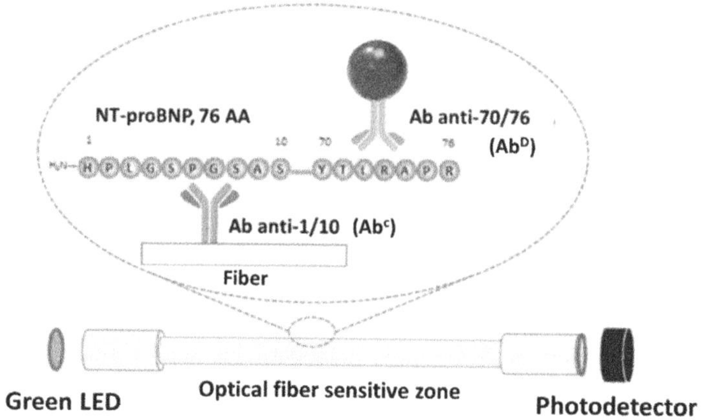

Figure 6.14. *Schematic diagram of the FONLISA (fiber optic nanogold-linked immunosorbent assay) strategy. The first antibody (Ab anti1/10, noted Ab^C) immobilized on the fiber captures the NT-proBNP, 76 AA protein. A second antibody (Ab anti-70/76, noted Ab^D) carrying AuNP recognizes the sequence 70-76 of the protein. The absorbance caused by the AuNP surface plasmons is "read" by the fiber at a wavelength of 527 nm. Adapted from Liu et al. (2022). For a color version of this figure, see www.iste.co.uk/lacaze/nanotechnologies.zip*

COMMENT ON FIGURE 6.14.– *The gold nanoparticles have a diameter of approximately 15 nm and are functionalized by the anti-NT-proBNP capture antibody via an alkylthiol layer. These functionalized nanoparticles are characterized by surface plasmon resonance at 527 nm. The NT-proBNP protein (denoted A) has 76 amino acids (AA). The antibody attached to the optical fiber* Ab^C *recognizes the peptide chain numbered from 1 to 10, while the antibody attached to the gold NP (*Ab^D*) recognizes the other end of the protein, numbered from 70 to 76. When the analyte A is sandwiched between the two antibodies (AuNP@*Ab^D*-A-*Ab^C*), the green light (532 nm) sent by the diode (LED) is partially absorbed by the plasmonic resonance of the gold NPs before reaching the photodetector and transmits a light intensity according to the protein concentration.*

The approach followed by these authors involves functionalizing an optical fiber with an antibody specific for a sequence of 10 amino acids located at one end of the NT-proBNP protein (Figure 6.14), which will capture the protein on the fiber surface. A second antibody, immobilized on the AuNPs, is specific for seven amino acids located at the other end of the protein. The presence of AuNPs near the optical fiber makes it possible to measure its optical density, which depends on the protein concentration.

The authors showed an excellent correlation between the concentrations estimated by their FONLISA technique and a reference commercial laboratory analysis technique (Roche Cobas e411 system, technology based on electrochemiluminescence). An excellent linear relationship is observed between the light intensity and the protein concentration for a range of concentrations between 0.5 and 5,000 pg/mL and a detection threshold of 0.058 pg/mL.

Other strategies using nanoparticles have been described elsewhere. We can cite, for example, the work by Wang et al. (2023), who developed a fiber optic sensor based on localized surface plasmon resonance (LSPR) to detect the cTnI protein. The fiber is functionalized by gold nanoparticles and cerium oxide (CeO_2) nanoparticles, the latter improving the detection threshold (LoD), established at 108 ng/mL (4.5 nM).

Vairaperumal et al. (2023) and Goryacheva et al. (2022) recently published two review articles discussing what already exists regarding optical POC sensors for detecting cardiac biomarkers.

Apart from the devices developed in research laboratories, several sensors are marketed and used by analysis laboratories today. We can cite, for example, a strip detection device marketed by 77 Elektronika (Hungary) (Figure 6.15). This strip-test type reader is based on a classic immunological principle. A monoclonal anti-human cTnI antibody conjugated to colloidal gold (according to the nanogold principle detailed previously for the detection of NT-proBNP) is deposited at the inlet of the test zone, and another anti-cTnI human monoclonal antibody is immobilized at another location on the strip. After deposition of the blood sample, the gold-labeled anti-cTnI binds to troponin I in the sample and forms a labeled complex. This complex is carried by capillarity and moves to a so-called test zone where it is captured by the immobilized anti-human cTnI monoclonal antibody, then forming a red-purplish line on the strip. The color intensity is proportional to the amount of cTnI in the sample and is read by the SmarTTester optical device shown in Figure 6.15.

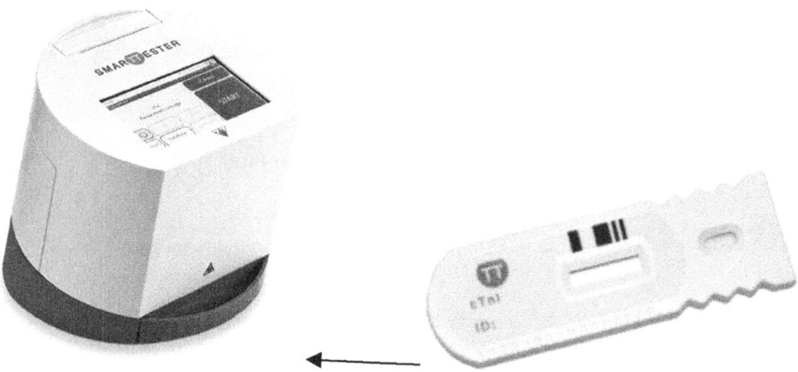

Figure 6.15. *Test strip reading device. On the left is the optical strip reader developed by the company 77 Elektronika (https://www.en.e77.hu/). On the right is the test strip based on the nanogold ELISA principle. The strip is inserted into the reader using a drawer in its lower part. For a color version of this figure, see www.iste.co.uk/lacaze/nanotechnologies.zip*

6.2.2.2. *Electrochemical and transistor biosensors*

Commercial POC devices, such as the Abbott i-STAT system, are aimed at hospital practitioners and are based on an electrochemical measurement. This device makes it possible to measure the cTnI protein in plasma for concentrations between 0 and 35 ng/mL; it also makes it possible to determine the BNP protein under the same measurement conditions for concentrations between 15 and 5,000 pg/mL. This is done using disposable "cassettes" dedicated to each analyte inserted into the reading device (Figure 6.16). The operating principle is based on ELISA. The antibody specific for cTnI is immobilized on one electrode. Another antibody is immobilized on another electrode and coupled to the alkaline phosphatase enzyme, which is specific for another part of the cTnI protein. When the sample is added, any cTnI present binds to the antibody labeled by the enzyme (this step takes 5–7 min).

The washing liquid, which is added at this moment, also contains an alkaline phosphatase substrate, which, when active, cleaves the antibody/antigen/antibody sandwich, releasing a phenolic electroactive substance. The measured current is proportional to the sample cTnI concentration. Abbott developed several other i-STAT cassettes for many other analytes (hCG, creatinine, lactate, etc.).

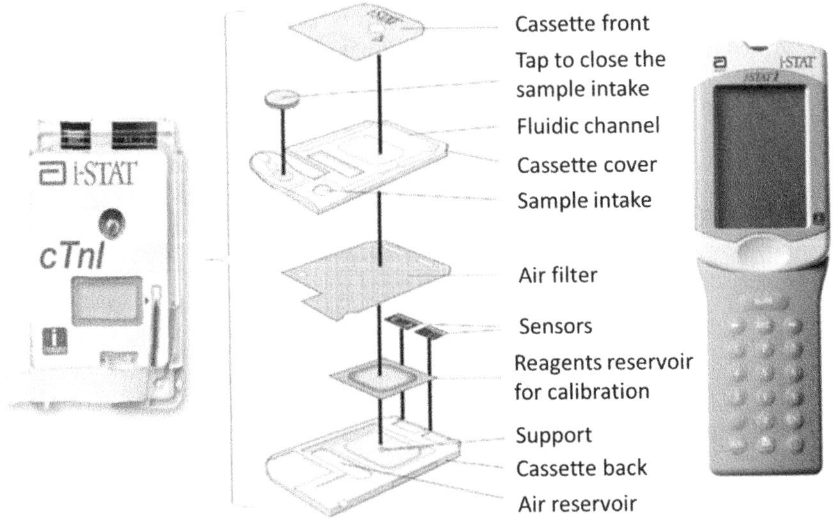

Figure 6.16. *i-STAT system developed by Abbott. (a) iSTAT cassette for cTnI protein analysis and its exploded view. The air filter and the air reservoir are needed to allow the sample to flow from the intake to the sensors. Visual extract from the i-STAT company presentation in 2002 at the UBS Warburg Global Life Sciences conference, October 2002. (b) i-STAT 1 reading device. For a color version of this figure, see www.iste.co.uk/lacaze/nanotechnologies.zip*

Similar to optical devices, many techniques have been used for the electrochemical detection of the most common cardiac biomarkers. As an illustration, we will describe the approach recently reported by R. Crooks and team to develop devices based on detection by galvanic exchange (Figure 6.17). It essentially involves the difference in redox potential of two different metals, such as gold and silver. Brought to a sufficiently close distance from each other, the electrochemical oxidation of gold releases Au^{3+} ions, which diffuse to the Ag nanoparticle, where galvanic exchange occurs, that is, the oxidation of Ag° into Ag^{+} ions, whereas the Au^{3+} ions are reduced to Au° on the nanoparticle. Then, the application of a potential corresponding to the reduction of Ag^{+} ions produces a current whose intensity correlates with the presence of the analyte. The detection limit, without optimization, reaches 0.6 nM (Pollok 2020)[5].

The same team recently improved the technique by replacing spherical silver nanoparticles with nanodiscs. Crystallographic effect made it possible to increase the dissolution rate on the disc sides relative to its crown, significantly increasing the detection sensitivity toward the NT-proBNP protein, ranging from 30 pM to 4.0 nM (Peng 2022).

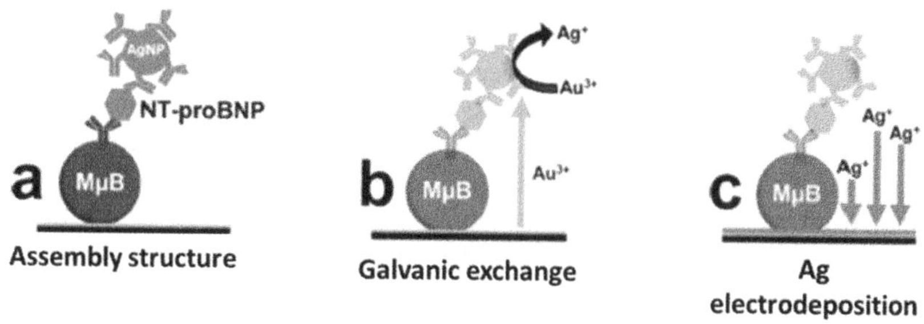

Figure 6.17. *Electrochemical detection of the NT-proBNP protein by sandwich assembly and galvanic exchange. (a) Construction of the assembly comprising a magnetic microbead (MμB) carrying an antibody specific for part of the NT-proBNP protein (when it is present in the sample) and an Ag nanoparticle (AgNP) functionalized by a second antibody specific for another part of the protein. (b) The production of Au^{3+} ions by applying a very brief positive potential to the gold electrode causes a galvanic exchange with the silver NP, resulting in the formation of Ag^{+} ions. c) The Ag+ ions produced are then electroreduced at the electrode. Such reactions are only possible because of their extreme proximity to the gold electrode surface and result from the sandwich assembly. According to Pollok et al. (2020). For a color version of this figure, see www.iste.co.uk/lacaze/nanotechnologies.zip*

A similar recognition device was also used by Kim et al. (2016) to detect Troponin I, a cardiovascular biomarker and preventive indicator of the risk of cardiac arrest, using a FET for transduction; the sensor channel is made up of Si nanowires functionalized by the troponin I antibody. According to the authors, a detection limit of 4 pg/mL is reached, well suited for diagnosing the risk of a heart attack, which generally appears at concentrations above 40 pg/mL.

Silicon, an essential element of conventional microelectronics, and in its nanostructured form, is the material of choice for developing innovative electronic devices. Its use for the manufacture of chemical and biological sensors is not novel

5 The electrochemical method used is known as adsorptive stripping voltammetry or inverse adsorption voltammetry.

and dates back to the early 2000s. Cui et al. (2001) were among the first to propose the use of Si nanowires to produce sensors for chemical and biological molecules based on conductivity measurements. Using a SiNW functionalized by a capture probe, any target protein that binds to this probe can be detected through the resulting variation in the nanowire conductivity. A variation in conductivity of 100 nS (easily measurable) was observed by these authors after adding 460 µg/mL of protein (i.e. a concentration of around 3 µM – very high but nevertheless promising at the time). Sensitivities improve when SiNWs are incorporated into FET-type devices, such as in the example of Kim et al. (2016). The sensitivity of detection devices is also better the smaller the diameter of the Si nanowire (diameters ≤ 20 nm maximize the surface/volume ratio) and if the silicon p or n doping level is not too high to preserve its semiconductor properties. Silicon nanowires (SiNW) are now obtained using proven techniques. They give rise to new families of sensors, inspired by the operation of field effect transistors (FETs) and benefiting from the one-dimensional contribution of Si nanowires. In addition to lithographic methods used in microelectronics, numerous methods combining chemistry and lithography make it possible to obtain SiNW networks at a better cost, which are now widely used by many laboratories. Leonardi et al. (2021) give an excellent description in a recent review article.

6.2.3. *Stress biomarkers*

Stress has a physiologically inhibitive effect, leading to depressive-type behaviors, which can affect the efficacy of the care provided to a patient.[6] We react to stress by releasing certain molecules (hormones and neurotransmitters), which in return have an important, immediate and long-term physiological effect. Hormones are released by the endocrine system while neurotransmitters are released by the nervous system. Hormones are generally released into the blood, from the endocrine glands to the target organ. Although less known, neurotransmitters can also be found in bodily fluids. Figure 6.18 gives a detailed view of the different neurotransmitters and hormones measurable in blood, sweat, saliva or urine, as well as their concentration ranges.

---

6 Whether in the sporting field to help an athlete manage stress during a competition, monitor the health of soldiers or police officers during a field operation, help control personnel stress in hospitals, in emergency medicine but also guide health professionals by helping them to treat the condition-related stress in patients; these are the problems that medical staff face. Also, a "connected watch" capable of providing its user with reliable stress indicators would have great success with the public (the indicator used today is essentially based on heart rate, spO_2 [pulsed oxygen saturation], and the movements by the individual, which are only very indirect measurements). There is, therefore, a very large market for devices capable of quantifying an individual's stress and determining its causes.

The methods for detecting these stress biomarkers (neurotransmitters are either high molecular weight proteins or simpler, low molecular weight molecules) do not differ from the methods already presented previously. For illustration purposes, we describe some techniques used for the detection of cortisol and dopamine.

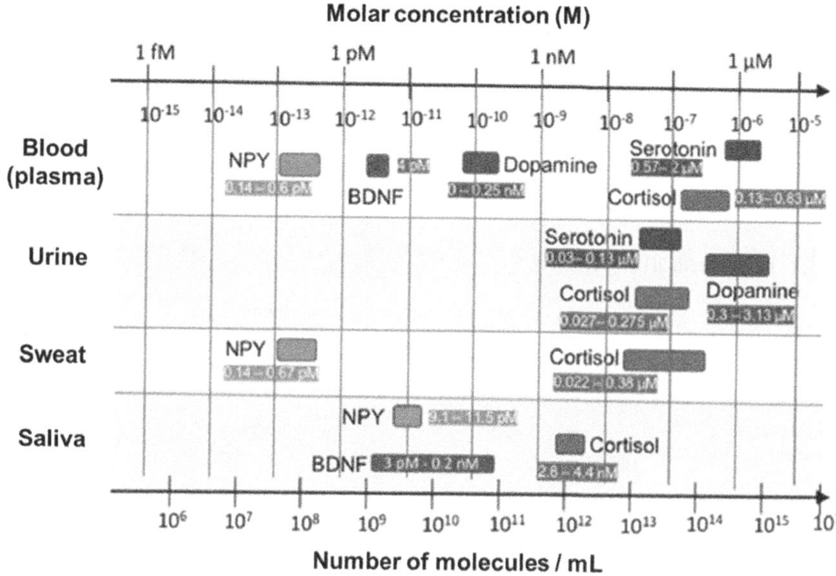

Figure 6.18. *The expected concentration ranges of stress biomarkers and the biological fluids in which these biomarkers are present and measured. NPY: neuropeptide Y; BDNF: brain-derived neurotrophic factor. From Steckl et al. (2018). For a color version of this figure, see www.iste.co.uk/lacaze/nanotechnologies.zip*

6.2.3.1. *Cortisol*

Cortisol is a steroid hormone that has many functions in humans. It is both anti-inflammatory and anti-stress. It plays a role in the release of glucose by the liver and muscles and enters into lipid storage regulatory loops. When chronically present in high levels, this hormone can promote weight gain, degrade the immune response, promote osteoporosis and generate heart or kidney problems. This justifies the need to develop cortisol sensors for therapeutic monitoring or screening outside the laboratory at the patient's bedside (POC).

Cortisol is present in most biological fluids. However, if its concentration is measured in µg/mL in the blood, this concentration drops to a few ng/mL in urine, saliva or sweat. This is because blood cortisol is composed of free cortisol (10%) and cortisol linked to glycoproteins (90%). Only free cortisol crosses cell barriers and enters bodily fluids. Although cortisol is present in low concentrations in these fluids, it strongly correlates with blood cortisol concentration and can be used for POC monitoring.

Although highly efficient, mass spectrometry is bulky and expensive to implement; therefore, most cortisol sensors described in the literature are based on immunodetection typically via the ELISA method, using specific cortisol antibodies as the capture probe (Zhang et al. 2023b); no enzymatic sensor has been developed for cortisol since this molecule is not a substrate or co-substrate of an enzyme. Sensors have also been developed based on molecularly imprinted polymers or DNA or RNA aptamers. We will see in the following examples that, in all cases, nanostructuring surfaces improve the sensitivity and the detection threshold.

6.2.3.1.1. Detection by ELISA

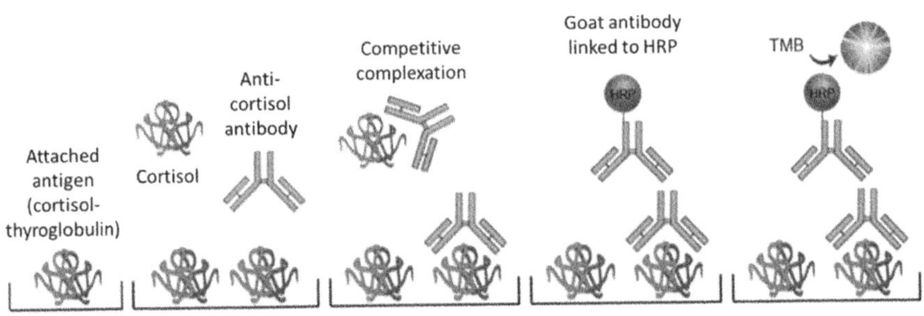

Figure 6.19. *Illustration of cortisol detection by competitive ELISA. Diagram taken from the LSBio company website (www.lsbio.com) for the description of its kit, ref. LS-F10024, dedicated to the detection of cortisol in humans. For a color version of this figure, see www.iste.co.uk/lacaze/nanotechnologies.zip*

COMMENT ON FIGURE 6.19.– *Cortisol is quantified competitively between the cortisol adsorbed in the well and the free cortisol, complexed by the anti-cortisol antibody in known and finite quantities. After washing the plates and removing the complexed cortisol in solution, the plates are subjected to a new antibody bound to HRP, which binds to the cortisol adsorbed in the well. The addition of TMB causes the green color, the intensity of which is directly proportional to the quantity of bound cortisol. By difference, we can deduce the concentration of free cortisol.*

Using the ELISA technique to detect blood cortisol is not new; Lewis and Elder described it in 1985 (Lewis and Elder 1985). In this work, the first published example, the authors used the ELISA technique to detect cortisol in plasma, with rabbit serum directly immunized against cortisol (the serum, therefore, contains rabbit "anti-cortisol" antibodies). They prepared cortisol linked to thyroglobulin as a capture complex, immobilized on the bottom of so-called “96-well” ELISA plates (the thyroglobulin is the “carrier” protein, correctly immobilizing the antibody at the bottom of the well, without denaturing the antibody). In the solution, both the free cortisol to be detected and the serum containing the rabbit anti-cortisol antibody are added. In this way, the antibody competitively binds to the free cortisol in solution or to cortisol immobilized at the bottom of the wells. The greater the cortisol concentration in the solution, the less the antibody is present at the bottom of the well. To detect and quantify cortisol, anti-goat IgG antibodies are added (an anti-IgG is an antibody that binds to other antibodies, in this case, rabbit antibodies) carrying the HRP enzyme. The signal is measured optically by adding the substrate TMB (3,3’,5,5’-tetramethylbenzidine), which HRP oxidizes into a green-colored molecule. Cortisol is quantified by measuring the optical density. The greater the initial cortisol concentration, the less intense the coloring. This so-called “competitive” ELISA approach is still common, with detection thresholds of around 50 pg/mL (Figure 6.19).

6.2.3.1.2. Use of magnetic microbeads

Modern methods are increasingly sensitive and require less preparation for faster analyses. An example is given by Evans et al. (2021) (Figure 6.20), who used an elegant technique based on the use of magnetic microbeads and the manipulation of microdroplets by microfluidics.

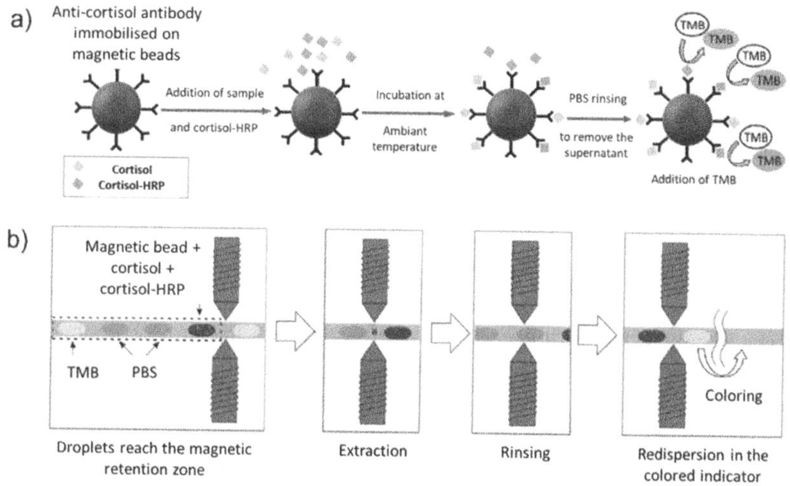

Figure 6.20. *Illustration of the competitive cortisol assay process by ELISA on magnetic beads by generating water microdroplets on oil in a microfluidic system. (a) The recognition principle involving magnetic microbeads. (b) Microfluidic processing involving aqueous microdroplets in an oil. According to Evans al. (2021). For a color version of this figure, see www.iste.co.uk/lacaze/nanotechnologies.zip*

COMMENT ON FIGURE 6.20.– *Trains of drops are generated in a set order: the first (brown) contains the magnetic beads, the sample and the HRP-coupled cortisol as the competitor; the next two contain only PBS buffer and are used for rinsing; the last contains the TMB. As the drop containing the sample passes, an electromagnet is activated to extract and immobilize the beads; while immobilized, these beads are washed by the two PBS microdroplets. As the droplet containing the TMB passes, the electromagnets are deactivated, releasing the microbeads into the droplet; the reaction produces a color read by a photodetector, from which the concentration of free cortisol is deduced indirectly.*

The classic immunological process is similar to that described in Figure 6.20(a). The magnetic beads (commercial) are functionalized by an anti-cortisol antibody coupled to the BSA carrier protein. These functionalized beads are then placed in contact with the sample containing cortisol and cortisol coupled to HRP. Antibody-antigen binding occurs competitively: in the presence of little free cortisol, the antibodies are saturated with cortisol-HRP; conversely, in the presence of a greater quantity of free cortisol in the sample, the density of bound cortisol-HRP will be lower. Adding TMB, oxidized into a colored form, gives a readable result. This very innovative reading process is based on the circulation of water-in-oil microdroplets with the reagents in the water droplets; the oil acts only as a vehicle (Figure 6.20(b)).

6.2.3.1.3. Use of nanoparticles

Another process, adapted from the ELISA technique, involves using gold nanoparticles (AuNP) and not enzymes, with photoabsorption properties alone on lateral flow immunoassay (LFIA) strips (in this case, 25 nm beads with surface plasmons resonant at 525 nm). The process is illustrated in Figure 6.21 (Di Nardo et al. 2019).

The sample is deposited onto an absorbent part and then migrates passing through an area on which AuNPs functionalized by anti-cortisol antibodies are deposited. These are carried to a "test" zone on which cortisol is immobilized. If the gold NPs are saturated with cortisol (in the case where the sample contains a high concentration of cortisol), they do not bind to the immobilized cortisol and reach the control zone, which is functionalized by secondary anti-cortisol antibodies IgG, which captures the nanoparticles in all cases. If the AuNPs are not saturated with cortisol (in the case where the sample contains a low concentration of cortisol), part of it may be captured in the "test" zone.

In this example, the LOD is around ng·mL^{-1}. The method is therefore not particularly sensitive but has the advantage of being very easy to use.

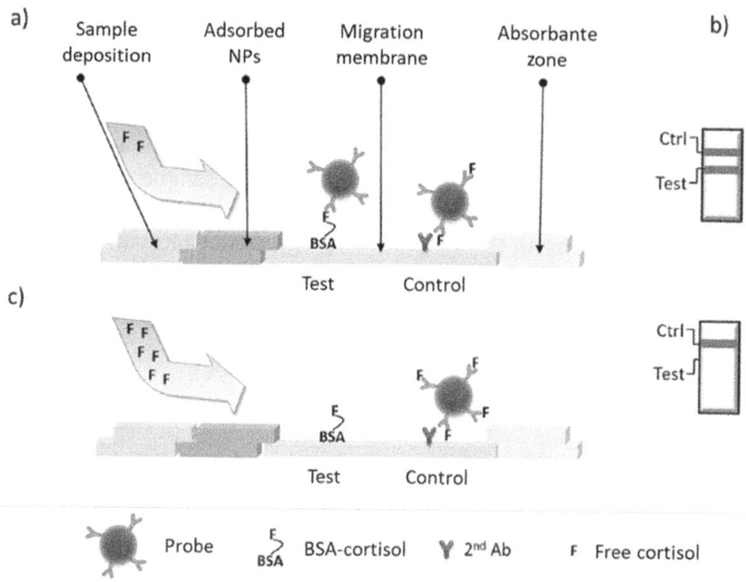

Figure 6.21. *Diagram of a lateral flow detection device for cortisol (F) using a competitive ELISA type mechanism. (a) For a sample containing little cortisol, the NPs are stopped by the test and control lines. (b) For a high concentration of cortisol, the NPs saturated with cortisol pass the test line and attach to the control line (Ctrl). Adapted from Di Nardo et al. (2019). For a color version of this figure, see www.iste.co.uk/lacaze/nanotechnologies.zip*

COMMENT ON FIGURE 6.21.– *The strip includes a test line and a control line (Ctrl) functionalized by the BSA-F chain and an anti-cortisol antibody (Y). The nanoparticles, functionalized by cortisol antibodies, initially adsorbed on the pad (pink zone), load with cortisol (F) depending on its concentration in the sample. Cortisol and NPs are carried by capillary action toward the test and control lines. For low amounts of cortisol, NPs are stopped by both the test and control lines, whereas for high cortisol concentrations, NPs (saturated with F) attach only to the control line.*

6.2.3.1.4. Molecularly imprinted polymers (MIPs)

Another method to capture and measure cortisol is to use an MIP (see Chapter 3). Wang and team demonstrated the relevance of this approach for measuring cortisol in sweat by the simple contact of a finger on an electrode functionalized by an MIP (Tang et al. 2021) (Figure 6.22).

The preparation of MIP in this case involved electropolymerizing pyrrole in the presence of cortisol and Prussian blue (PB), and then eluting the cortisol after polymerization. The authors showed that the redox kinetics of the PB electrochemical probe were slower when cortisol is trapped in the cavities, which allowed, through a preliminary calibration, to determine a relationship between the PB oxidation current and the cortisol concentration in the sample in contact with the MIP. Furthermore, a hydrogel layer is placed on top of the MIP, both to protect it mechanically and to act as an absorbent to store a greater quantity of sweat while the finger is in contact with the MIP.

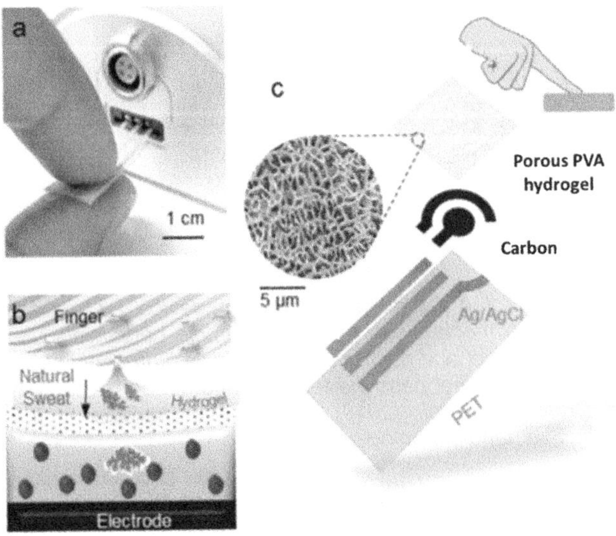

Figure 6.22. *Electrochemical device for detecting cortisol using molecular recognition by MIP followed by electrooxidation. (a) Photograph of the printed slide inserted into the measuring device. (b) Diagram showing the path of cortisol from the skin to the MIP. (c) Details of materials as per Tang et al. (2021). For a color version of this figure, see www.iste.co.uk/lacaze/nanotechnologies.zip*

In this system, 30 s are enough to saturate the hydrogel and obtain a stable electrochemical response. The detection threshold is between 10 and 100 nmol/L, that is, approximately 3–30 ng/mL.

6.2.3.1.5. Aptamers

Cortisol can also be recognized by aptamers (see Chapter 3). A particularly notable example was given by Wang et al. (2022). In this study, an EGFET (see Chapter 4) was manufactured by photolithography on a polyimide substrate, a layer of In_2O_3 as the semiconductor channel, placed between the Au/Ti source-drain contacts (Figure 6.23).

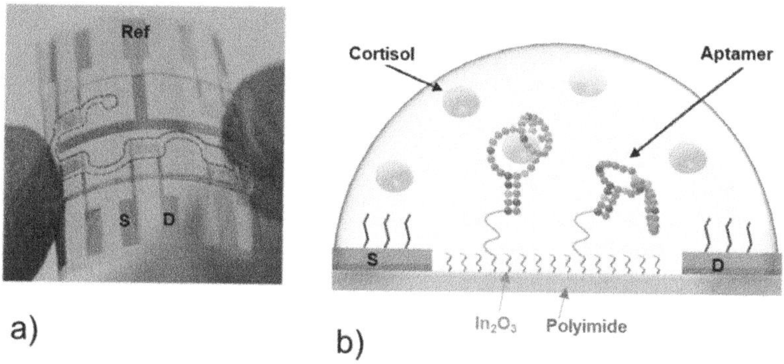

Figure 6.23. *Recognition of cortisol in sweat by an aptamer and transduction via electrolyte-gated transistors. (a) Photograph of five transistors and a coplanar Ag/AgCl gate on a polyimide sheet with the microfluidic circuit. (b) Schematic of the electrolyte-gated FET transistor and the microfluidic circuit. Adapted from Wang et al. (2022). For a color version of this figure, see www.iste.co.uk/lacaze/nanotechnologies.zip*

COMMENT ON FIGURE 6.23.– *The grid is a silver wire covered with silver chloride AgCl, serving as a pseudo-reference. Source and drain are deposited on an extremely thin (3 nm) indium oxide (In_2O_3) semiconductor layer, itself deposited on a very flexible polyimide sheet, 7-µm thick. The fluid containing cortisol is supplied through a microfluidic channel (black dotted lines). The aptamer strands, functionalized in the 5' position by a thiol group, are immobilized on the semiconductor surface by an aminosilane, then coupled with the aptamer activated by maleimidobenzoyl-N-hydroxysuccinimide ester; the source and drain are passivated by dodecanethiol adsorption. The transistors are in contact with the microfluidic channels cut by a laser on a double-sided self-adhesive sheet approximately 170-µm thick to provide a sufficient flow of sweat (these channels fill quite slowly, around 15 min). The transistors are controlled by dedicated electronics close by on a flexible printed circuit board (PCB). This circuit is capable of controlling and measuring current and voltage across the transistors via an analog input (AFE-analog front-end), processing the signals (via an analog-digital converter, ADC), transmitting them to a microcontroller (MCU), which sends data either to an LCD display or to a Bluetooth transmitter.*

The system was tested by measuring the transduction signal of an FET from sweat containing known quantities of cortisol and the variations in cortisol induced by a period of stress, according to a recognized medical protocol, the trier social stress test (TSST) (Figure 6.24).

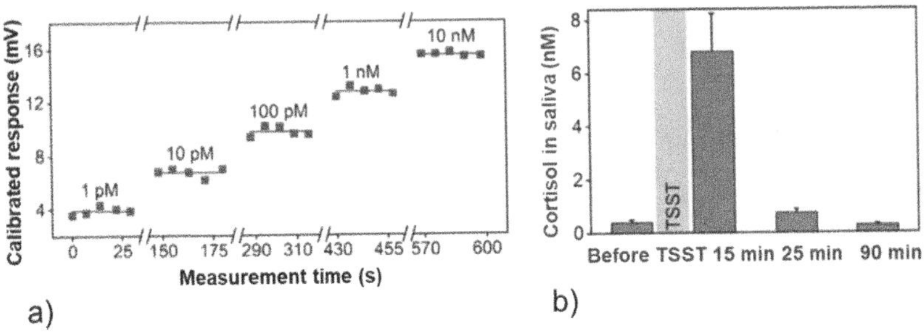

Figure 6.24. *Cortisol response of electrolyte-gated and aptamer-functionalized FETs. (a) Variations in drain current (measured in mV) as a function of cortisol content (calibrated) in sweat. (b) Variation in cortisol concentration as a function of time elapsed before and after the stress test (TSST) applied to an individual. Adapted from Wang et al. (2022). For a color version of this figure, see www.iste.co.uk/lacaze/nanotechnologies.zip*

6.2.3.1.6. Electrochemiluminescence

Electrochemiluminescence (see Chapter 4) has also been used to detect cortisol and developed sufficiently for commercial use, for example, Roche Diagnostic with its system Elecsys. This microfluidic system uses magnetic microbeads on which specific cortisol antibodies are adsorbed. A second antibody, capable of complexing cortisol, carries the phosphor $[Ru(bpy)]_3^{2+}$. A magnetic field is applied to accumulate the microbead/cortisol/luminophore assembly near a measuring electrode. Locally, tripropylamine is introduced into the microfluidic channel, and the classic principle of electrochemiluminescence is used for detection. The Elecsys system is sensitive enough to quantify cortisol in saliva, serum or urine for a concentration range between 0.5 ng/mL and 1.7 µg/mL.

In conclusion, the non-invasive detection of cortisol has been the subject of numerous studies and has resulted in the production of many autonomous and portable devices based on the use of various transduction phenomena. Several recent review articles provide detailed analyses. The detection of cortisol by POC type devices was described by B.D. Malhotra and colleagues (Kaur et al. 2023), the different electrochemical methods applied to the detection of cortisol were analyzed by Zea et al. (2020) and the use of molecularly imprinted polymers was reviewed by Yulianti et al. (2022). Finally, an overview of the currently marketed devices can be found in a report published by AFSSAPS (French Agency for Health Product Safety)[7].

6.2.3.2. *Dopamine biosensor*

Dopamine sensors have been the subject of publications for several decades; most of them are based on selective electrooxidation using enzymatic or metallic catalysts, the latter used in its original, meso-, micro-, or, more recently, nanostructured state. The review by Liu and Liu (2021) gives a complete vision of current processes.

One method, very different from previous methods, involves using the fluorescence properties of metal nanoclusters (MNCs), described in Chapter 1, also described for detecting cholesterol in this chapter. As before, it involves using the intrinsic fluorescence of AuNCs and their quenching by biological molecules. Halawa et al. (2018) carried out the one-pot synthesis of a fluorescent AuNC covered by β-CD molecules, leading to β-CD@AuNC core/shell type NCs. Dopamine in solution was complexed by β-CDs, which quenched the fluorescence due to electron transfer from the AuNCs to dopamine. The fluorescence intensity of β-CD(@)AuNCs decreases linearly as a function of dopamine concentration, ranging from 100 nM to 80 μM, with a detection limit of around 20 nM.

6.3. Pathogens

External molecules, of biological or non-biological origin, interacting with the body are considered potentially pathogenic. Viruses (e.g. RNA viruses), food pollutants (BPA, phthalates, etc.), mycotoxins and other toxins produced by bacteria, air pollutants such as volatile organic compounds, but also pollens, etc., are the main pathogens. In what follows, we will describe some of the most relevant examples.

6.3.1. *Virus detection*

The very recent global pandemic has given rise to many studies dedicated to the detection of RNA viruses, detected in a traditional way by DNA/RNA or RNA/RNA hybridization. It is a very effective virus detection technique, but requires relatively long analysis times, lasting several hours. This explains why, in recent years, new solutions for detecting the SARS-CoV-2 virus were proposed to replace this traditional detection technique with other more direct and faster technologies, including the use of thin-film transistors involving nanomaterials such as graphene,

7 Available at: https://archiveansm.integra.fr/content/download/1566/15221/version/3/file/rapport_controle_dmdiv_dosage_cortisol.pdf.

carbon nanotubes, 2D mineral nanosheets, as well as silicon semiconductor nanowires.

6.3.1.1. *CRISPR/Cas technology to detect SARS-CoV-2*

Coupled with FET transduction, Yu et al. (2022) recently proposed performing the SARS-CoV-2 (COVID) assay without an RT-PCR type preconcentration. Clustered regularly interspaced short palindromic repeats (CRISPR/Cas) is a specific RNA sequence (denoted crRNA) that binds to a Cas nuclease, activated when a complementary RNA sequence binds to the crRNA strand. Cas13a is a protein known to have a particular binding affinity for crRNA and combines this advantage with the ability to cleave RNA independently of its sequence. Thus, the nuclease cleaves not only the RNA linked to the crRNA, but also any strand of RNA or DNA in the vicinity. This cleavage property of Cas13 can be used to detect DNA and RNA/DNA viruses. It has already been described and proven very effective, particularly in viral diagnostics. It is very specific (able to discriminate single base mismatches) and works at low target DNA concentrations, thus avoiding the need for amplification steps such as PCR[8].

The device used by Yu et al. (2022) is an FET type transistor with the gate electrode used as a sensitive surface. It is first functionalized by a biopolymer layer on which the CRISPR/Cas13a is immobilized. The transistor comprising a graphene channel has standard geometry, with three contacts (top gate, source, drain)[9]. As soon as a target RNA strand is added to the electrolyte, it binds to the crRNA inside the Cas13a protein and CRISPR/Cas begins its cleavage activity, which induces a shift in the transfer curve of the transistor toward the more positive gate voltages. The authors attribute this shift to a decrease in the quantity of charges on the gate surface, resulting from the fact that the cleaved RNA is shorter and carries fewer charges.

In analytical terms, this device shows excellent sensitivity, with a target RNA at 100 fM leading to a shift in the transfer curve of approximately 100 mV (Figure 6.25(a)). The authors have also shown that their device can discriminate saliva samples from patients with COVID-19 compared to healthy patients (Figure 6.25(b)).

---

8 The use of CRISPR/Cas and a field effect transistor was described by Hajian et al. (2019) for the detection of DNA sequences linked to degenerative Duchenne disease.

9 The gold-gate electrode is first functionalized by a layer of thiol (11-sulfanylundecanol) to which dextran is attached. This is covalently linked to the previous thiol via epichlorohydrin and can bind covalently to proteins, in particular with the CRISPR/Cas13a complex. The transistor channel is composed of a graphene layer on which the source and drain electrodes are deposited.

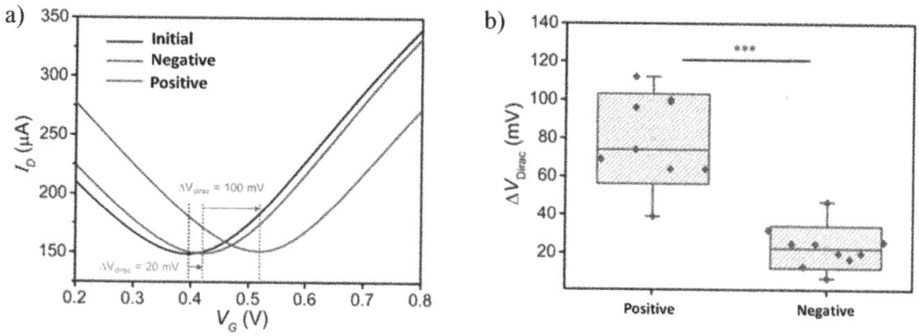

Figure 6.25. *Analytical results obtained on sRNA (COVID-19) from patient swabs. (a) Transfer curves of a graphene and an electrolyte-gated field effect transistor modified by CRISPR/Cas13a nuclease before addition of the target RNA (black curve), followed by incubation with a negative (blue curve) and positive (red curve) patient sample for which the target RNA concentration is estimated at approximately 100 fM. (b) Shifts of the minima (Dirac) of the transfer curves obtained for 15 samples (seven positive and eight negative), showing two distinct zones of potential variation. Adapted from Yu et al. (2022). For a color version of this figure, see www.iste.co.uk/lacaze/nanotechnologies.zip*

6.3.1.2. *SARS-CoV-2 assays via viral proteins*

SARs-CoV-2[10] is characterized by a core containing viral RNA surrounded by a crown membrane carrying different proteins, including "spike" S proteins (denoted SAg) and considered as antigens, membrane glycoproteins (M), an envelope protein (E), as well as a plasmid nucleoprotein (N), the latter coating the viral RNA (Figure 6.26(a)). The spike protein is one of the SARS-CoV-2 proteins that binds to human cells whose detection can be used as an indication of a positive test for SARS-CoV-2 (Figure 6.26(b)).

Numerous studies propose FETs with graphene (G) as a channel material to detect SARS-CoV-2, although graphene is not a semiconductor due to its zero band gap. However, given its two-dimensional nature, its charge carrier density is extremely sensitive to any variation in electric field, whether it comes from a gate or from nearby charged species. It is also electroactive and transparent, which has made it particularly attractive for many biomedical sensing applications, highlighted in several recent studies (Pumera 2011; Andronescu and Schuhmann 2017; Pena-Bahamonde et al. 2018; Dincer et al. 2019; Yildiz et al. 2021; Xue et al. 2022).

10 SARS-CoV-2 or severe acute respiratory syndrome corona virus.

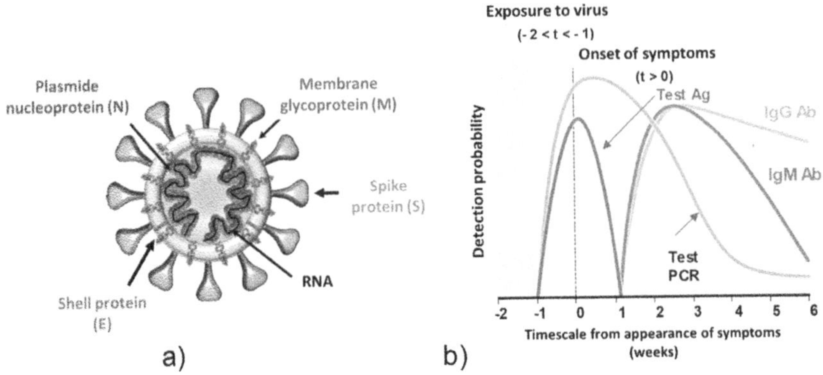

Figure 6.26. *Schematic representation of SARS-CoV-2 and the appearance of the different markers. (a) Location in the virus of the primary viral proteins (E, M, N and S) and the viral RNA combined with the nucleoprotein N. (b) Diagram indicating the appearance probability of viral proteins (Ag-antigens) and IgG-Ab and IgM-Ab antibodies after infection with the virus. We note that the appearance of the first symptoms of the disease only occurs 2 weeks (t > 0) after the first exposure to the virus. The concentrations of RNA (for a PCR test) and proteins (for an antigen test) only become detectable after a week of infection, whereas serological tests carried out on IgG-Ab and IgM-Ab antibodies only confirm infection after at least 3 weeks after the start of the infection. (a) Adapted from Ghodake et al. (2021). (b) Adapted from Guglielmi (2020). For a color version of this figure, see www.iste.co.uk/lacaze/nanotechnologies.zip*

An attractive characteristic of graphene is its wide commercial availability, in the form of Si/SiO_2/G/PMMA wafers, with diameters of 100 or 200 mm, easily transformable into Si/SiO_2/G to manufacture GFETs (graphene FETs) and other biosensors[11]. It is also available in the form of graphene oxide (GO), much easier to handle by wet processing, capable of reducing it to rGO, whose electronic properties are close to those of graphene.

A recent example of SARS-CoV-2 (COVID-19) assay using a GFET is an illustration of these properties and could favorably replace reference molecular

11 Si/SiO_2/G is in fact obtained after several transformation steps. The first step is the deposition of a G monolayer by evaporation of graphite on a copper film using a proven CVD technique, which guarantees high purity graphene. For handling purposes, the Cu/G film is covered with a layer of PMMA, then copper is disolved and the G/PMMA transferred onto a Si/SiO_2 wafer to produce the Si/SiO_2/G/PMMA multilayer structure. Finally, dissolving PMMA in acetone makes it possible to obtain a silicon wafer coated with good quality G, ready to be etched or cut for use in FET.

rRT-PCR[12] tests with a new type of antigen test. Seo et al. (2020) produced a device for detecting SARS-CoV-2 based on the use of a GFET by grafting the S (Spicule "Spike") protein antibody (SAb) into the channel. The S protein SAb antibody is attached to the graphene surface via a PBASE (1-pyrene butyric acid N-hydroxysuccinimide ester) coupling chain. It has a pyrene nucleus at its end, strongly adsorbing to the graphene due to its π-conjugated structure, while the SAb antibody reacts via its amine groups on the terminal succinimide group and binds covalently to butyric acid, giving PBASE-SAb (Figure 6.27).

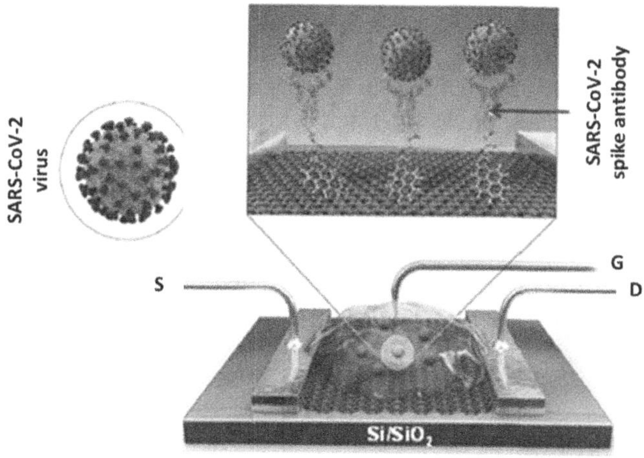

Figure 6.27. *Graphene FET used for SARS-CoV-2 assay. Adapted from Seo et al. (2020). For a color version of this figure, see www.iste.co.uk/lacaze/nanotechnologies.zip*

COMMENT ON FIGURE 6.27.– *The GFET is made from a $Si/SiO_2/G$ wafer on which the S and D electrodes are deposited by evaporation. The surface of the graphene is*

12 Currently, COVID-19 detection tests are carried out in real-time using the rRT-PCR (real-time reverse transcription-polymer chain reaction) technique. This is the most used technique (97% of all COVID-19 tests use rRT-PCR). Three distinct steps are necessary: (i) obtaining the viral sample with, for example, a nasopharyngeal swab; (ii) isolation of the virus RNA and conversion into cDNA using a particular enzyme (reverse transcriptase), which synthesizes the cDNA, complementary to the viral RNA, essential for PCR; (iii) amplification of the cDNA by PCR in the presence of a fluorescent marker and by carrying out several cDNA denaturation and recombination cycles, each cycle doubling the initial number of cDNA. Thus, after 35 PCR cycles the number of cDNAs obtained is 2^{35}, or more than 34 billion cDNA molecules. The fluorescence intensity threshold, characteristic of an infection, is reached quicker the greater the initial concentration of SARS-CoV-2.

functionalized by the SARS-CoV-2 "Spike" protein antibody. A fixed polarization is applied between the gate and the Si. After depositing a droplet containing the virus on the FET surface, the interaction between the antibody and the virus causes a variation in the I_{DS} current. The useful surface of the graphene constituting the GFET channel is delimited by the electrodes S and D, which corresponds to a 100 μm square. Adapted from Seo et al. (2020).

The specific interaction of the SARS-COV-2 "Spike" protein (SAg antigen) with the virus SAb antibody is responsible for the transduction signal. It produces a variation in channel conductivity of around 5%, characteristic of an infection, corresponding to a detection limit of 1 fg/mL in PBS (phosphate buffer saline solution) (Figure 6.28). However, the detection limit is only 100 fg/mL when SARS-CoV-2 is analyzed from clinical transport samples (virus transport medium [VTM]) containing various products (Hank's salts and bovine serum albumin [BSA]), likely to interfere with detection.

According to the authors, the LOD obtained using GFET-SAb is comparable to that obtained with rRT-PCR tests usually performed with a nasopharyngeal swab[13]. Unlike rRT-PCR tests, which require a preamplification step and an intervention of approximately 3 h, the test carried out using GFET-SAb is immediate and can be completed in 3 min.

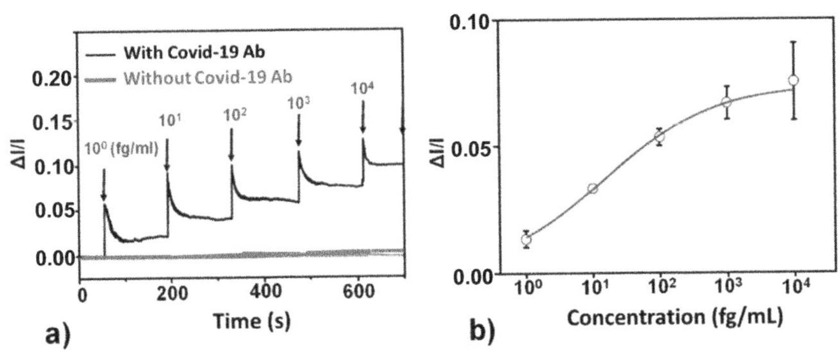

Figure 6.28. *FET-G-SAb response to COVID-19. (a) Relative variations in I_{DS} current as a function of time and for various concentrations of SARS-CoV-2 antibody (Ab). The blue curve indicates the absence of signal in the case of a sample without virus. About 100 s is necessary to reach pseudo-stability of the signal (V_{DS} = 0.01 V; V_{GS} = –1.5 V). (b) Relative I_{DS} variation curve as a function of the SARS-CoV-2 concentration. Adapted from Seo et al. (2020). For a color version of this figure, see www.iste.co.uk/lacaze/nanotechnologies.zip*

13 According to the authors, the LOD for their technique is 250 copies/mL, while that for current PCR tests is 50–100 copies/mL.

More recently, similar results were obtained using the same type of FET but replacing the graphene with single-wall semiconductor CNTs (SWCNT) (Shao et al. 2021; Zamzami et al. 2022). After deposition of the SWCNTs on the surface of Si/SiO_2, they are functionalized by SAb or N protein (NAb) antibodies with a covalent bond[14]. As with graphene, and for a fixed gate voltage (–0.5 V), the authors observed a linear variation in I_{DS} current as a function of the SAg concentration for a range of concentrations, albeit limited, between 5.5 fg/mL and 5.5 pg/mL in the case of FET-SWCNT-SAb (Shao et al. 2021) (Figure 6.29).

Compared to GFET-SAb, the sensitivity of FET-SWCNT-SAb is significantly improved. The increase in $\Delta I/I°$ in the linear part of the curve is around 0.25 for a variation of 10 fg/mL in SAg, while in the case of GFET it is only 0.025, or 10 times lower. This difference in sensitivity is partly due to the direct mode of grafting of the SAb antibody onto the SWCNTs, without an intermediate PBASE chain, which therefore does not attenuate the electrostatic effect resulting from the complexation of SAb and SAg. The increased relative sensitivity is, however, essentially due to the semiconductor nature of the SWCNTs, which provides the CNT-FET with a true on-off transition (conductor-insulator).

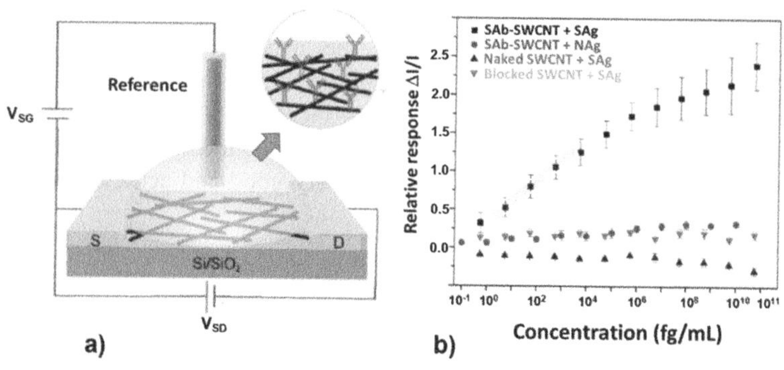

Figure 6.29. *FET-SWCNT for the SARS-CoV-2 assay. (a) Schematic of the device showing SWCNTs modified by protein S antibodies (SAbs) in green. (b) Relative variation of the I_{DS} current in response to SAg and NAg antigens on SWCNTs modified or not by the SAb antibody or with chemical blocking of the SWCNTs. The absence of a signal in the case of the NAg antigen indicates very good specificity of the FET-SWCNT/SAb sensor. The assays are carried out using calibrated samples; the linearity domain is indicated in light blue. Adapted from Shao et al. (2021). For a color version of this figure, see www.iste.co.uk/lacaze/nanotechnologies.zip*

14 The deposition of SWCNTs is carried out by dielectrophoresis by applying an alternating voltage (100 kHz) of around 100 V between the gold electrodes S and D of the FET. Functionalization involves a reaction between the residual COOH functional groups of the SWNTs and the amine functional groups of the SAg protein antibodies. Experimental details are described in the article by Shao et al. (2021).

Furthermore, the results obtained from nasopharyngeal samples in VTM, compared to certified NAAT (nucleic acid amplification test) assays, proved satisfactory. A concordance of 15 positive out of 18 positive NAATs and seven negative out of 10 negative NAATs is demonstrated; the duration of an assay is around 5 min and the execution cost is very low.

6.3.1.3. *Nanowire FETs for virus assays*

Using nanowires to produce FETs for analytic purposes is relatively long-standing and has been done with different materials (polymers and various semiconductors), a description of which can be found in the review article by Ahmad et al. (2018). Silicon is, however, the most used element and recent methods for manufacturing nanowires (SiNWs) have been described to obtain nanowires with excellent performance and reproducibility (Zafar 2018).

Using small diameter Si nanowires manufactured using a top-down technique, Nuzaihan et al. (2016) developed a transistor to detect the dengue virus. The "back gate" configuration transistor has a p-type Si nanowire as the channel onto which an oligonucleotide probe, complementary to dengue DNA, has been grafted[15]. This hybridization can be compared to a "molecular gate" effect; it adds to the gate polarization between the Si substrate and the nanowire and results from an accumulation of additional negative charges in the thin liquid layer in contact with the Si nanowire (Figure 6.30). This induces an increase in positive charges in the nanowire and therefore increases its conductivity and the intensity of the I_{DS} current passing through it. The biosensor sensitivity is very good and achieves an LOD of 2 fM.

Optimizing the device response essentially depends on three parameters: (i) the section of the Si nanowire, (ii) the pH of the solution containing the sample and (iii) the ionic strength of the solution containing the sample. The last two parameters are common to all FET-type devices.

Regarding the nanowire diameter, it has been shown that the sensitivity of a 20 nm diameter SiNW is more than 10 times greater than that obtained with a 1 µm diameter SiNW at constant lengths. This confirms that the greater the

15 The probe oligonucleotide (complementary to dengue DNA) – modified at its end by a NH_2 group – comprises 27 bases (5′-NH_2 C6-AAC AGC ATA TTG ACG CTG GCA GAG ACC-3′). It is attached to the surface of the silicon nanowire via amino propyl triethoxy siloxane $NH_2(CH_2)_3Si(OEt)_3$ (APTS) and glutaric aldehyde $OCH(CH_2)_3CHO$. In a basic medium, APTS is grafted directly onto the Si surface via the two groups Si-O-; the two glutaric aldehyde CHO groups react with the amino functional groups of the oligonucleotide and the APTS leading to the sequence $Si\text{-}O(CH_2)_3N{=}CH(CH_2)_3CH{=}N$-oligonucleotide.

surface/volume ratio of the SiNW, the greater the relative variation in the nanowire charge carrier concentration.

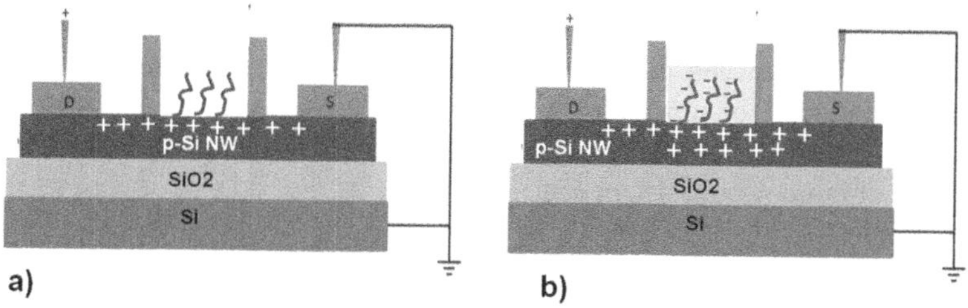

Figure 6.30. *Schematic of a silicon nanowire FET transistor for Dengue virus detection. (a) Grafting of the oligonucleotide complementary to the viral DNA (blue filaments) onto the Si nanowire (p-Si NW). (b) Addition of the solution containing the virus, leading to the appearance of additional negative charges (in red), responsible for an increase in positive charges in the nanowire, resulting in an increase in conductivity of the nanowire. Adapted from Nuzaihan et al. (2016). For a color version of this figure, see www.iste.co.uk/lacaze/nanotechnologies.zip*

The second critical point concerns the pH of the medium in which the detection of the charged species occurs. The variation in the charge carrier concentration in the SiNW is in fact linked to the presence of charges carried by the target analyte (in this case, viral DNA). It is therefore necessary to ensure that the pH of the medium corresponds to a charged form of the analyte and not to a neutral or even zwitterion form. In this case, the negative charges measured by Nuzaihan et al. came from PO_4^{3-} anionic groups from DNA strands.

The third parameter likely to negatively impact the measurement is the ionic strength of the solution. As indicated previously (Chapter 4), any electrolyte solution in contact with a polarized surface is characterized by a thin "diffuse" layer containing ions with the opposite sign to the surface polarization. This layer exerts a screening effect with respect to any charge outside of it, which means that any part of an electrically charged molecule (DNA strand) located outside of this layer will have no action on the transistor channel. In the present case, with the concentration of the ionic solution being around 0.1 M, the thickness λ_D of the diffuse layer is approximately 0.8 nm, or similar to the length of the DNA strand fixed on the electrode. Under these conditions, all the negative charges of the phosphate ions carried by the DNA strand include an electric field on the SiNWs. The Debye length must nevertheless be considered because a certain number of protein targets, including antibodies, exceed this length and are therefore not suitable for this type of transistor.

A final challenge, not mentioned previously, is the mechanical fragility of the Si wires suspended between electrodes S and D. This can be solved by adopting a network structure, for example, a honeycomb structure, which improves the mechanical resistance and increases the active surface area, favoring a stronger adsorption of biomolecules (Kim et al. 2016).

6.3.2. *Toxin detection*

Detecting toxins in food is generally carried out by applying ELISA-type immunoassay techniques. The use of transistors has recently been proposed for faster and lower cost analyses. White et al. (2016) designed a new type of transistor, adapted for toxin detection, illustrated using the ricin[16].

The transistor developed here is more complex than the one described previously. It consists of a control gate CG (covered by a buffer solution and acting as a reference electrode) and a floating gate FG, the latter divided into two parts. Part of this FG is functionalized by the aptamer and covered by the analyte solution (ricin). The other part is covered by an electrolyte gel and forms the control gate of a standard organic transistor with a P3HT channel. This transistor functions like a standard voltage inverter delivering a low voltage V_{out} for a high voltage V_{in} and vice versa (Figure 6.31). The presence of the analyte (ricin) causes a potential shift in the inverter, correlated with the concentration of ricin (Figure 6.32).

Another innovative part of this study was the powering of the circuit with an alternating voltage (at a frequency of 13 Hz). Figure 6.31(a) shows the values of ΔV_{out} for different concentrations of ricin (i.e. for V_{in} varying from –0.3 V to 0 (i.e. for 3 V and vice versa). It is interesting to note the asymmetry of the output signal and the increase in amplitude with ricin concentration.

Figure 6.32 shows the plots of ΔV_{out} according to the concentration of ricin for different media (water, milk and orange juice [OJ]) and for a "blank" sample. The device is capable of detecting ricin from 0.1 nM. This study opened the way to using active electronic components such as logic gates to produce sensors for comparison and decision (toggle of a potential from a given threshold).

16 Ricin is a protein toxin that can be extracted from the castor bean; its mass is approximately 60 kDa. It is frequently found as a toxic impurity in milk and orange juice (OJ).

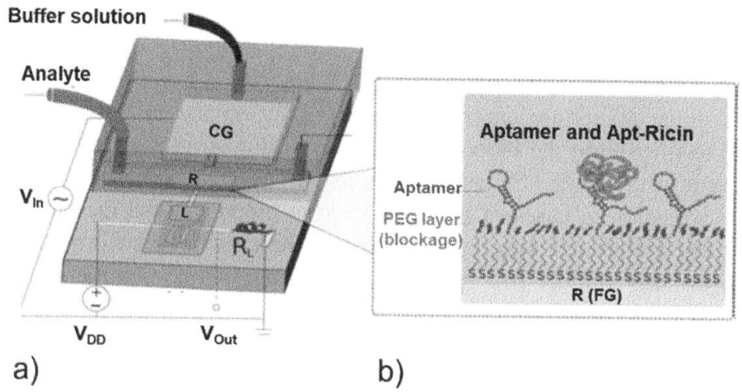

Figure 6.31. *Floating gate electrolyte-gated transistor developed for ricin detection. (a) Side view of the device showing a control gate (CG) and a floating gate (FG) of which a part (R) is immersed in the electrolyte containing the analyte; the other part (L) constituting the control gate is an FET transistor covered by an electrolyte gel. The FET channel is the P3HT semiconductor. (b) Part of the FG functionalized by the aptamer and the aptamer-ricin complex. Adapted from White et al. (2016). For a color version of this figure, see www.iste.co.uk/lacaze/nanotechnologies.zip*

COMMENT ON FIGURE 6.31.– *An alternating voltage V_{In} is applied between the CG and the FET source electrode, and it is the variations in electrode FG capacitance that induce the variations in the transistor output signal, correlated with the concentration of ricin.*

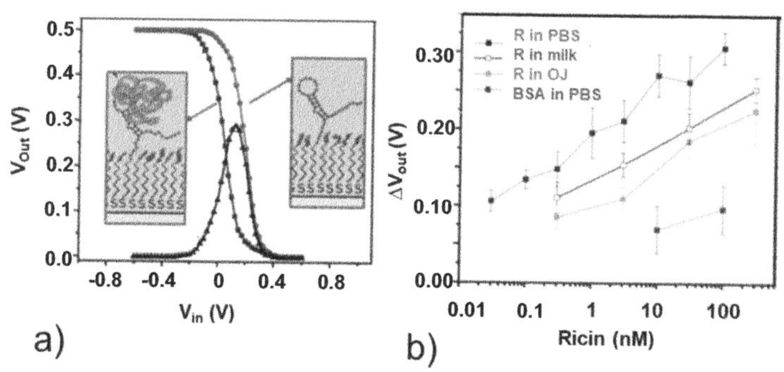

Figure 6.32. *Results obtained for the detection of ricin on the floating gate EGFET described above, with AC mode. (a) $V_{in}(t) = 0.3 \sin(\omega t)$ (in V), whereas $\Delta V_{out}(t)$ (black curve) presents a complex shape whose amplitude depends on the concentration of ricin. (b) Plots of the maximum value of ΔV_{out} for different concentrations of ricin (R). Adapted from White et al. (2016). For a color version of this figure, see www.iste.co.uk/lacaze/nanotechnologies.zip*

6.4. Conclusions

Health is one of the most studied areas of application because both societal and economic expectations are enormous. This is how, in the 1970s, the first sensor dedicated to measuring the blood oxygen level during surgery appeared. Economic development in the United States and then in Europe, generating new eating habits, led to an explosion in cases of diabetes, making blood glucose measuring devices economically viable, not only in health establishments but also in patient homes. Research on this subject has multiplied, allowing the development of portable devices accessible to all both in terms of ease of use and cost. Continuous blood glucose measuring devices are one of the most significant achievements of today.

Subsequently, new experimental developments concerning extremely varied biosensors both in terms of targets to be measured and in the techniques used have become essential for the monitoring of other molecules necessary for our metabolism. Equipping each patient with tools to continuously measure certain biomarkers, for predicting the appearance of symptoms of a disease, is a prospect that is no longer unrealistic. The scientific advances and technological developments made in recent years suggest that the global industry leaders and the WEB giants who have invested in this field will market very powerful analytical tools, connected to the Cloud and based on AI. Building on the techniques detailed in this chapter, such autonomous devices should be capable of carrying out real health assessments. Smart "connected watch", capable of measuring the oxygen saturation of our blood (spO_2), that produce an electrocardiogram equivalent to that provided by professional tools, is an example of this.

Conclusion

Increasingly stringent environmental regulations and the medical need for periodic monitoring of certain disease biomarkers have contributed to the development of increasingly sophisticated chemical and biochemical sensors and analytical methods. The oldest analytical techniques are separation techniques based on chromatography and electrophoresis. Considerable progress has been made in the automation of such analysis, with a huge diversity of techniques and commercial equipment. Unlike these techniques which often rely on bulky equipment, chemical and biological sensors are often of simpler design, portable and programmed to identify and dose a specific analyte. There are many required characteristics and they are sometimes difficult to achieve, the most important being sensitivity and selectivity. Sensitivity is the concentration range over which the sensor has a useful response with the lowest possible detection limit. Selectivity is the ability to detect a specific molecule in a mixture, to the exclusion of all others, with the added characteristic of being insensitive to possible interference from other interfering molecules. The other equally important qualities sought for such sensors depend on the technology used and concern response and recovery times, stability over time and sensor lifespan. These sensors are based on physicochemical transduction systems that enable miniaturization and provide access to portable devices for on-site testing (POC).

Throughout this work, we have witnessed significant advances in technology and materials that have profoundly changed the landscape of analyte detection and quantification. In particular, the synthesis and manipulation of new nanoscale organic and inorganic materials have played a key role in this progress. In the environmental field, the development of specific electrodes for ion analysis in solution is a prime example. The use of ionophores in their manufacture has made it possible to market ISEs dedicated to most metal ions. These electrochemical devices have gained widespread use due to their simplicity.

In the case of gases, most sensors, usually of the resistive type, are now designed to operate at room temperature and are unfortunately characterized by low selectivity. To overcome this drawback, one solution is to record the signals from multiple sensors, grouped together in a network, and processing the information by comparing it with that recorded in databases. This approach, akin to an electronic nose, can potentially enhance gas sensor performance.

The case of biological molecules, however, is very different. The sensor specificity is generally very high due to the nature of the recognition phenomenon involved between the solute and the sensor sensitive layer. Made up of molecules with a very high affinity for the analyte, this sensitive layer gives rise to associations similar to those encountered in immunological assays. Physical detection methods that combine electrochemistry and optics, such as electrochemiluminescence, have also been developed to provide speed and reliability in detecting specific biomolecules such as biomarkers and are now being adopted by analytical laboratories. Today, very low detection limits are achieved with optical devices based on luminescence signals, but also using transistors that deliver electrical signals directly usable by microcontrollers and even smartphones.

Finally, in addition to all of these scientific and technological advances, the potential contribution of artificial intelligence, the Internet of Things and statistical analysis coupled with databases is on the verge of revolutionizing analytical methods. This exciting prospect of future advancements, with new operating modes that are sometimes counterintuitive compared to those used until now, should inspire us all in our continued pursuit of knowledge and innovation.

References

Acquah, S.F.A., Penkova, A.V., Markelov, D.A., Semisalova, A.S., Leonhardt, B.E., Magi, J.M. (2017). Review–The beautiful molecule: 30 years of C60 and its derivatives. *ECS J. Solid State Sci. Techn.*, 6, M3155–M3162.

Adeel, M., Asif, K., Rahman, M.M., Daniele, S., Canzonieri, V., Rizzolio, F. (2021). Glucose detection devices and methods based on metal–organic frameworks and related materials. *Adv. Funct. Mater.*, 31, 2106023.

Ahmad, R., Mahmoudi, T., Ahn, M.-S., Hahn, Y.-B. (2018). Recent advances in nanowires-based field-effect transistors for biological sensor applications. *Biosensors and Bioelectronics*, 100, 312–325.

Akgönüllü, S. and Denizli, A. (2022). Recent advances in optical biosensing approaches for biomarkers detection. *Biosens. Bioelectron. X*, 12, 100269.

Alberty, R.A. (1950). Moving boundary systems formed by weak electrolytes. theory of simple systems formed by weak acids and bases. *J. Am. Chem. Soc.*, 72, 2361–2367.

Alemu, D., Wei, H.-Y., Hod, K.-C., Chu, C.-W. (2012). Highly conductive PEDOT:PSS electrode by simple film treatment with methanol for ITO-free polymer solar cells. *Energy Environ. Sci.*, 5, 9662–9671.

Ali, S.S., Hardt, J.I., Quick, K.L., Kim-Han, J.S., Erlanger, B.F., Huang, T.-T., Epstein, C.J., Dugan, L.L. (2004). A biologically effective fullerene (C_{60}) derivative with superoxide dismutase mimetic properties. *Free Rad. Biol. & Medicine*, 37, 1191–1202.

Alizadeh, A., Hsu, W.-L., Wang, M., Daiguji, H. (2021. Electroosmotic flow: From microfluidics to nanofluidics. *Electrophoresis*, 42, 834–868.

Allendorf, M.D., Bauer, C.A., Bhakta, R.K., Houk, R.J.T. (2009). Luminescent metal–organic frameworks. *Chem. Soc. Rev.*, 38, 1330–1352.

Alwarappan, S., Nesakumar, N., Sun, D., Hu, T.Y., Li, C.-Z. (2022). 2D metal carbides and nitrides (MXenes) for sensors and biosensors. *Biosensors and Bioelectronics*, 205, 113943.

Amendola, V., Pilot, R., Frasconi, M., Marago, O.M., Iati, M.A. (2017). Surface plasmon resonance in gold nanoparticles: A review. *J. Phys. Condens. Matter*, 29, 203002.

Amiri, M. and Arshi, S. (2020). An overview on electrochemical determination of cholesterol. *Electroanalysis*, 32, 1391.

Andersson, L., Sellergren, B., Mosbach, K. (1984). Imprinting of amino acid derivatives in macroporous polymers. *Tetrahedron Lett.*, 25, 5211–5214.

Andre, R.S., Facure, M.H.M., Mercante, L.A., Correa, D.S. (2022). Electronic nose based on hybrid free-standing nanofibrous mats for meat spoilage monitoring. *Sensors & Actuators: B. Chemical*, 353, 131114.

Andronescu, C. and Schuhmann, W. (2017). Review article: Graphene-based field effect transistors as biosensors. *Current Opinion in Electrochemistry*, 3, 11–17.

Anichini, C., Czepa, W., Pakulski, D., Aliprandi, A., Ciesielski, A., Samori, P. (2018). Chemical sensing with 2D materials. *Chem. Soc. Rev.*, 47, 4860–4908.

Asiala, S.M. and Schultz, Z.D. (2011). Characterization of hotspots in a highly enhancing SERS substrate. *Analyst*, 136, 4472–4479.

Askim, J.R., Mahmoudi, M., Suslick, K.S. (2013). Optical sensor arrays for chemical sensing: The optoelectronic nose. *Chem. Soc. Rev.*, 42, 8649–8682.

Axel, R. (1995). The molecular logic of smell. *Scientific American*, 273, 154–159.

Bachilo, S.M., Strano, M.S., Kittrell, C., Hauge, R.H., Smalley, R.E., Weisman, R.B. (2002). Structure-assigned optical spectra of single-walled carbon nanotubes. *Science*, 298, 2361–2368.

Bae, S., Kim, H., Lee, Y., Xu, X., Park, J.-S., Zheng, Y., Balakrishnan, J., Lei, T., Kim, H.R., Song, Y.I. et al. (2010). Roll-to-roll production of 30-inch graphene films for transparent electrodes. *Nat. Nanotechn.*, 5, 574–578.

Bag, A. and Lee, N.-E. (2019). Gas sensing with heterostructures based on two-dimensional nanostructured materials: A review. *J. Mater. Chem. C*, 7, 13367–13383.

Bagha, S.S. and Santiago, J.G. (2013). Coupling isotachophoresis and capillary electrophoresis: A review and comparison of methods. *Analyst*, 138, 735–754.

Bai, Y., Xu, T., Zhang, X. (2020). Graphene-based biosensors for detection of biomarkers. *Micromachines*, 11(1), 60.

Bakker, E. and Pretsch, E. (2005). Potentiometric sensors for trace-level analysis. *Trends Anal. Chem.*, 24(5), 199–207.

Bakker, E., Bühlmann, P., Pretsch, E. (1997). Carrier-based ion-selective electrodes and bulk optodes. 1. General Characteristics. *Chem. Rev.*, 97, 3083–3132.

Bakker, E., Diamond, D., Lewenstam, A., Pretsch, E. (1999). Ion sensors: Current limits and new trends, *Analytica Chimica Acta*, 393, 11–18.

Bakker, E., Bühlmann, P., Pretsch, E. (2004). The phase-boundary potential model. *Talanta*, 63, 3–20.

Bakry, R., Vallant, R.M., Najam-Ul-Haq, M., Rainer, M., Szabo, Z., Huck, C.W., Bonn, G.K. (2007). Medicinal applications of fullerenes. *Intern. J. Nanomedicine*, 2, 639–649.

Balch, A.L. and Olmstead, M.M. (1998). Reactions of transition metal complexes with fullerenes (C60, C70, etc.) and related materials. *Chem. Rev.*, 98, 2123–2165.

Bang, E., Choi, Y., Cho, J., Suh, Y.-H., Ban, H.W., Son, J.S., Park J. (2017). Large-scale synthesis of highly luminescent InP@ZnS quantum dots using elemental phosphorus precursor. *Chem. Mater.*, 29, 4236–4243.

Baratto, C., Comini, E., Faglia, G., Sberveglieri, G., Zha, M., Zappettini, A. (2005). Metal oxide nanocrystals for gas sensing. *Sensors and Actuators B*, 109(1), 2–6.

Barbante, G.J., Doeven, E.H., Kerr, E., Connell, T.U., Donnelly, P.S., White, J.M., Lopes, T., Laird, S., Wilson, D.J.D., Barnard, P.J. et al. (2014). Understanding electrogenerated chemiluminescence efficiency in blue-shifted iridium(iii)-complexes: An experimental and theoretical study. *Chem. Eur. J.*, 20(12), 3322–3332.

Bard, A.J. and Faulkner, L.R. (2001). *Electrochemical Methods: Fundamentals and Applications*, 2nd edition. John Wiley & Sons, New York.

Barsan, N. and Weimar, U. (2001). Conduction model of metal oxide gas sensors. *Journal of Electroceramics*, 7(3), 143–167.

Barsan, N., Schweizer-Berberich, M., Gopel, W. (1999). Fundamental and practical aspects in the design of nanoscaled SnO_2 gas sensors. A status report. *Fresenius J. Anal. Chem.*, 365, 287–304.

Barsan, N., Koziej, D., Weimar, U. (2007). Metal oxide-based gas sensor research: How to? *Sens. Actuators-B*, 121(1), 18–35.

Bastos-Soares, E.A., Sousa, R.M.O., Gómez A.F., Alfonso, J., Kayano, A.M., Zanchi, F.B., Funes-Huacca, M.E., Stábeli, R.G., Soares, A.M., Pereira, S.S. et al. (2020). Single domain antibodies in the development of immunosensors for diagnostics. *Int. J. Biol. Macromol.*, 165, 2244–2252.

Basu, S. and Bhattacharyya, P. (2012). Recent developments on graphene and graphene oxide based solid state gas sensors. *Sens. Actuators-B*, 173, 1–21.

Beckers, J.L. and Bocek P. (2000). Sample stacking in capillary zone electrophoresis: Principles, advantages and limitations. *Electrophoresis*, 21, 2747–2767.

Beckers, J.L., Gebauer, P., Bocek P. (2001). System zones in capillary zone electrophoresis. *Electrophoresis*, 22, 3648–3658.

BelBruno, J.J. (2019). Molecularly imprinted polymers. *Chem. Rev.* 119, 94–119.

Belyustin A.A. (2011). The centenary of glass electrode: From Max Cremer to F. G. K. Baucke. *J. Solid State Electrochem.*, 15, 47–65.

Benkovic, S.J. and Hammes-Schiffer, S. (2003). A perspective on enzyme catalysis. *Science* 301, 1196–1202.

Bera, M. and Maji, P.K. (2017). Graphene-based polymer nanocomposites: Materials for future revolution. *MOJ Poly Sci.*,1, 94–97.

Bergveld, P. (1970). Development of an ion-sensitive solid-state device for neurophysiological measurements. *IEEE Transactions on Biomedical Engineering*, 17(1), 70–71.

Bergveld, P. (1985). The impact of MOSFET-based sensors. *Sensors and Actuators*, 8(2), 109–127.

Berto, M., Diacci, C., Theuer, L., Di Lauro, M., Simon, D.T., Berggren, M., Biscarini, F., Beni, V., Bortolotti, C.A. (2018). Label free urea biosensor based on organic electrochemical transistors. *Flex. Print. Electron.* 3, 024001.

Bethune, D.S., Klang, C.H., De Vries, M.S., Gorman, G., Savoy, R., Vazquez, J., Beyers, R. (1993). Cobalt-catalysed growth of carbon nanotubes with single-atomic-layer walls. *Nature*, 363, 605–607.

Bhatia, P. and Gupta, B.D. (2012). Fabrication and characterization of a surface plasmon resonance based fiber optic urea sensor for biomedical applications. *Sens. Actuat. B*, 161, 434–438.

Bhattacharjee, S. (2016). DLS and zeta potential – What they are and what they are not? *J. of Controlled Release*, 235, 337–351.

Bhimwal, R., Rustandi, R.R., Payne, A., Dawod, M. (2022). Recent advances in capillary gel electrophoresis for the analysis of proteins. *Journal of Chromatography A*, 1682, 463453.

Bobacka, J. (2006). Conducting polymer-based solid-state ion-selective electrodes. *Electroanalysis*, 18(1), 7–18.

Bobacka, J., Ivaska, A., Lewenstam, A. (2008). Potentiometric ion sensors. *Chem. Rev.*, 108, 329–351.

Bolotin, K.I., Sikes, K.J., Jiang, Z., Klima, M., Fudenberg, G., Honec, J., Kim, P., Stormer, H.L. (2008a). Ultrahigh electron mobility in suspended graphene. *Solid State Comm.*, 146, 351–355.

Bolotin, K.I., Sikes, K.J., Hone, J., Stormer, H.L., Kim, P. (2008b). Temperature-dependent transport in suspended graphene. *Phys. Rev. Lett.*, 101, 096802.

Bonaccorso, F., Sun, Z., Hasan, T., Ferrari, A.C. (2010). Graphene photonics and optoelectronics. *Nat. Photonics*, 4, 611–622.

Botewad, S.N., Gaikwad, D.K., Girhe, N.B., Thorat, H.N., Pawar, P.P. (2023). Urea biosensors: A comprehensive review. *Biotechnology and Applied Biochemistry*, 70, 485–501.

Bourlinos, A.B., Bakandritsos, A., Kouloumpis, A., Gournis, D., Krysmann, M., Giannelis, E.P., Polakova, K., Safarova, K., Hola, K., Zboril, R. (2012). Gd(III)-doped carbon dots as a dual fluorescent-MRI probe. *J. Mater. Chem.*, 22, 23327–23330.

Breadmore, M.C., Dawod, M., Quirino, J.P. (2011). Recent advances in enhancing the sensitivity of electrophoresis and electrochromatography in capillaries and microchips (2008–2010). *Electrophoresis*, 32, 127–148.

Britz-McKibbin, P. and Chen, D.D.Y. (2000). Selective focusing of catecholamines and weakly acidic compounds by capillary electrophoresis using a dynamic pH junction. *Anal. Chem.*, 72, 1242–1252.

Bronikowski, M.J., Willis, P.A., Colbert, D.T., Smith, K.A., Smalley, R.E. (2001). Gas-phase production of carbon single-walled nanotubes from carbon monoxide via the HiPco process: A parametric study. *J. Vac. Sci. Technol. A*, 19, 1800–1805.

Brownson, D.A.C., Kampouris, D.K., Banks, C.E. (2011). An overview of graphene in energy production and storage applications. *J. Power Sources*, 196, 4873–4885.

Bruchez, M. Jr., Moronne, M., Gin, P., Weiss, S., Alivisatos, A.P. (1998). Semiconductor nanocrystals as fluorescent biological labels. *Science*, 281(5385), 2013–2016.

Buck, L.B. and Axel, R. (1991). A novel multigene family may encode odorant receptors: A molecular basis for odor recognition. *Cell*, 65(1), 175–187.

Bühlmann, P., Pretsch, E., Bakker, E. (1998). Carrier-based ion-selective electrodes and bulk optodes. 2. Ionophores for potentiometric and optical sensors. *Chem. Rev.*, 98, 1593–1687.

Buonanno, G., Brancaccio, A., Costanzo, S., Solimene, R. (2022). Spectral methods for response enhancement of microwave resonant sensors in continuous non-invasive blood glucose monitoring. *Bioengineering*, 9, 156.

Buth, F., Donner, A., Sachsenhauser, M., Stutzmann, M., Garrido, J.A. (2012). Biofunctional electrolyte-gated organic field-effect transistors. *Adv. Mat.*, 24, 4511–4517.

Cai, H.-H., Pi, J., Lin, X., Li, B., Li, A., Yang, P.-H., Cai, J. (2015). Gold nanoprobes-based resonance Rayleigh scattering assay platform: Sensitive cytosensing of breast cancer cells and facile monitoring of folate receptor expression. *Biosensors and Bioelectronics*, 74, 165–169.

Cao, L., Wang, X., Meziani, M.J., Lu, F., Wang, H., Luo, P.G., Lin, Y., Harruff, B.A., Veca, L.M., Murray, D. et al. (2007). Carbon dots for multiphoton bioimaging. *J. Am. Chem. Soc.*, 129, 11318–11319.

Carminati, M. (2017). Advances in high-resolution microscale impedance sensors. *Journal of Sensors*, 7638389.

Carne, A., Carbonell, C., Imaz, I., Maspoch, D. (2011). Nanoscale metal organic frameworks. *Chem. Soc. Rev.*, 40, 291–305.

Castro, E., Hernandez Garcia, A., Zavala, G., Echegoyen, L. (2017). Fullerenes in biology and medicine. *J. Mater. Chem. B Mater. Biol. Med.*, 5, 6523–6535.

Cattrall, R.W. (1971). Coated wire ion selective electrodes. *Anal. Chem.*, 43(13), 1905–1906.

Chai, Y., Guo, T., Jin, C., Haufler, R.E., Chibante, L.P.F., Pure, J., Wang, L., Alford, J.M., Smalley, R.E. (1991). Fullerenes with metals inside. *J. Phys. Chem.*, 95, 7564–7568.

Chambers, J.P., Arulanandam, B.P., Matta, L.L., Weis, A., Valdes J.J. (2008). Biosensor recognition elements. *Curr. Issues Mol. Biol.*, 10, 1–12.

Chang, Y.-R., Lee, H.-Y., Chen, K., Chang, C.-C., Tsai, D.-S., Fu, C.-C., Lim, T.-S., Tzeng, Y.K., Fang, C.-Y., Han, C.-C. et al. (2008). Mass production and dynamic imaging of fluorescent nanodiamonds. *Nat. Nanotechn.*, 3, 284–288.

Chen, J. and Yan, L. (2017). Recent advances in carbon nanotube-polymer composites. *Adv. Chem.*, 6, 129–148.

Chen, H., Müller, M.B., Gilmore, K.J., Wallace, G.G., Li, D. (2008). Mechanically strong, electrically conductive, and biocompatible graphene paper. *Adv. Mater.*, 20, 3557–3561.

Chen, X., Tian, X., Shin, I., Yoon, J. (2011). Fluorescent and luminescent probes for detection of reactive oxygen and nitrogen species. *Chem. Soc. Rev.*, 40, 4783–4804.

Chen, Y., Zhang, Y., Hu, Y., Kang, L., Zhang, S., Xie, H., Liu, D., Zhao, Q., Li, Q., Zhang, J. (2014a). State of the art of single-walled carbon nanotube synthesis on surfaces. *Adv. Mater.*, 26, 5898–5922.

Chen, L., Zeng, X., Si, P., Chen, Y., Chi, Y., Kim, D.-H., Chen, G. (2014b). Gold nanoparticle graphite-like C_3N_4 nanosheet nanohybrids used for electrochemiluminescent immunosensor. *Anal. Chem.*, 86, 4188–4195.

Chen, L.-Y., Wang, C.-W., Yuan, Z., Chang, H.-T. (2015). Fluorescent gold nanoclusters: Recent advances in sensing and imaging. *Anal. Chem.*, 87, 216–229.

Chen, Y., Lu, S., Zhang, S., Li, Y., Qu, Z., Chen, Y., Lu, B., Wang, X., Feng, X. (2017). Skin-like biosensor system via electrochemical channels for noninvasive blood glucose monitoring. *Sci. Adv.*, 3, e1701629.

Chen, X., Liu, Y., Ma, Q. (2018). Recent advances in quantum dot-based ECL sensors. *J. Mater. Chem. C*, 6, 942–959.

Chen, Y., Wang, Y., Zhu, S., Fu, K., Han, X., Wang, Y., Zhao, B., Li, T., Liu, B., Li, Y. et al. (2019). Nanomanufacturing of graphene nanosheets through nano-hole opening and closing. *Mater. Today*, 24, 26–32.

Chen, J., Fu, W., Jiang, F.-L., Liu, Y., Jiang, P. (2023). Recent advances in 2D metal carbides and nitrides (MXenes): Synthesis and biological application. *J. Mater. Chem. B*, 11, 702–715.

Cheong, W.J., Ali, F., Kim, Y.S., Lee, J.W. (2013). Comprehensive overview of recent preparation and application trends of various open tubular capillary columns in separation science. *J. Chromatogr. A*, 1308, 1–24.

Chhowalla, M., Shin, H.S., Eda, G., Li, L.-J., Loh, K.P., Zhang, H. (2013). The chemistry of two-dimensional layered transition metal dichalcogenide nanosheets. *Nature Chem.*, 5, 263–275.

Chhowalla, M., Liu, Z., Zhang, H. (2015). Two-dimensional transition metal dichalcogenide (TMD) nanosheets. *Chem. Soc. Rev.*, 44, 2584–2586.

Chien R.-L. and Burgi, D.S. (1991). Field amplified sample injection in capillary electrophoresis high-performance. *Journal of Chromatography*, 559, 141–152.

Chinnadayyala, S.R., Park, J., Le, H.T.N., Santhosh, M., Kadam, A.N., Cho, S. (2019). Recent advances in microfluidic paper-based electrochemiluminescence analytical devices for point-of-care testing applications. *Biosensors and Bioelectronics*, 126, 68–81.

Cho, S.-Y., Lee, Y., Koh, H.-J., Jung, H., Kim, J.-S., Yoo, H.-W., Kim, J., Jun, H.-T. (2016). Superior chemical sensing performance of black phosphorus: Comparison with MoS_2 and graphene. *Adv. Mater.*, 28, 7020–7028.

Cho, S.-Y., Koh, H.-J., Yoo, H.-W., Jung, H.-T. (2017). Tunable chemical sensing performance of black phosphorus by controlled functionalization with noble metals. *Chem. Mater.*, 29(17), 7197–7205.

Choi, W., Choudhary, N., Han, G.H., Park, J., Akinwande, D., Lee, Y.H. (2017a). Recent development of two-dimensional transition metal dichalcogenides and their applications. *Materials Today*, 20, 116–130.

Choi, H., Luzio, S., Beutler, J., Porch, A. (2017b). Microwave non-invasive blood glucose monitoring sensor: Human clinical trial results. *IEEE MTT-S International Microwave Symposium (IMS)*, 876–879.

Choi, J.H., Lee, J., Byeon, M., Hong, T.E., Park, H., Lee, C.Y. (2020). Graphene-based gas sensors with high sensitivity and minimal sensor-to-sensor variation. *ACS Appl. Nano Mater.*, 3(3), 2257–2265.

Chojnowska, S., Baran, T., Wilinska, I., Sienicka, P., Cabaj-Wiater, I., Knas, M. (2018). Human saliva as a diagnostic material. *Adv. Med. Sci.*, 63, 185–191.

Chumbimuni-Torres, K.Y., Dai, Z., Rubinova, N., Xiang, Y., Pretsch, E., Wang, J., Bakker, E. (2006). Potentiometric biosensing of proteins with ultrasensitive ion-selective microelectrodes and nanoparticle labels. *J. Am. Chem. Soc.*, 128, 13676–13677.

Clark, L.C. Jr. and Lyons, C. (1962). Electrode systems for continuous monitoring in cardiovascular surgery. *Annals of the N.Y. Acad. of Sci.*, 102, 29–45.

Clarke, W.L., Cox, D., Gonder-Frederick, L.A., Carter, W., Pohl, S.L. (1987). Evaluating clinical accuracy of systems for self-monitoring of blood glucose. *Diabetes Care*, 10, 622–628.

Conney, A.H., Lou, Y.R., Xie, J.G., Osawa, T., Newmark, H.L., Liu, Y., Chang, R.L., Huang, M.T. (1997). Some perspectives on dietary inhibition of carcinogenesis: Studies with curcumin and tea. *Proc. Soc. Exp. Bio. Med.*, 216(2), 234–245.

Crespo, G.A., Macho, S., Rius, F.X. (2008). Ion-selective electrodes using carbon nanotubes as ion-to-electron transducers. *Anal. Chem.*, 80, 1316–1322.

Cui, Y., Wei, Q., Park, H., Lieber, C.M. (2001). Nanowire nanosensors for highly sensitive and selective detection of biological and chemical species. *Science*, 293, 1289–1292.

Cui, H., Xu, Y., Zhang, Z.-F. (2004). Multichannel electrochemiluminescence of luminol in neutral and alkaline aqueous solutions on a gold nanoparticle self-assembled electrode. *Anal. Chem.*, 76, 4002–4010.

Cui, Y., Yue, Y., Qian, G., Chen, B. (2012). Luminescent functional metal-organic frameworks. *Chem. Rev.*, 112, 1126–1162.

Cui, Y., Yue, Y., Zhang, Y., Zhang, Z., Zhou, H.S. (2020). Advancing biosensors with machine learning. *ACS Sens.*, 5, 3346–3364.

Dai, B., Zhou, R., Ping, J., Ying, Y., Xie, L. (2022). Recent advances in carbon nanotube-based biosensors for biomolecular detection. *TrAC Trends in Analytical Chemistry*, 154, 116658.

Dak, P., Ebrahimi, A., Swaminathan, V., Duarte-Guevara, C., Bashir, R., Alam, M.A. (2016). Droplet-based biosensing for lab-on-a-chip, open microfluidics platforms. *Biosensors*, 6, 14.

Dan, Y.P., Lu, Y., Kybert, N.J., Luo, Z.T., Johnson, A.T.C. (2009). Intrinsic response of graphene vapor sensors. *Nano Lett.*, 9(4), 1472–1475.

Daniel, M.-C. and Astruc, D. (2004). Gold nanoparticles: Assembly, supramolecular chemistry, quantum-size-related properties, and applications toward biology, catalysis, and nanotechnology. *Chem. Rev.*, 104, 293–346.

Danilenko, V.V. (2004). On the history of the discovery of nanodiamond synthesis. *Physics of the Solid State*, 46, 595–599.

Dantism, S., Takenaga, S., Wagner, T., Wagner, P., Schöning, M.J. (2017). Differential imaging of the metabolism of bacteria and eukaryotic cells based on light-addressable potentiometric sensors. *Electrochimica Acta*, 246, 234–241.

Das, T.K. and Prusty, S. (2013). Graphene-based polymer composites and their applications. *Polymer-Plastics Techn. Engineering*, 52, 319–331.

De Volder, M.F.L., Tawfick, S.H., Baughman, R.H., Hart, A.J. (2013). Carbon nanotubes present and future commercial applications. *Science*, 239, 535–539.

Deamer, D., Akeson, M., Branton, D. (2016). Three decades of nanopore sequencing. *Nature Biotechn.*, 34, 518–524.

Degraff, J., Liang, R., Le, M.Q., Capsal, J.-F., Ganet, F., Cottinet, P.-J. (2017). Printable low-cost and flexible carbon nanotube buckypaper motion sensors. *Mat. & Design*, 133, 47–53.

Demchenko, P. and Dekaliuk, M.O. (2013). Novel fluorescent carbonic nanomaterials for sensing and imaging. *Methods and Applications in Fluorescence,* 1, 042001.

Deng, S. and Ju, H. (2013). Electrogenerated chemiluminescence of nanomaterials for bioanalysis. *Analyst*, 138, 43–61.

Dervisevic, E., Dervisevic, M., Nyangwebah, J.N., Şenel, M. (2017). Development of novel amperometric urea biosensor based on Fc-PAMAM and MWCNT bio-nanocomposite film. *Sensors and Actuators B: Chemical*, 246, 920–926.

Di Nardo, F., Cavalera, S., Baggiani, C., Giovannoli, C., Anfossi, L. (2019). Direct vs mediated coupling of antibodies to gold nanoparticles: The case of salivary cortisol detection by lateral flow immunoassay. *ACS Appl. Mater. Interfaces*, 11, 32758–32768.

Dikin, D.A., Stankovich, S., Zimney, E.J., Piner, R.D., Dommett, G.H.B., Evmenenko, G., Nguyen, S.T., Ruoff, R.S. (2007). Preparation and characterization of graphene oxide paper. *Nature*, 448, 457–460.

Diltemiz, S.E., Keçili, R., Ersöz, A., Say R. (2017). Molecular imprinting technology in quartz crystal microbalance (QCM) sensors. *Sensors*, 17, 454.

Dincer, C., Bruch, R., Costa-Rama, E., Fernández-Abedul, M.T., Merkoçi, A., Manz, A., Urban, G.A., Güder, F. (2019). Disposable sensors in diagnostics, food, and environmental monitoring. *Adv. Mater.*, 31, 1806739.

Dole, V.P. (1945). A theory of moving boundary systems formed by strong electrolytes. *J. Am. Chem. Soc.*, 67, 1119–1126.

Dong, Z., Jiang, C., Cheng, H., Zhao, Y., Shi, G., Jiang, L., Qu, L. (2012). Facile fabrication of light, flexible and multifunctional graphene fibers. *Adv. Mater.*, 24, 1856–1861.

Dong, H., Fan, Y., Zhang, W., Gu, N., Zhang, Y. (2019). Catalytic mechanisms of nanozymes and their applications in biomedicine. *Bioconjugate Chem.*, 30, 1273–1296.

Dorsey, J.G. and Dill, K.A. (1989). The molecular mechanism of retention in reversed-phase liquid chromatography. *Chem. Rev.*, 89, 331–346.

Dou, W.-T., Han, H.-H., Sedgwick, A.C., Zhu, G.-B., Zang, Y., Yang, X.-R., Yoon, J., James, T.D., Li, J., He, X.-P. (2022). Fluorescent probes for the detection of disease-associated biomarkers. *Science Bulletin*, 67, 853–878.

Douglas, A. and Pinta, C.L. (2017). Review–Electrochemical growth of carbon nanotubes and graphene from ambient carbon dioxide: Synergy with conventional gas-phase growth mechanisms. *ECS J. Solid State Sci. Techn.*, 6, M3084–M3089.

Dugan, L.L., Turetsky, D.M., Du, C., Lobner, D., Wheeler, M., Almli, C.R., Shen, C.K.-F., Luh, T.-Y, Choi, D.W., Lin, T.-S. (1997). Carboxyfullerenes as neuroprotective agents. *Proc. Natl. Acad. Sci. USA*, 94, 9434–9439.

Dungchai, W., Chailapakul, O., Henry, C.S. (2009). Electrochemical detection for paper-based microfluidics. *Anal. Chem.*,81, 5821–5826.

Dunn, M., Jimenez, R., Chaput, J. (2017). Analysis of aptamer discovery and technology. *Nat. Rev. Chem.*, 1, 76.

Dürkop, T., Getty, S.A., Cobas, E., Fuhre, M.S. (2004). Extraordinary mobility in semiconducting carbon nanotubes. *Nano Lett.*, 4, 35–39.

Eigler, S., Enzelberger-Heim, M., Grimm, S., Hofmann, P., Kroener, W., Geworski, A., Dotzer, C., Röckert, M., Xiao, J., Papp, C. et al. (2013). Wet chemical synthesis of graphene. *Adv. Mater.*, 25, 3583–3587.

Eisenman, G. (1962). Cation selective glass electrodes and their mode of operation. *Biophys. J.*, 2, 259–323.

Ellis, J.E. and Star, A. (2016). Carbon nanotube based gas sensors toward breath analysis. *ChemPlusChem*, 81, 1248–1265.

El-Schich, Z., Zhang, Y., Feith, M., Beyer, S., Sternbæk, L., Ohlsson, L., Stollenwerk, M., Gjörloff, A., Wingren, A.G. (2020). Molecularly imprinted polymers in biological applications. *BioTechnques*, 69, 407–419.

Emaminejad, S., Gao, W., Wu, E., Davies, Z.A., Nyein, H.Y.Y., Challa, S., Ryan, S.P., Fahad, H.M., Chen, K., Shahpar, Z. (2017). Autonomous sweat extraction and analysis applied to cystic fibrosis and glucose monitoring using a fully integrated wearable platform. *Proc. Natl. Acad. Sci. USA.*, 114, 4625–4630.

Emtsev, K.V., Bostwick, A., Horn, K., Jobst, J., Kellogg, G.L., Ley, L., Mcchesney, J.L., Ohta, T., Reshanov, S.A., Röhrl, J. et al. (2009). Towards wafer-size graphene layers by atmospheric pressure graphitization of silicon carbide. *Nat. Mat.*, 8, 403–407.

Etchegoin, P.G. and Le Ru, E.C. (2008). A perspective on single molecule SERS: Current status and future challenges. *Phys. Chem. Chem. Phys.*, 10, 6079–6089.

Eustis, S. and El-Sayed, M.A. (2006). Why gold nanoparticles are more precious than pretty gold: Noble metal surface plasmon resonance and its enhancement of the radiative and nonradiative properties of nanocrystals of different shapes. *Chem. Soc. Rev.*, 35, 209–217.

Evans, G.W.H., Bhuiyan, W.T., Pang, S., Warren, B., Makris, K., Coleman, S., Hassan, S., Niu, X. (2021). A portable droplet microfluidic device for cortisol measurements using a competitive heterogeneous assay. *Analyst*, 146(2021), 4535–4544.

Falcaro, P., Ricco, R., Doherty, C.M., Liang, K., Hill, A.J., Styles, M.J. (2014). MOF positioning technology and device fabrication. *Chem. Soc. Rev.*, 43, 5513–5560.

Fang, H., Tosun, M., Seol, G., Chang, T.C., Takei, K., Guo, J., Javey, A. (2013). Degenerate n-doping of few-layer transition metal Dichalcogenides by potassium. *Nano Lett.*, 13, 1991–1995.

Farha, O.K., Eryazici, I., Jeong, N.C., Hauser, B.G., Wilmer, C.E., Sarjeant, A.A., Snurr, R.Q., Nguyen, S.T., Yazaydin, A.Ö., Hupp, J.T. (2012). Metal-organic framework materials with ultrahigh surface areas: Is the sky the limit? *J. Am. Chem. Soc.*, 134, 15016–15021.

Fauzi, F., Rianjanu, A., Santoso, I., Triyana, K. (2021). Gas and humidity sensing with quartz crystal microbalance (QCM) coated with graphene-based materials – A mini review. *Sensors and Actuators A: Physical*, 330, 112837.

Feng, Q.-M., Shen, Y.-Z., Li, M.-X., Zhang, Z.-L., Zhao, W., Xu, J.-J., Chen, H.-Y. (2016a). Dual-wavelength electrochemiluminescence ratiometry based on resonance energy transfer between Au nanoparticles functionalized g-C3N4 nanosheet and $Ru(bpy)_3^{2+}$ for micro-RNA detection. *Anal. Chem.*, 88, 937–944.

Feng, Q., Li, X., Wang, J., Gaskov, A.M. (2016b). Reduced graphene oxide (rGO) encapsulated Co_3O_4 composite nanofibers for highly selective ammonia sensors. *Sens. Actuators B Chem.*, 222, 864–870.

Férey, G. (2008). Hybrid porous solids: Past, present, future. *Chem. Soc. Rev.*, 37, 191–214.

Fernando, K.A.S., Sahu, S., Liu, Y., Lewis, W.K., Guliants, E.A., Jafariyan, A., Wang, P., Bunker, C.E., Sun, Y.-P. (2015). Carbon quantum dots and applications in photocatalytic energy conversion. *ACS Appl. Mater. Interfaces*, 7, 8363–8376.

Ferrari, A.C., Meyer, J.C., Scardaci, V., Casiraghi, C., Lazzeri, M., Mauri, F., Piscanec, S., Jiang, D., Novoselov, K.S., Roth, S. et al. (2006). Raman spectrum of graphene and graphene layers. *Phys. Rev. Lett.*, 97, 187401.

Ferri, S., Kojima, K., Sode, K. (2011). Review of glucose oxidases and glucose dehydrogenases: A bird's eye view of glucose sensing enzymes. *J. Diabetes. Sci. Technol.*, 1068–1076.

Ferrier, D.C. and Honeychurch, K.C. (2021). Carbon nanotube (CNT)-based biosensors. *Biosensors*, 11, 486.

Fleischmann, M., Hendra, P.J., Mcquillan, A. (1974). Raman spectra of pyridine adsorbed at a silver electrode. *Chem. Physics Lett.*, 26, 163–166.

Frant, M.S. (1994). History of the early commercialization of ion-selective electrodes. *Analyst*, 11, 2293–2301.

Furukawa, H., Cordova, K.E., Keeffe, M.O., Yaghi, O.M. (2013). The chemistry and applications of metal-organic frameworks. *Science*, 341, 1230444.

Gaggin, H.K. and Januzzi, J.L. (2013). Biomarkers and diagnostics in heart failure. *Biochim. Biophys. Acta – Molecular Basis of Disease*, 1832, 2442–2450.

Gall, D. (2016). Electron mean free path in elemental metals. *J. Appl. Physics*, 119, 085101.

Gamero-Quijano, A., Molina-Osorio, A.F., Peljo, P., Scanlon, M.D. (2019). Closed bipolar electrochemistry in a four-electrode configuration. *Phys. Chem. Chem. Phys.*, 21, 9627.

Gao, N., Zhou, W., Jiang, X., Hong, G., Fu, T.-M., Lieber, C.M. (2015). General strategy for biodetection in high ionic strength solutions using transistor-based nanoelectronic sensors. *Nano Lett.*,15, 2143–2148.

Gao, X., Zheng, P., Kasani, S., Wu, S., Yang, F., Lewis, S., Nayeem, S., Engler-Chiurazzi, E.B., Wigginton, J.G., Simpkins, J.W. et al. (2017). Paper-based surface-enhanced Raman scattering lateral flow strip for detection of neuron-specific enolase in blood plasma. *Anal. Chem.*, 89, 10104–10110.

Gao W., Muzyka K., Ma X., Lou B. (2018). A single-electrode electrochemical system for multiplex electrochemiluminescence analysis based on a resistance induced potential difference. *Chem. Sci.*, 9, 3911–3916.

Garcia Yanez, M. (2013). Glucose meter fundamentals and design. Freescale Semiconductor, Application Note AN4364 [Online]. Available at: https://www.nxp.com/docs/en/application-note/AN4364.pdf [Accessed 26 March 2023].

Gardner, J.W. and Bartlett, P.N. (1994). A brief history of electronic noses. *Sensors and Actuators B*, 18–19, 211–220.

Gavgani, J.N., Dehsari, H.S., Hasani, A., Mahyari, M., Shalamzari, E.K., Salehic, A., Taromi, F.A. (2015). A room temperature volatile organic compound sensor with enhanced performance, fast response and recovery based on N-doped graphene quantum dots and poly(3,4-ethylenedioxythiophene) poly(styrenesulfonate) nanocomposite. *RSC Adv.*, 5, 57559–57567.

Geim, A.K. and Novoselov, K.S. (2007). The rise of graphene. *Nature Materials*, 6, 183–191.

Geng, K., He, T., Liu, R., Dalapati, S., Tan, K.T., Li, Z., Tao, S., Gong, Y., Jiang, Q., Jiang, D. (2020). Covalent organic frameworks: Design, synthesis, and functions. *Chem. Rev.*, 120, 8814–8933.

Georgakilas, V., Perman, J.A., Tucek, J., Zboril, R. (2015). Broad family of carbon nanoallotropes: Classification, chemistry, and applications of fullerenes, carbon dots, nanotubes, graphene, nanodiamonds, and combined superstructures. *Chem. Rev.*, 115, 4744–4822.

Gewers, F.L., Ferreira, G.R., De Arruda, H.F., Silva, F.N., Comin, C.H., Amancio, D.R., Costa, L.D.F. (2021). Principal component analysis: A natural approach to data exploration. *ACM Computing Surveys (CSUR)*, 54, 1–34.

Gharbi, N., Pressac, M., Hadchouel, M., Szwarc, H., Wilson, S.R., Moussa, F. (2005). [60]Fullerene is a powerful antioxidant in vivo with no acute or subacute toxicity. *Nano Lett.*, 5, 2578–2585.

Ghodake, G.S., Shinde, S.K., Kadam, A.A., Saratale, R.G., Saratale, G.D., Syed, A., Elgorban, A.M., Marraiki, N., Kim, D.-Y. (2021). Biological characteristics and biomarkers of novel SARS-CoV-2 facilitated rapid development and implementation of diagnostic tools and surveillance measures. *Biosensors and Bioelectronics*, 177, 112969.

Ghosh, R. and Giri, P.K. (2017). Silicon nanowire heterostructures for advanced energy and environmental applications: A review. *Nanotechnology*, 28, 012001.

Ghosh, R., Aslam, M., Kalita, H. (2022). Graphene derivatives for chemiresistive gas sensors: A review. *Mater. Today Commun.*, 30, 103182.

Gilroy, K.D., Ruditskiy, A., Peng, H.-C., Qin, D., Xia, Y. (2016). Bimetallic nanocrystals: Syntheses, properties, and applications. *Chem. Rev.*, 116, 10414–10472.

Glatz, Z. (2015). On-capillary derivatisation as an approach to enhancing sensitivity in capillary electrophoresis. *Electrophoresis*, 36, 744–763.

Goldoni, R., Farronato, M., Connelly, S.T., Tartaglia, G.M., Yeo, W.-H. (2021). Recent advances in graphene-based nanobiosensors for salivary biomarker detection. *Biosensors and Bioelectronics*, 171, 112723.

Gong, J., Li, Y.Z., Hu, Z., Zhou, Z., Deng, Y. (2010). Ultrasensitive NH_3 gas sensor from polyaniline nanograin enchased TiO_2 fibers. *J. Phys. Chem. C*, 114(21), 9970–9974.

Gore, J.P. and Sane, A. (2011). Flame synthesis of carbon nanotubes, synthesis, characterization, applications. *Intech Open*. doi: 10.5772/21012.

Gorg, A., Postel, W., Westermeier, R., Gianazza, E., Righetti, P.G. (1980). Gel gradient electrophoresis, isoelectric focusing and two-dimensional techniques in horizontal, ultrathin polyacrylamide layers. *J. Biochem. Biophys. Methods*, 3, 273–284.

Graf, D., Molitor, F., Ensslin, K., Stampfer, C., Jungen, A., Hierold, C., Wirtz, L. (2007). Spatially resolved Raman spectroscopy of single- and few-layer graphene. *Nano Lett.*, 7, 238–242.

Greiner, N.R., Phillips, D.S., Johnson, J.D., Volk, F. (1988). Diamonds in detonation soot. *Nature*, 333, 440–441.

Gremmels, H., Winkel, B.M.F., Schuurman, R., Rosingh, A., Rigter, N.A.M., Rodriguez, O., Ubijaan, J., Wensing, A.M.J., Bonten, M.J.M., Marije Hofstra, L. (2021). Real-life validation of the Panbio™ COVID-19 antigen rapid test(Abbott) in community-dwelling subjects with symptoms of potential SARS-CoV-2 infection. *E-Clinical Medicine*, 31, 100677.

Grobert, N. (2007). Carbon nanotubes – becoming clean. *Mat. Today*, 10, 28–35.

Gu, J., Zhang, X., Pang, A., Yang, J. (2016). Facile synthesis and photoluminescence characteristics of blue-emitting nitrogen-doped graphene quantum dots. *Nanotechnology*, 27, 165704.

Guglielmi, G. (2020). Fast coronavirus tests: What they can and can't do. *Nature*, 485, 496–498.

Gusmao, R., Sofer, Z., Pumera, M. (2017). Black phosphorus rediscovered: From bulk material tomonolayers. *Angew. Chem. Int. Ed.*, 56, 8052–8072.

Ha, N., Xu, K., Ren, G., Mitchell, A. Ou, J.Z. (2020). Machine learning-enabled smart. *Sensor Systems. Adv. Intell. Syst.*, 2, 2000063.

Hafeman, D.G., Parce, J.W., Mcconnell, H.M. (1988). Light-addressable potentiometric sensor for biochemical systems source. *Science*, 240, 1182–1185.

Hai, W., Goda, T., Takeuchi, H., Yamaoka, S., Horiguchi, Y., Matsumoto, A., Miyahara, Y. (2017). Specific recognition of human influenza virus with PEDOT bearing sialic acid-terminated trisaccharides. *ACS Appl. Mater. Interfaces*, 9, 14162–14170.

Hajba, L., Jeong, S., Chung, D.S., Guttman, A. (2023). Capillary gel electrophoresis of proteins: Historical overview and recent advances. *Trends in Anal. Chem.*, 162, 117024.

Hajian, R., Balderston, S., Tran, T., deBoer, T., Etienne, J., Sandhu, M., Wauford, N.A., Chung, J.Y., Nokes, J., Athaiya, M. et al. (2019). Detection of unamplified target genes via CRISPR–Cas9 immobilized on a graphene field-effect transistor. *Nat. Biomed. Eng.*, 3, 427.

Halawa, M.I., Lai, J., Xu, G. (2018a). Gold nanoclusters: Synthetic strategies and recent advances in fluorescent sensing. *Materials Today Nano*, 3, 9–27.

Halawa, M.I., Wu, F., Fereja, T.H., Lou, B., Xu, G. (2018b). One-pot green synthesis of supramolecular β-cyclodextrin functionalized gold nanoclusters and their application for highly selective and sensitive fluorescent detection of dopamine. *Sens. Actuat. B: Chemical*, 254, 1017–1024.

Han, X., Liu, Y., Yin, Y. (2014). Colorimetric stress memory sensor based on disassembly of gold nanoparticle chains. *Nano Lett.*, 14, 2466–2470.

Han, Y., Huang, D., Ma, Y., He, G., Hu, J., Zhang, J., Hu, N., Su, Y., Zhou, Z., Zhang, Y. et al. (2018). Design of hetero-nanostructures on MoS_2 nanosheets to boost NO_2 room-temperature sensing. *ACS Appl. Mater. Interfaces*, 10, 22640–22649.

Hancu, G., Simon, B., Rusu, A., Mircia, E., Gyére, Á. (2013). Principles of micellar electrokinetic capillary chromatography applied in pharmaceutical analysis. *Adv. Pharmaceutical Bull.*, 3(1), 1–8.

Hassoun, A., Sahar, A., Lakhal, L., Aït-Kaddour, A. (2019). Fluorescence spectroscopy as a rapid and non-destructive method for monitoring quality and authenticity of fish and meat products: Impact of different preservation conditions. *LWT – Food Sci. and Technol.*, 103, 279–292.

Haupt, K. and Mosbach, K. (1998). Plastic antibodies: Developments and applications. *Trends Biotechnol.*, 16, 468–475.

Haupt, K., Rangel, P.X.M., Tse Sum Bui, B. (2020). Molecularly imprinted polymers: Antibody mimics for bioimaging and therapy. *Chem. Rev.*, 120, 9554–9582.

Haynes, C.L. and Van Duyne, R.P. (2001). Nanosphere lithography: A versatile nanofabrication tool for studies of size-dependent nanoparticle optics. *J. Phys. Chem. B*, 105, 5599–5611.

Haynes, C.L. and Van Duyne, R.P. (2003). Plasmon-sampled surface-enhanced Raman excitation spectroscopy. *J. Phys. Chem. B*, 107, 7426–7433.

He, Q., Zeng, Z., Yin, Z., Li, H., Wu, S., Huang, X., Zhang, H. (2012). Fabrication of flexible MoS_2 thin-film transistor arrays for practical gas-sensing applications. *Small*, 8, 2994–2999.

He, X., ElNaggar, M., Ostrowski, M.A., Guttman, A., Gentalen, E., Sperry, J. (2022). Evaluation of an icIEF-MS system for comparable charge variant analysis of biotherapeutics with rapid peak identification by mass spectrometry. *Electrophoresis*, 43, 1215–1222.

Heath, J.R., O'Brien, S.C., Zhang, Q., Liu, Y., Curl, R.F., Kroto, H.W., Tittel, F.K., Smalley, R.E. (1985). Lanthanum complexes of spheroidal carbon shells. *J. Am. Chem. Soc.*, 107, 7779–7780.

Hercules, D.M. (1964). Chemiluminescence resulting from electrochemically generated species. *Science*, 143, 308–309.

Hersam, M.C. (2008). Progress towards monodisperse single-walled carbon nanotubes. *Nature Nanotechn.*, 3, 387–394.

Hetemi, D., Noël, V., Pinson, J. (2020). Grafting of diazonium salts on surfaces: Application to biosensors. *Biosensors*, 10, 4.

Hjerten, S. (1983). High-performance electrophoresis: The electrophoretic counterpart of high-performance liquid chromatography. *Journal of Chromatography*, 270, 1–6.

Hjerten, S. and Zhu, M.-D. (1985). Adaptation of the equipment for high-performance electrophoresis to isoelectric focusing. *Journal of Chromatography*, 346, 265–270.

Horcajada, P., Chalati, T., Serre, C., Gillet, B., Sebrie, C., Baati, T., Eubank, J.F., Heurtaux, D., Clayette, P., Kreuz, C. et al. (2010). Porous metal-organic-framework nanoscale carriers as a potential platform for drug delivery and imaging. *Nat. Mat.*, 9, 172–178.

Hsiao, W.W-W., Hui, Y., Ytsai, P.-C., Chang, H.-C. (2016). Fluorescent nanodiamond: A versatile tool for long-term cell tracking, super-resolution imaging, and nanoscale temperature sensing. *Acc. Chem. Res.*, 49, 400–407.

Hu, C. (2010). *Modern Semiconductor Devices for Integrated Circuits*. Pearson/Prentice Hall, New Jersey.

Hu, S. and Dovichi, N.J. (2002). Capillary electrophoresis for the analysis of biopolymers. *Anal. Chem.*, 74, 2833–2850.

Hu, Z., Deibert, B.J., Li, J. (2014b). Luminescent metal-organic frameworks for chemical sensing and explosive detection. *Chem. Soc. Rev.*, 43, 5815–5840.

Hu, K., Kulkarni, D.D., Choi, I., Tsukruk, V.V. (2014a). Graphene-polymer nanocomposites for structural and functional applications. *Progr. Polymer Sci.*, 39, 1934–1972.

Hu, J., Stein, A., Bühlmann, P. (2016). Rational design of all-solid-state ion-selective electrodes and reference electrodes. *Trends in Anal. Chem.*, 76, 102–114.

Hu, C., Li, M., Qiu, J., Sun, Y.-P. (2019a). Design and fabrication of carbon dots for energy conversion and storage. *Chem. Soc. Rev.*, 48, 2315–2337.

Hu, W., Wan, L., Jian, Y., Ren, C., Jin, K., Su, X., Bai, X., Haick, H., Yao, M., Wu, W. (2019b). Electronic noses: From advanced materials to sensors aided with data processing. *Adv. Mater. Technol.*, 4, 1800488.

Hu, L.-F., Yin, S.-J., Zhang, H., Yang, F.-Q. (2020). Recent developments of monolithic and open-tubular capillary electrochromatography (2017–2019). *J. Sep. Sci.*, 43, 1942–1966.

Huang, J. and Wan, Q. (2009). Gas sensors based on semiconducting metal oxide one-dimensional nanostructures. *Sensors*, 9, 9903–9924.

Huang, T., Mi, J.-Q., Zhang, X.-X. (2006). Capillary electrochromatography of amino acids with a protein-bonded porous-layer open-tubular column. *J. Sep. Sci.*, 29, 277–281.

Huang, C.P., Li, Y.K., Chen, T.M. (2007). A highly sensitive system for urea detection by using CdSe/ZnS core-shell quantum dots. *Biosens. Bioelectron.*, 22, 1835–1838.

Huang, X., Neretina, S., El-Sayed, M.A. (2009). Gold nanorods: From synthesis and properties to biological and biomedical applications. *Adv. Mater.*, 21, 4880–4910.

Huang, X., Zeng, Z., Zhang, H. (2013). Metal dichalcogenide nanosheets: Preparation, properties and applications. *Chem. Soc. Rev.*, 42, 1934–1946.

Huang, Y., Ren, J., Qu, X. (2019). Nanozymes: Classification, catalytic mechanisms, activity regulation, and applications. *Chem. Rev.*, 19, 4357–4412.

Hummers, W.S. Jr. and Offeman, R.E. (1958). Preparation of graphite oxide. *J. Am. Chem. Soc.*, 80, 1339–1339.

Huo, X.-L., Lu, H.-J., Xu, J.-J., Zhou, H., Chen, H.-Y. (2019). Recent advances of ratiometric electrochemiluminescence biosensors. *J. Mater. Chem. B*, 7, 6469–6475.

Iijima, S. (1991). Helical microtubules of graphitic carbon. *Nature*, 354, 56–58.

Ishida, Y., Corpuz, R.D., Yonezawa, T. (2017). Matrix sputtering method: A novel physical approach for photoluminescent noble metal nanoclusters. *Acc. Chem. Res.*, 50, 2986–2995.

Issaq, H.J. and Barr, E.W. (1977). Instrumentation. Recent developments in thin-layer chromatography. *Anal. Chem.*, 49(1), 84A–96A.

Ji, J., Zhou, Z., Zhao, X., Sun, J., Sun, X. (2015). Electrochemical sensor based on molecularly imprinted film at Au nanoparticles-carbon nanotubes modified electrode for determination of cholesterol. *Biosens. Bioelectron.*, 66, 590–595.

Jimenez-Jorquera, C., Orozco, J., Baldi, A. (2010). ISFET based microsensors for environmental monitoring. *Sensors*, 10, 61–83.

Jin, R., Zeng, C., Zhou, M., Chen, Y. (2016). Atomically precise colloidal metal nanoclusters and nanoparticles: Fundamentals and opportunities. *Chem. Rev.*, 116, 10346–10413.

Johnson, R.D. and Bachas, L.G. (2003). Ionophore-based ion-selective potentiometric and optical sensors. *Anal. Bioanal. Chem.*, 376, 328–341.

Jorgenson, J.W. and Lukacs, K.D. (1981). Zone electrophoresis in open-tubular glass capillaries. *Anal. Chem.*, 53(8), 1298–1302.

Jovčevska, I. and Muyldermans, S. (2020). The therapeutic potential of nanobodies. *Bio Drugs.*, 34, 11–26.

Jozanovic, M., Pukles, I., Sakac, N., Carrilho, E., Kilar, A., Matasovic, B., Samardzic, M., Budetic, M., Kilar, F. (2023). Nanomaterials in microchip electrophoresis–A review. *Trends in Analyt. Chem.*, 165, 117111.

Jung, B., Bharadwaj, R., Santiago, J.G. (2006). On-chip millionfold sample stacking using transient isotachophoresis. *Anal. Chem.*, 78, 2319–2327.

Kang, S., Araya-Secchi, R., Wang, D., Wang, B., Huynh, T., Zhou, R. (2014). Dual inhibitory pathways of metallofullerenol Gd@$C_{82}(OH)_{22}$ on matrix metalloproteinase-2: Molecular insight into drug-like nanomedicine. *Sci. Rep.*, 4, 4775.

Karunakaran, C., Bhargava, K., Benjamin, R. (2015). Introduction to biosensors. In *Biosensors and Bioelectronics*, Karunakaran, C., Bhargava, K., Benjamin, R. (eds). Elsevier, Amsterdam, 1–68.

Kataura, H., Kumazawa, Y., Maniwa, Y., Umezu, I., Suzuki, S., Ohtsuka, Y., Achiba, Y. (1999). Optical properties of single-wall carbon nanotubes. *Synth. Met.*, 103, 2555–2558.

Kaur, M., Kaur, M., Sharma, V.K. (2018). Nitrogen-doped graphene and graphene quantum dots: A review on synthesis and applications in energy, sensors and environment. *Adv. in Colloids and Interface Sci.*, 259, 44–64.

Kaur, S., Gupta, N., Malhotra, B.D. (2023). Recent developments in wearable & non-wearable point-of-care biosensors for cortisol detection. *Expert Review of Molecular Diagnostics*, 23, 217–230.

Kausar, A.S.M.Z., Reza, A.W., Saleh, M.U., Ramiah, H. (2014). Energizing wireless sensor networks by energy harvesting systems: Scopes, challenges and approaches. *Renew. Sustain. Energy Rev.*, 38, 973–989.

Kergoat, L., Piro, B., Berggren, M., Horowitz, G., Pham, M.-C. (2012a). Advances in organic transistor-based biosensors: From organic electrochemical transistors to electrolyte-gated organic field-effect transistors. *Anal. Bioanal. Chem.*, 402, 1813–1826.

Kergoat, L., Piro, B., Berggren, M., Pham, M.-C., Yassar, A., Horowitz, G. (2012b). DNA detection with a water-gated organic field-effect transistor. *Org. Electronics*, 13, 1–6.

Kesler, V., Murmann, B., Soh, H.T. (2020). Going beyond the Debye length: Overcoming charge screening limitations in next-generation bioelectronic sensors. *ACS Nano*, 14, 16194–16201.

Khalafi, L. and Rafiee, M. (2017). Cyclic voltammetry. In *Encyclopedia of Physical Organic Chemistry*, 1st edition, Wang, Z. (ed.). John Wiley & Sons, Hoboken, 3437–3478.

Khan, R. and Andreescu, S. (2020). MXenes-based bioanalytical sensors: Design, characterization, and applications. *Sensors*, 20, 5434.

Khan, R.R.M. and Kang, S.-W. (2015). Highly sensitive multi-channel IDC sensor array for low concentration taste detection. *Sensors*, 15, 13201–13221. doi: 10.3390/s150613201.

Khodagholy, D., Rivnay, J., Sessolo, M., Gurfinkel, M., Leleux, P., Jimison, L.H., Stavrinidou, E., Herve, T., Sanaur, S., Owens, R.M. et al. (2013). High transconductance organic electrochemical transistors. *Nature Commun.*, 4, 2133.

Kim, K.S., Zhao, Y., Jang, H., Lee, S.Y., Kim, J.M., Kim, K.S., Ahn, J.-H., Kim, P., Choi, J.-Y., Hong, B.H. (2009). Large-scale pattern growth of graphene films for stretchable transparent electrodes. *Nature*, 457, 706–710.

Kim, D.H., Viventi, J., Amsden, J.J., Xiao, J., Vigeland, L., Kim, Y.-S., Blanco, J.A., Panilaitis, B., Frechette, E.S., Contreras, D. et al. (2010). Dissolvable films of silk fibroin for ultrathin conformal bio-integrated electronics. *Nat. Mater.*, 9, 511–517.

Kim, S.H., Choi, J.B., Nguyen, Q.N., Lee, J.M., Park, S., Chung, T.D., Byun, J.Y. (2013). Nanoporous platinum thin films synthesized by electrochemical dealloying for nonenzymatic glucose detection. *Phys. Chem. Chem. Phys.*, 15, 5782–5787.

Kim, K., Park, C., Kwon, D., Meyyappan, M., Jeon, S., Lee, J.-S. (2016). Silicon nanowire biosensors for detection of cardiac troponin I (cTnI) with high sensitivity. *Biosensors and Bioelectronics*, 77, 695–701.

Kim J., Campbell A.S., Esteban-Fernandez De Ávila B., Wang J. (2019). Wearable biosensors for healthcare monitoring. *Nature Biotechnol.*, 37, 389–406.

Kisiel, A., Marcisz, H., Michalska, A., Maksymiuk, K. (2005). All-solid-state reference electrodes based on conducting polymers. *Analyst*, 130, 1655–1662.

Knox, J.H. (1994). Terminology and nomenclature in capillary electroseparation systems. *J. of Chromatography A*, 680, 3–13.

Kobayashi, T., Bando, M., Kimura, N., Shimizu, K., Kadono, K., Umezu, N., Miyahara, K., Hayazaki, S., Nagai, S., Mizuguchi, Y. (2013). Production of a 100-m-long high-quality graphene transparent conductive film by roll-to-roll chemical vapor deposition and transfer process. *Appl. Phys. Lett.*, 102, 023112.

Kolmakov, A., Zhang, Y., Cheng, G., Moskovits, M. (2003). Detection of CO and O_2 using Tin oxide nanowire sensors. *Adv. Mat.*, 15, 997–1000.

Komatsu, N. and Wang, F. (2010). A comprehensive review on separation methods and techniques for single-walled carbon nanotubes. *Materials*, 3, 3818–3844.

Kondo, T., Chen, W.J., Schlau-Cohen, G.S. (2017). Single-molecule fluorescence spectroscopy of photosynthetic systems. *Chem. Rev.*, 117, 860–898.

Kortan, R., Hull, R., Opila, R., Bawendi, M., Steigerwald, M., Carroll, P., Brus, L.J. (1990). Nucleation and growth of CdSe on ZnS quantum crystallite seeds, and vice versa, in inverse micelle media. *J. Am. Chem. Soc.*, 112, 1327–1332.

Kortshagen, U.R., Sankaran, R.M., Pereira, R.N, Girshick, S.L., Wu, J.J., Aydil, E.S. (2016). Non thermal plasma synthesis of nanocrystals: Fundamental principles, materials, and applications. *Chem. Rev.*, 116, 11061–11127.

Kovalenko, M.V., Manna, L., Cabot, A., Hens, Z., Talapin, D.V., Kagan, C.R., Klimov, V.I., Rogach, A.L., Reiss, P., Milliron, D.J. et al. (2015). Prospects of nanoscience with nanocrystals. *ACS Nano*, 9, 1012–1057.

Koziol, K.K., Janas, D., Brown, E., Hao, L. (2017). Thermal properties of continuously spun carbon nanotube fibres. *Physica E: Low-dimensional Syst. and Nanostructures*, 88, 104–108.

Kraut, J. (1988). How do enzymes work? *Science*, 242, 533–540.

Krishnan, U., Kaur, M., Singh, K., Kumar, M., Kumar, A. (2019). A synoptic review of MoS_2: Synthesis to applications. *Superlattices and Microstructures*, 128, 274–297.

Kristoff, C.J., Bwanali, L., Veltri, L.M., Gautam, G.P., Rutto, P.K., Newton, E.O., Holland, L.A. (2020). Challenging bioanalyses with capillary electrophoresis. *Anal. Chem.*, 92, 49–66.

Kroto, H.W., Heath, J.R., O'Brien, S.C., Curl, R.F., Smalley, R.E. (1985). C60: Buckminsterfullerene. *Nature*, 318, 162–163.

Krylov, S.N. and Dovichi, N.J. (2000). Capillary electrophoresis for the analysis of biopolymers. *Anal. Chem.*, 72, 111R–128R.

Kuban, P. and Kuban, P. (2023). Novel developments in capillary electrophoresis miniaturization, sampling, detection and portability: An overview of the last decade. *Trends Anal. Chem.*, 159, 116941.

Kudelski, A. (2008). Analytical applications of Raman spectroscopy. *Talanta*, 76, 1–8.

Kumar, S., Gandhi, K.S., Kumar, R. (2006). Modeling of formation of gold nanoparticles by citrate method. *Industrial & Engineering Chemistry Research*, *46*, 3128–3136.

Kumar, S., Bukkitgar, S.D., Singh, S., Pratibha, Singh, V., Reddy, K.R., Shetti, N.P., Reddy, C.V., Shuad, V., Naveen, S. (2019). Electrochemical sensors and biosensors based on graphene functionalized with metal oxide nanostructures for healthcare applications. *Chem. Select*, 4, 5322–5337.

Kumar N., Wang, W., Ortiz-Marquez, J.C., Catalano, M., Gray, M., Biglari, N., Hikari, K., Ling, X., Gao, J., van Opijnen, T. et al. (2020). Dielectrophoresis assisted rapid, selective and single cell detection of antibiotic resistant bacteria with G-FETs. *Biosensors and Bioelectronics*, 156, 112123.

Laborda, E., Gonzalez, J., Molina, Á. (2014). Recent advances on the theory of pulse techniques: A mini review. *Electrochemistry Commun.*, 43, 25–30.

Lacaze, P.C. and Lacroix J.-C. (2021). *Nanotechnology and Nanomaterials for Energy*. ISTE Ltd, London, and John Wiley & Sons, New York.

Lai, C.-Z., Fierke, M.A., Stein, A., Bühlmann, P. (2007). Ion-selective electrodes with three-dimensionally ordered macroporous carbon as the solid contact. *Anal. Chem.*, 79, 4621–4626.

Lakshmi, K.R., Nelakurthi, H., Kumar, A.S., Rudraraju, A. (2017). Oral fluid-based biosensors: A novel method for rapid and noninvasive diagnosis. *Indian J. Dent. Sci.*, 9, 60–66.

Lalwani, G. and Sitharaman, B. (2013). Multifunctional fullerene and metallo fullerene-based nanobiomaterials. *Nano Life*, 3, 1342003.

Lan, A., Li, K., Wu, H., Olson, D.H., Emge, T.J., Ki, W., Hong, M., Li, J. (2009). A luminescent microporous metal–organic framework for the fast and reversible detection of high explosives. *Angew. Chem. Int. Ed.*, 48, 2334–2338.

Lan, W.-J., Zou, X.U., Hamedi, M.M., Hu, J., Parolo, C., Maxwell, E.J., Bühlmann, P., Whitesides, G.M. (2014). Paper-based potentiometric ion sensing. *Anal. Chem.*, 86(19), 9548–9553.

Länge, K. (2019). Bulk and surface acoustic wave sensor arrays for multi-analyte detection: A review. *Sensors*, 19, 5382.

Late, D.J., Huang, Y.-K., Liu, B., Acharya, J., Shirodkar, S.N., Luo, J., Yan, A., Charles, D., Waghmare, U.V., Dravid, V.P., Rao, C.N.R. (2013). Sensing behavior of atomically thin-layered MoS_2 transistors. *ACS Nano*, 7(6), 4879–4891.

Latif, U., Can, S., Hayden, O., Grillberger, P., Dicker, F.L. (2013). Sauerbrey and anti-Sauerbrey behavioral studies in QCM sensors–Detection of bioanalytes. *Sensors and Actuators B*, 176, 825–830.

Le, T.-H., Kim, Y., Yo, H. (2017). Electrical and electrochemical properties of conducting polymers. *Polymers*, 9, 150.

Lee, H., Song, C., Hong, Y.S., Kim, M., Cho, H.R., Kang, T., Shin, K., Choi, S.H., Hyeon, T., Kim, D.H. (2017). Wearable/disposable sweat-based glucose monitoring device with multistage transdermal drug delivery module. *Science Advances*, 3, e1601314.

Lee, Y., Howe, C., Mishra, S., Lee, D.S., Mahmood, M., Piper, M., Kim, Y., Tieu, K., Byun, H.-S., Coffey, J.P. et al. (2018). Wireless, intraoral hybrid electronics for real-time quantification of sodium intake toward hypertension management. *Proc. Nation. Acad. Sci.*, 115, 5377–5382.

Leonardi, A.A., Lo Faro, M.J., Irrera, A. (2021). Silicon nanowires synthesis by metal-assisted chemical etching: A review. *Nanomaterials*, 11, 383.

Lewis, J.G. and Elder, P.A. (1985). An enzyme-linked immunosorbent assay (ELISA) for plasma cortisol. *J. Steroid Biochem.*, 22, 673–676.

Li, H. and Steckl, A.J. (2019). Paper microfluidics for point-of-care blood-based analysis and diagnostics. *Anal. Chem.*, 91, 352–371.

Li, X.Z. and Suslick, K.S. (2021). The optoelectronic nose. *Acc. Chem. Res.*, 54, 950–960.

Li, H., Eddaoudi, M., O'Keeffe, M., Yaghi, O.M. (1999). Design and synthesis of an exceptionally stable and highly porous metal-organic framework. *Nature*, 402, 276–279.

Li, W., Moore, M.J., Vasiliev, A.N., Sui, J., Wong, S.K., Berne, M.A., Somasundaran, M., Sullivan, J.L., Luzuriaga, K., Greenough, T.C. et al. (2003). Angiotensin-converting enzyme 2 is a functional receptor for the SARS coronavirus. *Nature*, 426, 450–454.

Li, X., Cai, W., An, J., Kim, S., Nah, J., Yang, D., Piner, R., Velamakanni, A., Jung, I., Tutuc, E. et al. (2009). Large-area synthesis of high-quality and uniform graphene films on copper foils. *Science*, 324, 1312–1314.

Li, H., Yin, Z., He, Q., Li, H., Huang, X., Lu, G., Fam, D.W.H., Tok, A.I.Y., Zhang, Q. (2012). Fabrication of single- and multilayer MoS_2 film-based field-effect transistors for sensing NO at room temperature. *Small*, 8(1), 63–67.

Li, R., Liu, Y., Li, Z., Shen, J., Yang, Y., Cui, X., Yang, G. (2016a). Bottom-up fabrication of single-layered nitrogen-doped graphene quantum dots through intermolecular carbonization arrayed in a 2d plane. *Chem. Eur. J.*, 22, 272–278.

Li, L., Long, Y., Gao, J.M., Song, K., Yang, G. (2016b). Label-free and pH-sensitive colorimetric materials for the sensing of urea. *Nanoscale*, 8, 4458–4462.

Li, L., Chen, Y., Zhu, J.-J. (2017). Recent advances in electrochemiluminescence analysis. *Anal. Chem.*, 89, 358–371.

Li, S., Liu, A, Yang, Z., He, J., Wang, J., Liu, F., Lu, H., Yan, X., Sun, P., Liang, X. et al. (2019a). Room temperature gas sensor based on tin dioxide@ polyaniline nanocomposite assembled on flexible substrate: ppb-level detection of NH_3. *Sensors & Actuators: B*, 299, 126970.

Li, Y., Xie, M., Zhang, X., Liu, Q., Lin, D., Xu, C., Xie, F., Sun, X. (2019b). Co-MOF nanosheet array: A high-performance electrochemical sensor for non-enzymatic glucose detection. *Sens. Actuat. B: Chem.*, 278, 126–132.

Li, X., Ma, Q., Zheng, X., Chen, Q., Sun, X. (2023). Recent applications and chiral separation development based on stationary phases in open tubular capillary electrochromatography (2019-2022). *Journal of Pharmaceutical Analysis*, 13, 323-339.

Liang, M. and Yan, X. (2019). Nanozymes: From new concepts, mechanisms, and standards to applications. *Acc. Chem. Res.*, 52, 2190–2200.

Liang, Y., Ernst, M., Brings, F., Kireev, D., Maybeck, V., Offenhäusser, A., Mayer, D. (2018). High performance flexible organic electrochemical transistors for monitoring cardiac action potential. *Adv. Healthcare Mater.*,7, 1800304.

Liang, T., Qiu, Y., Gan, Y., Sun, J., Zhou, S., Wan, H., Wang, P. (2019). Recent developments of high-resolution chemical imaging systems based on light-addressable potentiometric sensors (LAPSs). *Sensors*, 19, 4294.

Liang X., Lin Z., Li L., Tang D., Kong J. (2022). Ratiometric fluorescence enzyme-linked immunosorbent assay based on carbon dots@SiO_2@CdTe quantum dots with dual functionalities for alpha-fetoprotein. *Analyst*, 147, 2851–2858.

Liao, J., Si, H., Zhang, X., Lin, S. (2019). Functional sensing interfaces of PEDOT:PSS organic electrochemical transistors for chemical and biological sensors: A mini review. *Sensors*, 19, 218.

Ligtenberg, J.J., Borggreve, H.F., Stemerdink, A., Zijlstra, J.G. (2010). Subcutaneous glucose measurements and glucose regulation. *Intensive Care Med.*, 36, 897–898.

Lim, S., Ciuparu, D., Pak, C., Dobek, F., Chen, Y., Harding, D., Pfefferle, L., Haller, G. (2003). Synthesis and characterization of highly ordered Co–MCM-41 for production of aligned single walled carbon nanotubes (SWNT). *J. Phys. Chem. B*, 107, 11048–11056.

Lim, S.Y., Shen, W., Gao, Z. (2015). Carbon quantum dots and their applications. *Chem. Soc. Rev.*, 44, 362–381.

Lin, S., Feng, W., Miao, X., Zhang, X., Chen, S., Chen, Y., Wang, W., Zhang, Y. (2018). A flexible and highly sensitive nonenzymatic glucose sensor based on DVD-laser scribed graphene substrate. *Biosensors and Bioelectronics*, 10, 89–96.

Lin, Y.-T., Darvishi, S., Preet, A., Huang, T.-Y., Lin, S.-H., Girault, H.H, Wang, L., Lin, T.-E. (2020). A review: Electrochemical biosensors for oral cancer. *Chemosensors*, 8, 54–68.

Linares, M., Perez-Tanoira, R., Carrero, A., Romanyk, J., Perez-Garcia, F., Gomez-Herruz, P., Arroyo, T., Cuadros, J. (2020). Panbio antigen rapid test is reliable to diagnose SARS-CoV-2 infection in the first 7 days after the onset of symptoms. *Journal of Clinical Virology*, 133, 104659.

Lindner, E. and Pendley, B.D. (2013). A tutorial on the application of ion-selective electrode potentiometry: An analytical method with unique qualities, unexplored opportunities and potential pitfalls; tutorial. *Anal. Chim. Acta*, 762, 1–13.

Liu, X. and Liu, J. (2021). Biosensors and sensors for dopamine detection. *VIEW*, 2, 20200102.

Liu, K. and Zhang, C. (2021). Volatile organic compounds gas sensor based on quartz crystal microbalance for fruit freshness detection: A review. *Food Chem.*, 334, 27615.

Liu, X., Sozic, Z., Krull, I.S. (1996). Capillary isoelectric focusing as a tool in the examination of antibodies, peptides and proteins of pharmaceutical interest. *Journal of Chromatography A*, 735, 165–190.

Liu, S., Shi, Y., Ja, W.W., Mathies, R.A. (1999). Optimization of high-speed DNA sequencing on microfabricated capillary electrophoresis channels. *Anal. Chem.*, 71, 566–573.

Liu X., Jiang H., Lei J., Ju H. (2007). Anodic electrochemiluminescence of CdTe quantum dots and its energy transfer for detection of catechol derivatives. *Anal. Chem.*, 79, 8055–8060.

Liu, Y., Bryantsev, V.S., Diallo, M.S., Goddard, W.A. (2009). PAMAM dendrimers undergo pH responsive conformational changes without swelling. *J. Am. Chem. Soc.*, 131, 2798–2799.

Liu, B., Chen, L., Liu, G., Abbas, A.N., Fathi, M., Zhou, C. (2014a). High-performance chemical sensing using Schottky-contacted chemical vapor deposition grown monolayer MoS_2 transistors. *ACS Nano*, 8(5), 5304–5314.

Liu, H., Neal, A.T., Zhu, Z., Luo, Z., Xu, X., Tomanek, D., Ye, P.D. (2014b). Phosphorene: An unexplored 2D semiconductor with a high hole mobility. *ACS Nano*, 8(4), 4033–4041.

Liu, Z., Qi, W., Xu, G. (2015). Recent advances in electrochemiluminescence. *Chem. Soc. Rev.*, 44, 3117–3142.

Liu, Y., Xu, Z., Zhan, J., Li, P., Gao, C. (2016). Superb electrically conductive graphene fibers via doping strategy. *Adv. Mater.*, 28, 7941–7947.

Liu, H., Ren, W., Zong, L., Zhang, J., Wang, Y. (2019a). Characterization of recombinant monoclonal antibody charge variants using WCX chromatography, icIEF and LC-MS/MS. *Anal. Biochemistry*, 564–565, 1–12.

Liu J., Chen X., Wang Q., Xiao M., Zhong D., Sun W., Zhang G., Zhang Z. (2019b). Ultrasensitive monolayer MoS_2 field-effect transistor based DNA sensors for screening of Down syndrome. *Nano Lett.*, 19, 1437−1444.

Liu Y., Zheng H., Zhao L., Liu S., Yao K., Li D., Yiu C., Gao S., Avila R., Pakpong C. et al. (2020a). Electronic skin from high-throughput fabrication of intrinsically stretchable lead zirconate titanate elastomer. *Research*, 1085417.

Liu, J., Siavash Moakhar, R., Sudalaiyadum Perumal, A., Nicolae Roman, H., Mahshid, S., Wachsmann-Hogiu, S. (2020b). An AgNP-deposited commercial electrochemistry test strip as a platform for urea detection. *Sci. Rep.*, 10, 9527.

Liu, Q., Zhang, A., Wang, R., Zhang, Q., Cui, D. (2021). A review on metal- and metal oxide-based nanozymes: Properties, mechanisms, and applications. *Nano-Micro Lett.*, 13, 154.

Liu, F., Du, F., Yuan, F., Quan, S., Guan, Y., Xu, G. (2022a). Electrochemiluminescence bioassays based on carbon nitride nanomaterials and 2D transition metal carbides. *Current Opinion in Electrochemistry*, 34, 100981.

Liu, Y., Zhu, P., Liu, S., Chen, Y., Liang, D., Wang, M., Du, L., Wu, C. (2022b). The light-addressable potentiometric sensor and its application in biomedicine towards chemical and biological sensing. *Chemosensors*, 10, 156.

Liu, H.L., Tseng, Y.T., Lai, M.C., Chau, L.K. (2022c). Ultrasensitive and rapid detection of N-terminal pro-B-type natriuretic peptide (NT-proBNP) using fiber optic nanogold-linked immunosorbent assay. *Biosensors*, 12, 746.

Longsworth, L.G. (1945). Moving boundary studies on salt mixtures. *J. Am. Chem. Soc.*, 67, 1109–1119.

Lu, G., Crihfield, C.L., Gattu, S., Veltri, L.M., Holland, L.A. (2018). Capillary electrophoresis separations of glycans. *Chem. Rev.*, 118, 7867−7885.

Luo, Z., Yuan, X., Yu, Y., Zhang, Q., Leong, D.T., Lee, J.Y., Xie, J. (2012). From aggregation-induced emission of Au(I)–thiolate complexes to ultrabright Au(0)@Au(I)–thiolate core–shell nanoclusters. *J. Am. Chem. Soc.*, 134, 16662–16670.

Luo, Z., Qi, G., Chen, K., Zou, M., Yuwen, L., Zhang, X., Huang, W., Wang, L. (2016). Microwave-assisted preparation of white fluorescent graphene quantum dots as a novel phosphor for enhanced white-light-emitting diodes. *Adv. Funct. Mat.*, 26, 2739–2744.

Luong, D.X., Bets, K.V., Algozeeb, W.A., Stanford, M.G., Kittrell, C., Chen, W., Salvatierra, R.V., Ren, M., McHugh, E.A., Advincula, P.A. et al. (2020). Gram-scale bottom-up flash graphene synthesis. *Nature*, 577, 647–653.

Lv, R., Robinson, J.A., Schaak, R.E., Sun, D., Sun, Y., Mallouk, T.E., Terrones, M. (2015). Transition metal dichalcogenides and beyond: Synthesis, properties, and applications of single- and few-layer nanosheets. *Acc. Chem. Res.*, 48, 56–64.

MacDonald, D.D. (2006). Reflections on the history of electrochemical impedance spectroscopy. *Electrochimica Acta*, 51, 1376–1388.

Mack, S., Cruzado-Park, I., Chapman, J., Ratnayake, C., Vigh, G. (2009). A systematic study in CIEF: Defining and optimizing experimental parameters critical to method reproducibility and robustness. *Electrophoresis*, 30, 4049–4058.

Mack, S., Arnold, D., Bogdan, G., Bousse, L., Danan, L., Dolnik, V., Ducusin, M.A., Gwerder, E., Herring, C., Jensen, M. et al. (2019). A novel microchip-based imaged CIEF-MS system for comprehensive characterization and identification of biopharmaceutical charge variants. *Electrophoresis*, 40, 3084–3091.

Maduraiveeran, G. (2022). Nanomaterials-based portable electrochemical sensing and biosensing systems for clinical and biomedical applications. *J. Anal. Sci. & Technology*, 13, 35.

Magar, H.S., Hassan, R.Y.A., Mulchandani, A. (2021). Electrochemical impedance spectroscopy (EIS): Principles, construction, and biosensing applications. *Sensors*, 21, 6578.

Mahler, B., Spinicelli, P., Buil, S., Quelin, X., Hermier, J.-P., Dubertret, B. (2008). Towards non-blinking colloidal quantum dots. *Nat. Mat.*, 7, 661–664.

Malucelli, G. (2017). Graphene-based polymer nanocomposites: Recent advances and still open challenges. *Curr. Graphene Sci.*, 1, 16–25.

Manabe, T., Iwasaki, A., Miyamoto, H. (1997). Separation of human plasma/serum proteins by capillary isoelectric focusing in the absence of denaturing agents. *Electrophoresis*, 18, 1159–1165.

Mandal, D. and Banerjee, S. (2022). Surface acoustic wave (SAW) sensors: Physics, materials and applications. *Sensors*, 22, 820.

Mannoor, M.S., Tao, H., Clayton, J.D., Sengupta, A., Kaplan, D.L., Naik, R.R., Verma, N., Omenetto, F.G., Mcalpine, M.C. (2012). Graphene-based wireless bacteria detection on tooth enamel. *Nature Communications*, 3, 763.

Mansouri Majd, S., Mirzapour, F., Shamsipur, M., Manouchehri, I., Babaee, E., Pashabadi, A., Moradian, R. (2023). Design of a novel aptamer/molecularly imprinted polymer hybrid modified Ag–Au@Insulin nanoclusters/Au-gate-based MoS_2 nanosheet field-effect transistor for attomolar detection of BRCA1 gene. *Talanta*, 257, 124394.

Mao, Z. and Chen, Z. (2019). Advances in capillary electro-chromatography. A review. *Journal of Pharmaceutical Analysis*, 9, 227–237.

March, C., García, J.V., Sánchez, Á., Arnau, A., Jiménez, Y., García, P., Manclús, J.J., Montoya, Á. (2015). High-frequency phase shift measurement greatly enhances the sensitivity of QCM immunosensors. *Biosensors and Bioelectronics*, 65, 1–8.

Marczewska, B. and Marczewski, K. (2010). Glass electrode and its creators F. Haber and Z. Klemensiewicz – On 100th anniversary. *Z. Phys. Chem.*, 224, 795–799.

Martinez, A.W., Phillips, S.T., Butte, M.J., Whitesides, G.M. (2007). Patterned paper as a platform for inexpensive, low-volume, portable bioassays. *Angew. Chem. Int. Ed.*, 46, 1318–1320.

Martinez, A.W., Phillips, S.T., Carrilho, E., Thomas, S.W., Sindi, H., Whitesides, G.M. (2008). Simple telemedicine for developing regions: Camera phones and paper-based microfluidic devices for real-time, off-site diagnosis. *Anal. Chem.*, 80, 3699–3707.

Maultzsch, J., Telg, H., Reich, S., Thomsen, C. (2005). Radial breathing mode of single-walled carbon nanotubes: Optical transition energies and chiral-index assignment. *Phys. Rev. B*, 72, 205438.

Maxam, A.M. and Gilbert, W. (1977). A new method for sequencing DNA. *Proc. Natl. Acad. Sci. USA*, 74, 560–564.

McCord, P. and Bard, A.J. (1991). Electrogenerated chemiluminescence: Part 54. Electrogenerated chemiluminescence of ruthenium (II) 4,4'-diphenyl-2,2'-bipyridine and ruthenium (II) 4,7-diphenyl-l,l0-phenanthroline systems in aqueous and acetonitrile solutions. *J. Electroanal. Chem.*, 318, 91–99.

Meyyappan, M. (2016). Carbon nanotube-based chemical sensors. *Small*, 12, 2118–2129.

Miao, W. (2008). Electrogenerated chemiluminescence and its biorelated applications. *Chem. Rev.*, 108, 2506–2553.

Miao, P., Zhang, L., Zhang, J., Ma, M., Du, Y., Gan, J., Yang, J. (2022). Metal organic framework-modified monolithic column immobilized with pepsin for enantioseparation in capillary electrochromatography. *Journal of Chromatography B*, 1203, 123306.

Michaelis, L. and Menten, M.L. (1913). Die kinetik der invertinwirkung. *Biochem. Z.*, 49, 333.

Michalska, A. (2012). All-solid-state ion selective and all-solid-state reference electrodes. *Electroanalysis*, 24, 1253–1265.

Middya, S., Bhattacharjee, M., Bandyopadhyay, D. (2019). Reusable nano-BG-FET for point-of-care estimation of ammonia and urea in human urine. *Nanotechnology*, 30, 145502.

Milone, A., Monteduro, A.G., Rizzato, S., Leo, A., Maruccio, G. (2021). Gas sensing technologies – status, trends, perspectives and novel applications. arXiv:2110.13637.

Milone, A., Monteduro, A.G., Rizzato, S., Leo, A., Di Natale, C., Kim, S.S., Maruccio, G. (2023). Advances in materials and technologies for gas sensing from environmental and food monitoring to breath analysis. *Adv. Sustainable Syst.*, 7, 2200083.

Molaei, M.J. (2020). Principles, mechanisms, and application of carbon quantum dots in sensors: a review. *Anal. Methods*, 12, 1266–1287.

Morales, M.A. and Halpern, J.M. (2018). Guide to selecting a biorecognition element for biosensors. *Bioconjug. Chem.*, 29, 3231–3239.

Moser, N., Lande, T.S., Toumazou, C., Georgiou, P. (2016). ISFETs in CMOS and emergent trends in instrumentation: A review. *IEEE Sensors J.*, 16(17), 6496–6514.

Mousavi, Z., Granholm, K., Sokalski, T., Lewenstam, A. (2013). An analytical quality solid-state composite reference electrode. *Analyst*, 138, 5216–5220.

Mujahid, A. and Dickert, F.L. (2017). Surface acoustic wave (SAW) for chemical sensing applications of recognition layers. *Sensors*, 17, 2716.

Murauer, A., Bakry, R., Partl, G., Huck, C.W., Ganzera, M. (2019). Optimization of an innovative vinylimidazole-based monolithic stationary phase and its use for pressured capillary electrochromatography. *J. of Pharmaceutical and Biomedical Analysis*, 162, 117–123.

Murray, C.B., Norris, D.J., Bawendi, M.G. (1993). Synthesis and characterization of nearly monodisperse CdE (E = S, Se, Te) semiconductor nanocrystallites. *J. Am. Chem. Soc.*, 115, 8706–8715.

Nägele, M., Bakker, E., Pretsch, E. (1999). General description of the simultaneous response of potentiometric ionophore-based sensors to ions of different charge. *Anal. Chem.*, 71, 1041–1048.

Naguib, M., Kurtoglu, M., Presser, V., Lu, J., Niu, J., Heon, M., Hultman, L., Gogotsi, Y., Barsoum, M.W. (2011). Two-dimensional nanocrystals produced by exfoliation of Ti_3AlC_2. *Adv. Mat.*, 23(37), 4248–4253.

Naguib, M., Mashtalir, O., Carle, J., Presser, V., Lu, J., Hultman, L., Gogotsi, Y., Barsoum, M.W. (2012). Two-dimensional transition metal carbides. *ACS Nano*, 6(2), 1322–1331.

Naguib, M., Mochalin, V.N., Barsoum, M.W., Gogotsi, Y. (2014). 25th anniversary article: MXenes: A new family of two-dimensional materials. *Adv. Mater.*, 26(7), 992–1005.

Nakhleh, M.K., Amal, H., Jeries, R., Broza, Y.Y., Aboud, M., Gharra, A., Ivgi, H., Khatib, S., Badarneh, S., Har-Shai, L., et al. (2017). Diagnosis and classification of 17 diseases from 1404 subjects via pattern analysis of exhaled molecules. *ACS Nano*, 11, 112–125.

Nallon, E.C., Schnee, V.P., Bright, C.J., Polcha, M.P., Li, Q. (2016). Chemical discrimination with an unmodified graphene chemical sensor. *ACS Sens.*, 1, 26–31.

Narsaiah, K., Jha, S.N., Bhardwaj, R., Sharma, R., Kumar, R. (2012). Optical biosensors for food quality and safety assurance–A review. *Journal of Food Sci & Technology*, 49, 383–406.

Narwal, V., Deswal, R., Batra, B., Kalra, V., Hooda, R., Sharma, M., Rana, J.S. (2019). Cholesterol biosensors: A review. *Steroids*, 143, 6–17.

Nazri, N.A.A., Azeman, N.H., Luo, Y., Ashrif, A., Bakar, A. (2021). Carbon quantum dots for optical sensor applications: A review. *Optics & Laser Technology*, 139, 106928.

Nguyen, V.Q., Ai, Y., Martin, P., Lacroix, J.C. (2017). Plasmon-induced nanolocalized reduction of diazonium salts. *ACS Omega*, 2, 1947–1955.

Nguyen, D.K., Hoang, D.-Q., Hoat, D.M. (2022). Exploring a silicene monolayer as a promising sensor platform to detect and capture NO and CO gas. *RSC Adv.*, 12, 9828–9835.

Nilghaz, A., Guan, L, Tan, W., Shen, W. (2016). Advances of paper-based microfluidics for diagnostics – The original motivation and current status. *ACS Sens.*, 1, 1382–1393.

Nirmal, M., Dabbousi, B., Bawendi, M., Macklin, J., Trautman, J., Harris, T., Brus, L. (1996). Fluorescence intermittency in single cadmium selenide nanocrystals. *Nature*, 383, 802–804.

Niu, X., Lan, M., Zhao, H., Chen. C. (2013). Highly sensitive and selective nonenzymatic detection of glucose using three-dimensional porous nickel nanostructures. *Anal. Chem.*, 85, 3561–3569.

Niu, X., Li, Y., Tang, J., Hu, Y., Zhao, H., Lan, M. (2014). Electrochemical sensing interfaces with tunable porosity for nonenzymatic glucose detection: A Cu foam case. *Biosens. Bioelectron.*, 51, 22–28.

Novell, M., Guinovart, T., Blondeau, P., Rius, X.F., Andrade, F.J. (2014). A paper-based potentiometric cell for decentralized monitoring of Li levels in whole blood. *Lab Chip*, 14, 1308–1314.

Novoselov, K.S., Geim, A.K., Morozov, S.V., Jiang, D., Zhang, Y., Dubonos, S.V., Grigorieva, I.V., Firsov, A.A. (2004). Electric field effect in atomically thin carbon films. *Science*, 306, 666–669.

Novoselov, K.S., Jiang, D., Schedin, F., Booth, T.J., Khotkevich, V.V., Morozov, S.V., Geim, A.K. (2005a). Two-dimensional atomic crystals. *Proc. Nat. Acad. Sci. USA*, 10, 10451–10453.

Novoselov, K.S., Geim, A. K., Morozov, S.V., Jiang, D., Katsnelson, M.I., Grigorieva, I.V., Dubonos, S.V., Firsov, A.A. (2005b). Two-dimensional gas of massless Dirac fermions in graphene. *Nature*, 438, 197–200.

Numnuam, A., Chumbimuni-Torres, K.Y., Xiang, Y., Bash, R., Thavarungkul, P., Kanatharana, P., Pretsch, E., Wang, J., Bakker, E. (2008). Aptamer-based potentiometric measurements of proteins using ion-selective microelectrodes. *Anal. Chem.*, 80, 707–712.

Nuzaihan, M.N., Hashim, U.M., Arshad, M.K., Kasjoo, S.R., Rahman, S.F.A., Ruslinda, A.R., Fathil, M.F.M., Adzhri, R., Shahimin, M.M. (2016). Electrical detection of dengue virus (DENV) DNA oligomer using silicon nanowire biosensor with novel molecular gate control. *Biosens. Bioelectron.*, 83, 106–114.

Odom, T.W., Huang, J.-L., Kim, P., Lieber, C.M. (1998). Atomic structure and electronic properties of single-walled carbon nanotubes. *Nat.*, 391, 62–64.

O'Farrell, P.H. (1975). High resolution two-dimensional electrophoresis of proteins. *J. Biological Chem.*, 250, 4007–4021.

Ohnsorg, M.L., Beaudoin, C.K., Anderson, M.E. (2015). Fundamentals of MOF thin film growth via liquid-phase epitaxy: Investigating the initiation of deposition and the influence of temperature. *Langmuir*, 31, 6114–6121.

Olczuk, D. and Priefer, R. (2018). A history of continuous glucose monitors (CGMs) in self-monitoring of diabetes mellitus. *Diabetes & Metabolic Syndrome: Clin. Res. & Rev.*, 12, 181–187.

Osbourn, D.M., Weiss, D.J., Lunte, C.E. (2000). On-line preconcentration methods for capillary electrophoresis. *Electrophoresis*, 21, 2768–2779.

O'Sullivan, C.K. and Guilbault, G.G. (1999). Commercial quartz crystal microbalances – Theory and applications. *Biosens. Bioelectron.*, 14, 663–670.

Owen, J. and Brus, L. (2017). Chemical synthesis and luminescence applications of colloidal semiconductor quantum dots. *J. Am. Chem. Soc.*, 139, 10939–10943.

Ozkan, S.A., Bakirhan, N.K., Mollarasouli, F. (eds) (2021). *The Detection of Biomarkers: Past, Present and the Future Prospects*. Elsevier, The Netherlands.

Palermo, G., Rippa, M., Conti, Y., Vestri, A., Castagna, R., Fusco, G., Suffredini, E., Zhou, J., Zyss, J., De Luca, A., Petti, L. (2021). Plasmonic metasurfaces based on pyramidal nanoholes for high-efficiency SERS biosensing. *ACS Appl. Mater. Interfaces*, 13, 43715–43725.

Palmer, J., Munro, N.J., Landers, J.P. (1999). A Universal concept for stacking neutral analytes in micellar capillary electrophoresis. *Anal. Chem.*, 71, 1679–1687.

Parolo, C., Sena-Torralba, A., Bergua, J.F., Calucho, E., Fuentes-Chust, C., Hu, L., Rivas, L., Álvarez-Diduk, R., Nguyen, E.P., Cinti, S. et al. (2020). Tutorial: Design and fabrication of nanoparticle based lateral-flow immunoassays. *Nature Protocols*, 15, 3788–3816.

Pasternack, R.F. and Collings, P.J. (1995). Resonance light scattering: A new technique for studying chromophore aggregation. *Science*, 269, 935–939.

Patel, V.D., Shamsi, S.A., Sutherland, K. (2021). Capillary electromigration techniques coupled to mass spectrometry: Applications to food analysis. *Trends in Analyt. Chem.*, 139, 116240.

Pena-Bahamonde, J., Nguyen, H.N., Fanourakis, S.K., Ro, D.F. (2018). Recent advances in graphene-based biosensor technology with applications in life sciences. *J Nanobiotechnol.*, 16, 75.

Peng, Y., Raj, N., Strasser, J.W., Crooks, R.M. (2022). Paper biosensor for the detection of NT-proBNP using silver nanodisks as electrochemical labels. *Nanomaterials*, 30, 2254.

Petitjean, J., Tanguy, J., Lacroix, J-.C., Chane-Ching, K.I., Aeiyach, S., Delamar, M., Lacaze, P.C. (2005). Interpretation of the ultra-fast electropolymerization of pyrrole in aqueous media on zinc in a one-step process: The specific role of the salicylate salt investigated by x-ray photoelectron spectroscopy (XPS) and by electrochemical quartz crystal microbalance (EQCM). *J. Electroanal. Chem.*, 581, 111–121.

Peto, T. (2021). COVID-19: Rapid antigen detection for SARS-CoV-2 by lateral flow assay: A national systematic evaluation of sensitivity and specificity for mass-testing. *EClinicalMedicine*, 36, 100924.

Pham, K., Temerov, F., Saarinen, J.J. (2020). Multicompound inverse opal structures with gold nanoparticles for visible light photocatalytic activity. *Mater. Des.*, 194, 108886.

Piccinini, E., Alberti, S., Longo, G.S., Berninger, T., Breu, J., Dostalek, J., Azzaroni, O., Knoll, W. (2018). Pushing the boundaries of interfacial sensitivity in Graphene FET sensors: Polyelectrolyte multilayers strongly increase the Debye screening length. *J. Phys. Chem. C*, 122, 10181–10188.

Poghossian, A., Yoshinobu, T., Simonis, A., Ecken, H., Lüth, H., Schöning, M.J. (2001). Penicillin detection by means of field effect based-sensors: EnFET, capacitive EIS sensors or LAPs? *Sensors and Actuators*, 78, 237–242.

Poller, A.M., Spieker, E., Lieberzeit, P.A., Preininger, C. (2017). Surface imprints: Advantageous application of ready-to-use materials for bacterial quartz-crystal microbalance sensors. *ACS Appl. Mater. Interfaces*, 9, 1129–1135.

Pollok, N.E., Rabin, C., Walgama, C.T., Smith, L., Richards, I., Crooks, R.M. (2020). Electrochemical detection of NT-proBNP using a metalloimmunoassay on a paper electrode platform. *ACS Sens.*, 27, 853–860.

Ponomarenko, L.A., Schedin, F., Katsnelson, M.I., Yang, R., Hill, E.H., Novoselov, K.S., Geim, A.K. (2008). Chaotic Dirac billiard in graphene quantum dots. *Science*, 320(5874), 356–358.

Pop, E., Mann, D., Wang, Q., Goodson, K., Dai, H. (2006). Thermal conductance of an individual single-wall carbon nanotube above room temperature. *Nano Lett.*, 6, 96–100.

Portehault, D., Delacroix, S., Gouget, G., Grosjean, R., Chan-Chang, T.-H.-C. (2018). Beyond the compositional threshold of nanoparticle-based materials. *Acc Chem. Res.*, 51, 930–939.

Potts, R.O., Tamada, J.A., Tierney, M.J. (2002). Glucose monitoring by reverse iontophoresis. *Diabetes Metab. Res. Rev.*, 18, S49–S53.

Potts, J.R., Dreyer, D.R., Bielawski, C.W., Ruoff, R.S. (2011). Graphene-based polymer nanocomposites. *Polymer*, 52, 5–25.

Poznyak, S.K., Talapin, D.V., Shevchenko, E.V., Weller, H. (2004). Quantum dot chemiluminescence. *Nano Lett.*, 4, 693–698.

Preethichandra, D.M.G., Gholami, M.D., Izake, E.L., O'Mullane, A.P, Sonar, P. (2023). Conducting polymer based ammonia and hydrogen sulfide chemical sensors and their suitability for detecting food spoilage. *Adv. Mater. Technol.*, 8, 2200841.

Pretorius, V., Hopkins, B.J., Schieke, J.D. (1974). Electro-osmosis. A new concept for high- speed liquid chromatography. *J. Chromatography A*, 99, 23–30.

Pretsch, E. (2007). The new wave of ion-selective electrodes. *Trends Anal. Chem.*, 26(1), 46–51.

Pu, Y., Cai, F., Wang, D., Wang, J.-X., Chen, J.-F. (2018a). Colloidal synthesis of semiconductor quantum dots toward large-scale production: A review. *Ind. Eng. Chem. Res.*, 57, 1790–1802.

Pu, X., Hu, W., Wang, Z.L. (2018b). Toward wearable self-charging power systems: The integration of energy-harvesting and storage devices. *Small*, 14, 1702817.

Pumera, M. (2011). Graphene in biosensing. *Mat. Today*, 14, 308–315.

Qi, H., Li, M., Dong, M., Ruan, S., Gao, Q., Zhang, C. (2014). Electrogenerated chemiluminescence peptide-based biosensor for the determination of prostate-specific antigen based on target-induced cleavage of peptide. *Anal. Chem.*, 86, 1372–1379.

Qin, D., Meng, S., Wu, Y., Mo, G., Jiang, X., Deng, B. (2021). Design of a dual wavelength ratiometric electrochemiluminescence immunosensor for sensitive detection of amyloid-β protein in human serum. *ACS Sust. Chem. Eng.*, 9, 7541–7549.

Qiu, Z., Shu, J., Liu, J., Tang, D. (2019). Dual-channel photoelectrochemical ratiometric aptasensor with up-converting nanocrystals using spatial-resolved technique on homemade 3D printed device. *Anal. Chem.*, 91, 1260–1268.

Qu, B., Ma, C., Ji, G., Xu, C., Xu, J., Meng, Y.S., Wang, T., Lee, J.Y. (2014). Layered SnS_2-reduced graphene oxide composite – A high-capacity, high-rate, and long-cycle life sodium-ion battery anode material. *Adv. Mat.*, 26, 3854–3859.

Quigley, W.W.C. and Dovichi, N.J. (2004). Capillary electrophoresis for the analysis of biopolymers. *Anal. Chem.*, 76, 4645–4658.

Quirino, J.P., Terabe, S., Bocek, P. (2000). Sweeping of neutral analytes in electrokinetic chromatography with high-salt-containing matrixes. *Anal. Chem.*, 72, 1934–1940.

Qureshi, A., Gurbuz, Y., Niazi, J.H. (2012). Biosensors for cardiac biomarkers detection: A review. *Sensors and Actuators B*, 171–172, 62–76.

Ra, H.-W., Choi, K.-S., Kim, J.-H., Hahn, Y.-B., Im, Y.-H. (2008). Fabrication of ZnO nanowires using nanoscale spacer lithography for gas sensors. *Small*, 4, 1105–1109.

Radisavljevic, B., Radenovic, A., Brivio, J., Giacometti, V., Kis, A. (2011). Single-layer MoS2 transistors. *Nat. Nanotechn.*, 6, 147–150.

Radu, A., Radu, T., Mcgraw, C., Dillingham, P., Anastasova-Ivanova, S., Diamond, D. (2013). Ion selective electrodes in environmental analysis. *J. Serb. Chem. Soc.*, 78, 1729–1761.

Radziuk, D. and Moehwald, H. (2015). Prospects for plasmonic hot spots in single molecule SERS towards the chemical imaging of live cells. *Phys. Chem. Chem. Phys.*, 17, 21072–21093.

Raj, V.B., Singh, H., Nimal, A.T., Sharma, M.U., Gupta, V. (2013). Oxide thin films (ZnO, TeO_2, SnO_2, and TiO_2) based surface acoustic wave (SAW) E-nose for the detection of chemical warfare agents. *Sensors and Actuators B*, 178, 636–647.

Rakow, N.A. and Suslick, K.S. (2000). A colorimetric sensor array for odour visualization. *Nature*, 406, 710–713.

Ramachandran, A. and Santiago, J.G. (2022). Isotachophoresis: Theory and microfluidic applications. *Chem. Rev.*, 122, 12904–12976.

Ramayah, T., Ahmad, N.H., Halim, H.A., Zainal, S.R.M., Lo, M.-C. (2010). Discriminant analysis: An illustrated example. *African J. Business Management*, 4, 1654–1667.

Rejšek, J., Vrkoslav, V., Vaikkinen, A., Haapala, M., Kauppila, T.J., Kostiainen, R., Cvačka, J. (2016). Thin-layer chromatography/desorption atmospheric pressure photoionization orbitrap mass spectrometry of lipids. *Anal. Chem.*, 88, 12279–12286.

Relyea, R.A. (2009). A cocktail of contaminants: How mixtures of pesticides at low concentrations affect aquatic communities. *Oecologia*, 159, 363–376.

Rengaraj, A., Haldorai, Y., Kwak, C.H., Ahn, S., Jeon, K.J., Park, S.H., Han, Y.K., Huh, Y.S. (2015). Electrodeposition of flower-like nickel oxide on CVD-grown graphene to develop an electrochemical non-enzymatic biosensor. *J. Mater. Chem. B*, 3, 6301–6309.

Richter, M.M. (2004). Electrochemiluminescence (ECL). *Chem. Rev.* 104, 3003–3036.

Righetti, P.G., Sebastiano, R., Citterio, A. (2013). Capillary electrophoresis and isoelectric focusing in peptide and protein analysis. *Proteomics*, 13, 325–340.

Rius-Ruiz, F.X., Crespo, G.A., Bejarano-Nosas, D., Blondeau, P., Riu, J., Rius, F.X. (2011a). Potentiometric strip cell based on carbon nanotubes as transducer layer: Toward low-cost decentralized measurements. *Anal. Chem.*, 83(22), 8810–8815.

Rius-Ruiz, F.X., Bejarano-Nosas, D., Blondeau, P., Riu, J., Rius, F.X. (2011b). Disposable planar reference electrode based on carbon nanotubes and polyacrylate membrane. *Anal. Chem.*, 83, 5783–5788.

Rivnay, J., Inal, S., Salleo, A., Owens, R. M., Berggren, M., Malliaras, G.G. (2018). Organic electrochemical transistors. *Nature Rev. Mat.*, 3, 17086.

Ruecha, N., Chailapakul, O., Suzuki, K., Citterio, D. (2017). Fully inkjet-printed paper-based potentiometric ion-sensing devices. *Anal. Chem.*, 89,10608–10616.

Russier-Antoine, I., Bertorelle, F., Vojkovic, M., Rayane, D., Salmon, E., Jonin, C., Dugourd, P., Antoine, R., Brevet, P.-F. (2014). Non-linear optical properties of gold quantum clusters. The smaller the better. *Nanoscale*, 6, 13572–13578.

Saha, K., Agasti, S.S., Kim, C., Li, X., Rotello, V.M. (2012). Gold nanoparticles in chemical and biological sensing. *Chemical Reviews*, 112, 2739–2779.

Sanchez, C., Arribart, H., Giraud Guille, M.M. (2005). Biomimetism and bioinspiration as tools for the design of innovative materials and systems. *Nature Materials*, 4, 277–288.

Sander, L.C. and Wise, S.A. (1987). Effect of phase length on column selectivity for the separation of polycyclic aromatic hydrocarbons by reversed-phase liquid chromatography. *Anal. Chem.*, 59, 2309–2313.

Sanger, F., Nicklen, S., Coulson, A.R. (1977). DNA sequencing with chain-terminating inhibitors (DNA polymerase/nucleotide sequences/bacteriophage 4X174). *Proc. Nat. Acad. Sci. USA*, 74(12), 5463–5467.

Sänger-van de Griend, E. (2019). CE-SDS method development, validation, and best practice–An overview. *Electrophoresis*, 40, 2361–2374.

Sau, T.K. and Murphy, C.J. (2004). Seeded high yield synthesis of short Au nanorods in aqueous solution. *Langmuir*, 20, 6414–6420.

Schedin, F., Geim, A.K., Morozov, S.V., Hill, E.W., Blake, P., Katsnelson, M.I., Novoselov, K.S. (2007). Detection of individual gas molecules adsorbed on graphene. *Nature. Mat.*, 6, 652–655.

Schirhagl, R., Chang, K., Loretz, M., Degen, C.L. (2014). Nitrogen-vacancy centers in diamond: Nanoscale sensors for physics and biology. *Annu. Rev. Phys. Chem.*, 65, 83–105.

Schmalzing, D., Tsao, N., Koutny, L., Chisholm, D., Srivastava, A., Adourian, A., Linton, L., McEwan, P., Matsudaira, P., Ehrlich, D. (1999). Toward real-world sequencing by microdevice electrophoresis. *Genome Res.*, 9, 853–858.

Schnorr, J.M. and Swager, T.M. (2011). Emerging applications of carbon nanotubes. *Chem. Mat.*, 23, 646–657.

Schöning, M.J. and Poghossian, A. (2002). Recent advances in biologically sensitive field-effect transistors (BioFETs). *Analyst*, 127, 1137–1151.

Schramm, L.L., Stasiuk, E.N., Marangoni, D.G. (2003). Surfactants and their applications. *Annu. Rep. Prog. Chem., Sect. C*, 99, 3–48.

Schwierz, F. (2010). Graphene transistors. *Nat. Nanotechnol.*, 5, 487–496.

Seabold, J.A., Shankar, K., Wilke, R.H., Paulose, M., Varghese, O.K., Grimes, C.A., Choi, K.S. (2008). Photoelectrochemical properties of heterojunction CdTe/TiO2 electrodes constructed using highly ordered TiO2 nanotube arrays. *Chem. Mater.*, 20, 5266–5273.

Sekar, R.B. and Periasamy, A. (2003). Fluorescence resonance energy transfer (FRET) microscopy imaging of live cell protein localizations. *Journal of Cell Biology*, 160, 629–633.

Senillou, A., Jaffrezic-Renault, N., Martelet, C., Cosnier. S. (1999). A miniaturized urea sensor based on the integration of both ammonium-based urea enzyme field effect transistor and a reference field effect transistor in a single chip. *Talanta*, 50, 219–226.

Seo, D., Yoo, C., Chung, I.S., Park, S.M., Ryu, S., Song, H. (2008). Shape adjustment between multiply twinned and single-crystalline polyhedral gold nanocrystals: Decahedra, icosahedra, and truncated tetrahedral. *J. Phys. Chem. C*, 112, 2469–2475.

Seo, D., Park, J.H., Jung, J., Park, S.M., Ryu, S., Kwak, J., Song, H. (2009). One-dimensional gold nanostructures through directed anisotropic overgrowth from gold decahedrons. *J. Phys. Chem. C*, 113, 3449–3454.

Seo, G., Lee, G., Kim, M.J., Baek, S.-H., Choi, M., Ku, K.B., Lee, C.-S., Jun, S., Park, D., Kim, H.G. et al. (2020). Rapid detection of COVID-19 causative virus (SARS-CoV-2) in human nasopharyngeal swab specimens using field-effect transistor-based biosensor. *ACS Nano*, 14, 5135–5142.

Shao, Y., Wang, J., Wu, H., Liu, J., Aksay, I. A., Lin, Y. (2010). Graphene based electrochemical sensors and biosensors: A review. *Electroanalysis*, 22, 1027–1036.

Shao, W., Shurin, M.R., Wheeler, S.E., He, X., Star, A. (2021). Rapid detection of SARS-CoV-2 antigens using high-purity semiconducting single-walled carbon nanotube-based field-effect transistors. *ACS Appl. Mater. Interfaces*, 13, 10321–10327.

Sharma, S. and Madou, M. (2012). A new approach to gas sensing with nanotechnology. *Phil. Trans. R. Soc. A*, 370, 2448–2473.

Shekhah, O., Wang, H., Paradinas, M., Ocal, C., Schüpbach, B., Terfort, A., Zacher, D., Fischer, R.A., Wöll, C. (2009). Controlling interpenetration in metal–organic frameworks by liquid-phase epitaxy. *Nat. Mat.*, 8, 481–484.

Shen, G., Chen, P.C., Ryu, K., Zhou, C.W. (2009). Devices and chemical sensing applications of metal oxide nanowires. *J. Mater. Chem.*, 19, 828–839.

Shen, J., Zhou, T., Huang, R. (2019). Recent advances in electrochemiluminescence sensors for pathogenic bacteria detection. *Micromachines*,10, 532–549.

Shendure, J. and Ji, H. (2008). Next-generation DNA sequencing. Review. *Nature Biotechn.*, 26, 1135–1145.

Sherman, L.R. and Goodrich, J.A. (1985). The historical development of sodium dodecyl sulphate-polyacrylamide gel electrophoresis. *Chem. Soc. Rev.*, 14, 225–236.

Shi, Y., Li, H., Li, L.-J. (2015). Recent advances in controlled synthesis of two-dimensional transition metal dichalcogenides via vapour deposition techniques. *Chem. Soc. Rev.*, 44, 2744–2756.

Shipp, D.W, Sinjab, F., Notingher, I. (2017). Raman spectroscopy: Techniques and applications in the life sciences. *Advances in Optics and Photonics*, 9, 315–390.

Simon, D.T., Gabrielsson, E.O., Tybrandt, K., Berggren, M. (2016). Organic bioelectronics: Bridging the signaling gap between biology and technology. *Chem. Rev.*, 116, 13009–13041.

Singhal, N., Kumar, M., Kanaujia, P., Virdi, J. (2015). MALDI-TOF mass spectrometry: An emerging technology for microbial identification and diagnosis. *Frontiers in Microbiology*, 6, 791.

Smejkal, P., Bottenus, D., Breadmore, M.C., Guijt, R.M., Ivory, C.F., Foret, F., Macka, M. (2013). Microfluidic isotachophoresis: A review. *Electrophoresis*, 34, 1493–1509.

Smith, E. and Dent, G. (2005). *Modern Raman Spectroscopy. A Practical Approach*. Wiley, Chichester.

Sokalski, T., Ceresa, A., Zwickl, T., Pretsch, E. (1997). Large improvement of the lower detection limit of ion-selective polymer membrane electrodes. *J. Am. Chem. Soc.*, 119, 11347–11348.

Sokalski, T., Ceresa, A., Fibbioli, M., Zwickl, T., Bakker, E., Pretsch, E. (1999). Lowering the detection limit of solvent polymeric ion-selective membrane electrodes. 2. Influence of composition of sample and internal electrolyte solution. *Anal. Chem.*, 71, 1210–1214.

Song, M.S., Choi, S.P., Lee, J., Kwon, Y.J., Sim, S.J. (2013). Real-time, sensitive, and specific detection of promoter-polymerase interactions in gene transcription using a nanoplasmonic sensor. *Adv. Mater.*, 25, 1265–1269.

Stankovich, S., Dikin, D.A., Dommett, G.H.B., Kohlhaas, K.M., Zimney, E.J., Stach, E.A., Piner, R.D., Nguyen, S.T., Ruoff, R.S. (2006). Graphene-based composite materials. *Nature*, 442, 282–286.

Stanton, S.G., Pecora, R., Hudson, B.S. (1981). Resonance enhanced dynamic Rayleigh scattering. *J. Chem. Physics*, 75(12), 5615–5621.

Stern, E., Wagner, R., Sigworth, F., Breaker, R., Fahmy, T., Reed, M. (2007). Importance of the Debye screening length on nanowire field effect transistor sensors. *Nano Lett.*, 7, 3405–3409.

Su, Y., Kravets, V.G., Wong, S.L., Waters, J., Geim, A.K., Nair, R.R. (2014). Impermeable barrier films and protective coatings based on reduced graphene oxide. *Nat. Commun.*, 5, 4843.

Su, G., Dang, L., Liu, G., Feng, T., Wang, W., Wang, C., Wei, H. (2022). MOF-derived hierarchical porous 3D ZnO/Ag nanostructure as a reproducible SERS substrate for ultrasensitive detection of multiple environmental pollutants. *Spectrochimica Acta Part A: Mol. Biomol. Spectroscopy*, 270, 120818.

Sun, Y.-P., Zhou, B., Lin, Y., Wang, W., Fernando, K.A.S., Pathak, P., Meziani, M.J., Harruff, B.A., Wang, X., Wang, H. et al. (2006). Quantum-sized carbon dots for bright and colorful photoluminescence. *J. Am. Chem. Soc.*, 128, 7756–7757.

Sun, X., Sun, H., Li, H., Peng, H. (2013). Developing polymer composite materials: Carbon nanotubes or graphene? *Adv. Mater.*, 25, 5153–5176.

Sun, B., Lv, H., Liu, Z., Wang, J., Bai, X., Zhang, Y., Chen, J., Kan, K., Shi, K. (2021). Co_3O_4@PEI/Ti_3C_2Tx MXene nanocomposites for a highly sensitive NOx gas sensor with a low detection limit. *J. Mater. Chem. A*, 9, 6335–6344.

Sung, D. and Koo, J. (2021). A review of BioFET's basic principles and materials for biomedical applications. *Biomedical Engineering Letters*, 11, 85–96.

Sunney, X. (2013). Enzyme kinetics, past and present. *Science*, 342, 1457–1459.

Suslick, K.S., Rakow, N.A., Sen, A. (2004). Colorimetric sensor arrays for molecular recognition. *Tetrahedron*, 60, 11133–11138.

Sutter, J., Radu, A., Peper, S., Bakker, E., Pretsch, E. (2004). Solid-contact polymeric membrane electrodes with detection limits in the subnanomolar range. *Analyt. Chimica Acta*, 523, 53–59.

Svec, F. (2005). Recent developments in the field of monolithic stationary phases for capillary Electrochromatography. *J. Sep. Sci.*, 28, 729–745.

Szigeti, Z., Vigassy, T., Bakker, E., Pretsch, E. (2006). Approaches to improving the lower detection limit of polymeric membrane ion-selective electrodes. *Electroanalysis*, 18(13–14), 1254–1265.

Szuber, J., Uljanowb, J., Karczewska-Buczek, T., Jakubik, W., Waczynski, K., Kwoka, M., Konczak, S. (2005). On the correlation between morphology and gas sensing properties of RGTO SnO_2 thin films. *Thin Solid Films*, 490, 54–58.

Szunerits, S. and Boukherroub, R. (2018). Graphene-based biosensors. *Interface Focus*, 8, 20160132. doi: 10.1098/rsfs.2016.0132.

Tahir, M.A., Zhang, X., Cheng, H., Xu, D., Feng, Y., Sui, G., Fu, H., Valev, V.K., Zhang, L., Chen, J. (2020). Klarite as a label-free SERS-based assay: A promising approach for atmospheric bioaerosol detection. *Analyst*, 145, 277–285.

Tai, H., Zhen, Y., Liu, C., Ye, Z., Xie, G., Du, X., Jiang, Y. (2016). Facile development of high performance QCM humidity sensor based on protonated polyethylenimine-graphene oxide nanocomposite thin film. *Sens. Actuators, B Chem.*, 230, 501–509.

Tan, C. and Zhang, H. (2015). Two-dimensional transition metal dichalcogenide nanosheet-based composites. *Chem. Soc. Rev.*, 44, 2713–2731.

Tang, W., Yin, L., Sempionatto, J.R., Moon, J.M., Teymourian, H., Wang, J. (2021). Touch-based stressless cortisol sensing. *Advanced Materials*, 33, 2008465.

Tarongoy, F.M. Jr., Haddad, P.R., Boysen, R.I., Hearn, M.T.W., Quirino, J.P. (2016). Open tubular-capillary electrochromatography: Developments and applications from 2013 to 2015. *Electrophoresis*, 37, 66–85.

Taylor, R. and Walton, D.R.M. (1993). The chemistry of fullerenes. *Nature*, 363, 685–693.

Terabe, S. (2004). Micellar electrokinetic chromatography. *Anal. Chem.*, 241A–246A.

Terabe, S. Otsuka, K., Ichikawa, K., Tsuchiya, A., Ando, T. (1984). Electrokinetic separations with micellar solutions and open-tubular capillaries. *Anal. Chem.*, 56, 113–116.

Tian, P., Tang, L., Teng, K.S., Lau, S.P. (2018). Graphene quantum dots from chemistry to applications. *Mat. Today Chem.*, 10, 221–258.

Tierney, M., Tamada, J., Potts, R., Jovanovic, L., Garg, S. (2001). Clinical evaluation of the GlucoWatch biographer: A continual, non-invasive glucose monitor for patients with diabetes. *Biosens. Bioelectron.*, 16, 621–629.

Tokel, N.E. and Bard, A.J. (1972). Electrogenerated chemiluminescence. IX. Electrochemistry and emission from systems containing tris(2,2'-bipyridine)ruthenium(II) dichloride. *J. Am. Chem. Soc.*, 94, 2862–2863.

Toghill, K.E. and Compton, R.G. (2010). Electrochemical non-enzymatic glucose sensors. *Int. J. Electrochem. Sci.*, 5, 1246–1301.

Tosun, M., Chuang, S., Fang, H., Sachid, A.B., Hettick, M., Lin, Y., Zeng, Y., Javey, A. (2014). High-gain inverters based on WSe2 complementary field-effect transistors. *ACS Nano*, 8, 4948–4953.

Tran, H.V., Piro, B., Reisberg, S., Tran, L.D., Duc, H.T., Pham, M.C. (2013). Label-free and reagentless electrochemical detection of microRNAs using a conducting polymer nanostructured by carbon nanotubes: Application to prostate cancer biomarker miR-141. *Biosensors and Bioelectronics*, 49, 164–169.

Truong, P.L., Choi, S.P., Sim, S.J. (2013). Amplification of resonant Rayleigh light scattering response using immunogold colloids for detection of lysozyme. *Small*, 9(20), 3485–3492.

Turkevich, J., Stevenson, P.C., Hillier, J. (1951). A study of the nucleation and growth processes in the synthesis of colloidal gold. *Discussions of the Faraday Soc.*, 11, 55–75.

Umezawa, Y., Umezawa, K., Sato, H. (1995). Selectivity coefficients for ion-selective electrodes: Recommended methods for reporting K_{AB}^{pot} values. *Pure & Appl. Chem.*, 67, 507–518.

VahidMohammadi, A., Rosen, J., Gogotsi, Y. (2021). The world of two-dimensional carbides and nitrides (MXenes). *Science*, 372(6547), eabf1581.

Valappil, M.O., Pillai, V.K., Alwarappan S. (2017). Spotlighting graphene quantum dots and beyond: Synthesis, properties and sensing applications. *Appl. Mat. Today*, 9, 350–371.

Vlatakis, G., Andersson, L.I., Muller, R., Mosbach, K. (1993). Drug assay using antibody mimics made by molecular imprinting. *Nature*, 361, 645–647.

Voeten, R.L.C., Ventouri, I.K., Haselberg, R., Somsen, G.W. (2018). Capillary electrophoresis: Trends and recent advances. *Anal. Chem.*, 90, 1464–1481.

Wagner, T., Werner, C.F., Miyamoto, K., Schöning, M.J., Yoshinobu, T. (2012). Development and characterisation of a compact light-addressable potentiometric sensor (LAPS) based on the digital light processing (DLP) technology for flexible chemical imaging. *Sensors and Actuators B*, 170, 34–39.

Wagner, T., Vornholt, W., Werner, C.F., Yoshinobu, T., Miyamoto, K.-I., Keusgen, M., Schöning, M.J. (2016). Light-addressable potentiometric sensor (LAPS) combined with magnetic beads for pharmaceutical screening. *Physics in Medicine*, 1, 2–7.

Wall, M. (2011). The Raman spectroscopy of graphene and the determination of layer thickness. Application Note 52252. Thermo Fisher Scientific, Madison, WI.

Wang, J. (2008). Electrochemical glucose sensors. *Chem. Rev.*, 108, 814–825.

Wang, Y. and Hu, A. (2014). Carbon quantum dots: synthesis, properties and applications. *J. Mater. Chem. C*, 2, 6921–6939.

Wang, Q.H., Kalantar-Zadeh, K., Kis, A., Coleman, J.N., Strano, M.S. (2012). Electronics and optoelectronics of two-dimensional transition metal dichalcogenides. *Nat. Nanotechn.*, 7, 699–712.

Wang, H., Yuan, Y., Wei, L., Goh, K., Yu, D., Chen, Y. (2015). Catalysts for chirality selective synthesis of single-walled carbon nanotubes. *Carbon*, 81, 1–19.

Wang, T., Huang, D., Yang, Z., Xu, S., He, G., Li, X., Hu, N., Yin, G., He, D., Zhang, L. (2016a). A review on graphene-based gas/vapor sensors with unique properties and potential applications. *Nano-Micro Lett.*, 8(2), 95–119.

Wang, D., Noël, V., Piro, B. (2016b). Electrolytic gated organic field-effect transistors for application in biosensors–A review. *Electronics*, 5, 1–24.

Wang, Y.-Z., Zhao, W., Dai, P.-P., Lu, H.-J., Xu, J.-J., Pan, J., Chen, H.-Y. (2016c). Spatial-resolved electrochemiluminescence ratiometry based on bipolar electrode for bioanalysis. *Biosensors and Bioelectronics*, 86, 683–689.

Wang, K., Zhang, R., Sun, N., Li, X., Wang, J., Cao, Y., Pei, R. (2016d). Near-infrared light-driven photoelectrochemical aptasensor based on the upconversion nanoparticles and TiO_2/CdTe heterostructure for detection of cancer cells. *ACS Appl. Mater. Interfaces*, 8, 25834–25839.

Wang, J., Yu, J., Wang, T., Li, C., Wei, Y., Deng, X., Chen, X. (2020a). Emerging intraoral biosensors. *J. Mater. Chem. B*, 8, 3341–3356.

Wang, B., Cao, J.-T., Liu, Y.-M. (2020b). Recent progress of heterostructure-based photoelectrodes in photoelectrochemical biosensing: A mini review. *Analyst*, 145, 1121–1128.

Wang, B., Zhao, C., Wang, Z., Yang, K.A., Cheng, X., Liu, W., Yu, W., Lin, S., Zhao, Y., Cheung K.M. et al. (2022). Wearable aptamer-field-effect transistor sensing system for noninvasive cortisol monitoring. *Science Advances*, 8, eabk0967.

Warner, M. (1994). Migrating toward capillary electrophoresis. *Anal. Chem.*, 66(22), 1137A–1141A.

Waterhouse, G.I., Chen W.-T., Chan, A., Jin, H., Sun-Waterhouse, D., Cowie, B.C.C. (2015). Structural, optical, and catalytic support properties of γ-Al_2O_3 inverse opals. *J. Phys. Chem. C*, 119, 6647–6659.

Wei, B.Q., Vajtai, R., Ajayan, P.M. (2001). Reliability and current carrying capacity of carbon nanotubes. *Appl. Phys. Lett.*, 79, 1172–1174.

Weinberger, R. (ed.) (2000). Capillary isoelectric focusing. In *Practical Capillary Electrophoresis*, 2nd edition. Academic Press, San Diego, CA, 209–243.

Westermeier, R. (2014). Looking at proteins from two dimensions: A review on five decades of 2D electrophoresis. *Arch. Physiol. Biochem.*, 120, 168–172.

White, H.S., Kittlesen, G.P., Wrighton, M.S. (1984). Chemical derivatization of an array of 3 gold microelectrodes with polypyrrole – Fabrication of a molecule-based transistor. *J. Am. Chem. Soc.*, 106, 5375–5377.

White, S.P., Sreevatsan, S., Frisbie, C.D., Dorfman, K.D. (2016). Rapid, selective, label-free aptameric capture and detection of ricin in potable liquids using a printed floating gate transistor. *ACS Sensors*, 1, 1213.

Wilson, D.A. (2018). Applications of electronic-nose technologies for noninvasive early detection of plant, animal and human diseases. *Chemosensors*, 6, 45.

Wu, J. and Pawliszyn, J. (1992). Universal detection for capillary isoelectric focusing without mobilization using a concentration gradient imaging system. *Anal. Chem.*, 64, 224–227.

Wu, M.-S., Yuan, D.-J., Xu, J.-J., Chen, H.-Y. (2013). Electrochemiluminescence on bipolar electrodes for visual bioanalysis. *Chem. Sci.*, 4, 1182–1188.

Wu, P., Hou, X., Xu, J.J., Chen, H.Y. (2014). Electrochemically generated versus photoexcited luminescence from semiconductor nanomaterials: Bridging the valley between two worlds. *Chem. Rev.*, 114, 11027–11059.

Wu, J., McElroy, W., Pawliszyn, J., Heger, C.D. (2022). Imaged capillary isoelectric focusing: Applications in the pharmaceutical industry and recent innovations of the technology. *Trends in Analytical Chemistry*, 150, 116567.

Xia, Y., Gilroy, K.D., Peng, H.-C., Xia, X. (2017). Seed-mediated growth of colloidal metal nanocrystals. *Angew. Chem. Int. Ed.*, 56, 60–95.

Xia, Q., Fan, Y., Li, S., Zhou, A., Shinde, N., Mane, R.S. (2023). MXene-based chemical gas sensors: Recent developments and challenges. *Diamond & Related Materials*, 131, 109557.

Xiao, Y., Li, Y., Ying, J., Tian, Y., Xiao, Y., Mei, Z. (2015). Determination of alditols by capillary electrophoresis with indirect laser-induced fluorescence detection. *Food Chemistry*, 174, 233–239.

Xie, Q., Pérez-Cordero, E., Echegoyen, L. (1992). Electrochemical detection of C_{60}^{6-} and C_{70}^{6-}: Enhanced stability of fullerides in solution. *J. Am. Chem. Soc.*, 114, 3978–3980.

Xin, H. and Li, W. (2018). A review on high throughput roll-to-roll manufacturing chemical vapor deposition graphene. *Appl. Phys. Rev.*, 5, 031105.

Xiong, P., Ma, R., Wang, G., Sasaki, T. (2019). Progress and perspective on two-dimensional unilamellar metal oxide nanosheets and tailored nanostructures from them for electrochemical energy storage. *Energy Storage Materials*, 19, 281–298.

Xu, X.Y., Ray, R., Gu, Y.L., Ploehn, H.J., Gearheart, L., Raker, K., Scrivens, W.A. (2004). Electrophoretic analysis and purification of fluorescent single-walled carbon nanotube fragments. *J. Am. Chem. Soc.*, 126, 12736–12737.

Xu, M., Liang, T., Shi, M., Chen, H. (2013). Graphene-like two-dimensional materials. *Chem. Rev.*, 113(5), 3766–3798.

Xu, Z., Liu, Y., Zhao, X., Peng, L., Sun, H., Xu, Y., Ren, X., Jin, C., Xu, P., Wang, M., Gao, C. (2016). Ultrastiff and strong graphene fibers via full-scale synergetic defect engineering. *Adv. Mater.*, 28, 6449–6456.

Xu, S.J., Mo, R.Z., Jin, C., Cui, X.Q., Bai, R.H., Ji, Y.B. (2017). Mesoporous silica nanoparticles incorporated hybrid monolithic stationary phase immobilized with pepsin for enantioseparation by capillary electrochromatography. *J Pharm. Biomed. Anal.*, 140, 190–198.

Xu, S., Zhang, G., Fang, B., Xiong, Q., Duan, H., Lai, W. (2019). Lateral flow immunoassay based on polydopamine-coated gold nanoparticles for the sensitive detection of zearalenone in maize. *ACS Appl. Mater. Interfaces*, 11, 31283–31290.

Xue, M., Mackin, C., Weng, W.-H., Zhu, J., Luo, Y., Luo, S.-X. L., Lu, A.-Y., Hempel, M., McVay, E., Kong, J. et al. (2022). Integrated biosensor platform based on graphene transistor arrays for real-time high-accuracy ion sensing. *Nat. Commun.* 13, 5064.

Yaghi, O.M., O'Keeffe, M., Ockwig, N.W., Chae, H.K., Eddaoudi, M., Kim, J. (2003). Reticular synthesis and the design of new materials. *Nature*, 423, 705–714.

Yamazoe, N. (1991). New approaches for improving semiconductor gas sensors. *Sensors and Actuators B*, 5, 7–19.

Yamazoe, N. (2005). Toward innovations of gas sensor technology. *Sens. Actuators, B*, 108, 2–14.

Yan, Y., Yang, G., Xu, J.-L., Zhang, M., Kuo, C.-C., Wang S.-D. (2020). Conducting polymer-inorganic nanocomposite-based gas sensors: A review. *Sci. Technology Adv. Materials*, 21(1), 768–786.

Yang, Y.A., Wu, H., Williams, K.R., Cao, Y.C. (2005). Synthesis of CdSe and CdTe nanocrystals without precursor injection. *Angew. Chem. Int. Ed.*, 44, 6712–6715.

Yang, L., Zhao, H., Fan, S., Zhao, G., Ran, X., Li, C.P. (2015). Electrochemical detection of cholesterol based on competitive host–guest recognition using a β-cyclodextrin/poly (N-acetylaniline)/graphene-modified electrode. *RSC Advances*, 5, 64146–64155.

Yang, A., Wang, D., Wang, X., Zhang, D., Koratkar, N., Rong, M. (2018). Recent advances in phosphorene as a sensing material. *Nano Today*, 20, 13–32.

Yang, W., Yang, X., Zhu, L., Chu, H., Li, X., Xu, W. (2021). Nanozymes: Activity origin, catalytic mechanism, and biological application. *Coordination Chemistry Rev.*, 448, 214170.

Yang, E., Zhang, Y., Shen, Y. (2022). Quantum dots for electrochemiluminescence bioanalysis – A review. *Analytica Chimica Acta*, 1209, 339140.

Yao, J., Yang, M., Duan, Y. (2014). Chemistry, biology, and medicine of fluorescent nanomaterials and related systems: New insights into biosensing, bioimaging, genomics, diagnostics, and therapy. *Chem. Rev.*, 114, 6130–6178.

Ye, S., Zhou, T., Cheng, K., Chen, M., Wang, Y., Jiang, Y., Yang, P. (2015). Carboxylic acid fullerene (C60) derivatives attenuated neuroinflammatory responses by modulating mitochondrial dynamics. *Nanoscale Res. Lett.*, 10, 246.

Ye, F., Zhao, Y., El-Sayed, R., Muhammed, M., Hassan, M. (2018). Advances in nanotechnology for cancer biomarkers. *Nano Today*, 18, 103–123.

Yildiz, G., Bolton-Warberg, M., Awaja, F. (2021). Graphene and graphene oxide for biosensing: General properties and the effects of graphene ripples. *Acta Biomaterialia*, 131, 62–79.

Yoshinobu, T. and Schöning, M.J. (2021). Light-addressable potentiometric sensors for cell monitoring and biosensing. *Current Opinion in Electrochemistry*, 28, 100727.

Yoshinobu, T., Iwasaki, H., Ui, Y., Furuichi, K., Ermolenko, Y., Mourzina, Y., Wagner, T., Näther, N., Schöning, M.J. (2005). The light-addressable potentiometric sensor for multi-ion sensing and imaging. *Methods*, 37, 94–102.

Yu, H., Zhang, H., Li, J., Zhao, Z., Deng, M., Ren, Z., Li, Z., Xue, C., Li, M.G., Chen, Z. (2022). Rapid and unamplified detection of SARS-COV-2 RNA via CRISPR-Cas13a modified solution-gated graphene transistors. *ACS Sensors*, 7, 3923.

Yuan, W.J. and Shi, G.Q. (2013). Graphene-based gas sensors. *J. Mater. Chem. A*, 1(35), 10078–10091.

Yulianti, E.S., Rahman, S.F., Whulanza, Y. (2022). Molecularly imprinted polymer-based sensor for electrochemical detection of cortisol. *Biosensors*, 12, 1090.

Zafar, S., D'Emic, C., Jagtiani, A., Kratschmer, E., Miao, X., Zhu, Y., Mo, R., Sosa, N., Hamann, H., Shahidi, G., Riel, H. (2018). Silicon nanowire field effect transistor sensors with minimal sensor-to-sensor variations and enhanced sensing characteristics. *ACS Nano.*, 12, 6577–6587.

Zamzami, M.A., Rabbani, G., Ahmad, A., Basalah, A.A., Al-Sabban, W.H., Ahn, S.N., Choudhry, H. (2022). Carbon nanotube field-effect transistor (CNT-FET)-based biosensor for rapid detection of SARS-CoV-2 (COVID-19) surface spike protein S1. *Bioelectrochemistry*, 143, 107982.

Zarad, W., Shawky, A., Ali, A., Aboulella, Y., Kamal, M., Masujima, T., Emara, S., El-Gendy, H. (2021). Field amplified sample stacking and in-capillary derivatization for forensic analysis of morphine and morphine-6-glucuronide in human urine by capillary electrophoresis. *Talanta Open*, 3, 100041.

Zaumseil, J. and Sirringhaus, H. (2007). Electron and ambipolar transport in organic field-effect transistors. *Chem. Rev.*, 107, 1296–1323.

Zdrachek, E. and Bakker, E. (2019). Potentiometric sensing. *Anal. Chem.*, 91, 2–26.

Zea, M., Bellagambi, F.G., Ben Halima, H., Zine, N., Jaffrezic-Renault, N., Villa, R., Gabriel, G., Errachid, A. (2020). Electrochemical sensors for cortisol detections: Almost there. *TrAC Trends in Analytical Chemistry*, 132, 116058.

Zeng, M., Tan, L., Wang, J., Chen, L., Rümmeli, M.H., Fu, L. (2014). Liquid metal: An innovative solution to uniform graphene films. *Chem. Mater.*, 26, 3637–3643.

Zhang, L. and Wang, E. (2014). Metal nanoclusters: New fluorescent probes for sensors and bioimaging. *Nano Today*, 9, 132–157.

Zhang, Y., Zhang, L., Zhou, C. (2013a). Review of chemical vapor deposition of graphene and related applications. *Acc. Chem. Res.*, 46, 2329–2339.

Zhang, H.-R., Xu, J.-J., Chen, H.-Y. (2013b). Electrochemiluminescence ratiometry: A new approach to DNA biosensing. *Anal. Chem.*, 85, 5321–5325.

Zhang, Q., Jie, J., Diao, S., Shao, Z., Zhang, Q., Wang, L., Deng, W., Hu, W., Xia, H., Yuan, X., Lee, S.-T. (2015). Solution-processed graphene quantum dot deep-UV photodetectors. *ACS Nano*, 9, 1561–1570.

Zhang, K., Feng, Y., Wang, F., Yang, Z., Wang, J. (2017a). Two dimensional hexagonal boron nitride (2D-hBN): Synthesis, properties and applications. *J. Mater. Chem. C*, 5, 11992–12022.

Zhang, Z.J., Zhang, X.H., Liu, B.W., Liu, J.W. (2017b). Molecular imprinting on inorganic nanozymes for hundred-fold enzyme specificity. *J. Am. Chem. Soc.*, 139, 5412–5419.

Zhang, Y., Zhang, J., Liu, Q. (2017c). Gas sensors based on molecular imprinting technology. *Sensors*, 17, 1567.

Zhang, H., Zhang, C., Liu, D., Zuo, F., Chen, S., Yuan, R., Xu, W. (2018a). A ratiometric electrochemiluminescent biosensor for Con A detecting based on competition of dissolved oxygen. *Biosensors and Bioelectronics*, 120, 40–46.

Zhang, B., Cheng, M., Liu, G., Gao, Y., Zhao, L., Li, S., Wang, Y., Liu, F., Liang, X., Zhang, T. (2018b). Room temperature NO_2 gas sensor based on porous Co_3O_4 slices/reduced graphene oxide hybrid. *Sens. Actuators B Chem.*, 263, 387–399.

Zhang, J., Shin, H., Lu, W. (2019a). Highly ambient-stable few-layer black phosphorene by pulsed laser exfoliation and HEMM. *Chem. Commun.*, 55, 2601–2604.

Zhang, F., Wang, M., Zhang, L., Su, X. (2019b). Ratiometric fluorescence system for pH sensing and urea detection based on MoS2 quantum dots and 2,3-diaminophenazine. *Anal. Chim. Acta*, 1077, 200–207.

Zhang, X., Chen, T., Li, V., Bo, T., Du, M., Huang, T. (2023a). Cutting-edge mass spectrometry strategy based on imaged capillary isoelectric focusing (icIEF) technology for characterizing charge heterogeneity of monoclonal antibody. *Analyt. Biochemistry*, 660, 114961.

Zhang, Y., Lai, Q., Chen, W., Zhang, C., Mo, L., Liu, Z. (2023b). Recent advance in cortisol immunosensing technologies and devices. *Chemosensors*, 11, 90.

Zhao, W.-W., Wang, J., Zhu, Y.-C., Xu, J.-J., Chen, H.-Y. (2015). Quantum dots: Electrochemiluminescent and photoelectrochemical bioanalysis. *Anal. Chem.*, 87, 9520–9531.

Zheng, J., Zhang, C., Dickson, R.M. (2004). Highly fluorescent, water-soluble, size-tunable gold quantum dots. *Phys. Rev. Lett.*, 93, 077402.

Zheng, G., Patolsky, F., Cui, Y., Wang, W.U., Lieber, C.M. (2005). Multiplexed electrical detection of cancer markers with nanowire sensor arrays. *Nat. Biotechnol*, 23, 1294–1301.

Zheng, J., Zhou, C., Yu, M., Liu, J. (2012). Different sized luminescent gold nanoparticles. *Nanoscale*, 4, 4073–4083.

Zheng, M., Liu, S., Li, J., Qu, D., Zhao, H., Guan, X., Hu, X., Xie, Z., Jing, X., Sun, Z. (2014a). Integrating oxaliplatin with highly luminescent carbon dots: An unprecedented theranostic agent for personalized medicine. *Adv. Mater.*, *26*, 3554–3560.

Zheng, Y., Zhong, X., Li, Z., Xia, Y. (2014b). Successive, seed-mediated growth for the synthesis of single-crystal gold nanospheres with uniform diameters controlled in the range of 5–150 nm. *Particle & Part. Syst. Charact.*, 31, 266–273.

Zhou, Y., Liu, B., Yang, R., Liu, J. (2017). Filling in the gaps between nanozymes and enzymes: Challenges and opportunities. *Bioconjugate Chem.*, 28, 2903–2909.

Zhu, Z., Lu, J.J., Liu, S. (2012). Protein separation by capillary gel electrophoresis: A review. *Anal. Chim. Acta*, 709, 21–31.

Zhu, S.J., Meng, Q.N., Wang, L., Zhang, J.H., Song, Y.B., Jin, H., Zhang, K., Sun, H.C., Wang, H.Y., Yang, B. (2013). Highly photoluminescent carbon dots for multicolor patterning, sensors, and bioimaging. *Angew. Chem., Int. Ed.*, 52, 3953–3957.

Zhu, S., Wang, L., Li, B., Song, Y., Zhao, X., Zhang, G., Zhang, S., Lu, S., Zhang, J., Wang, H. et al. (2014a). Investigation of photoluminescence mechanism of graphene quantum dots and evaluation of their assembly into polymer dots. *Carbon*, 77, 462–472.

Zhu, G., Sun, L., Yan, X., Dovichi, N.J. (2014b). Bottom-up proteomics of *Escherichia coli* using dynamic pH junction preconcentration and capillary zone electrophoresis electrospray ionization-tandem mass spectrometry. *Anal. Chem.*, 86, 6331–6336.

Zhu, H., Zhang, H., Xu, Y., Lassakova, S., Korabecna, M., Neuzil, P. (2020). PCR past, present and future. *BioTechniques*, 69, 317–325.

Zijlstra, P., Chon, J.W.M., Gu, M. (2009). Five-dimensional optical recording mediated by surface plasmons in gold nanorods. *Nature*, 409, 410–413.

Zou, R., Teng, X., Lin, Y., Lu, C. (2020). Graphitic carbon nitride-based nanocomposites electrochemiluminescence systems and their applications in biosensors. *Trends in Anal. Chem.*, 132, 116054.

Zuo, W., Tang, L., Xiang, J., Ji, R., Luo, L., Rogee, L., Lau, S.P. (2017). Functionalization of graphene quantum dots by fluorine: Preparation, properties, application, and their mechanisms. *Appl. Phys. Lett.*, 110, 221901.

Index

A, B

C, D

E, F

G

I, L

M

N, O

P, Q

R, S

T, V, Z

Other titles from

ISTE

in

Nanoscience and Nanotechnology

2022

LACAZE Pierre Camille, LACROIX Jean-Christophe
Nanotechnology and Nanomaterials for Energy

2017

JALABERT Denis, VICKRIDGE Ian, CHABLI Amal
Swift Ion Beam Analysis in Nanosciences

2016

BASSET Philippe, BLOKHINA Elena, GALAYKO Dimitri
Electrostatic Kinetic Energy Harvesting
(Nanotechnologies for Energy Recovery Set – Volume 3)

LAFONTAINE Eric, COMET Marc
Nanothermites

TRIOZON François, DOLLFUS Philippe
Simulation of Transport in Nanodevices

2015

LEPRINCE-WANG Yamin

Piezoelectric ZnO Nanostructure for Energy Harvesting
(Nanotechnologies for Energy Recovery Set – Volume 1)

ROSSI Carole
Al-based Energetic Nanomaterials
(Nanotechnologies for Energy Recovery Set – Volume 2)

2014

BALESTRA Francis
Beyond CMOS Nanodevices 1
Beyond CMOS Nanodevices 2

LACAZE Pierre Camille, LACROIX Jean-Christophe
Non-volatile Memories

LATU-ROMAIN Laurence, OLLIVIER Maelig
Silicon Carbide One-dimensional Nanostructures

NICU Liviu, LEÏCHLE Thierry
Micro- and Nanoelectromechanical Biosensors

SENEZ Vincent, THOMY Vincent, DUFOUR Renaud
Nanotechnologies for Synthetic Super Non-wetting Surfaces

TRIPON-CANSELIET Charlotte, CHAZELAS Jean
Nanoscale Microwave Engineering: Optical Control of Nanodevices

2012

BELLEVILLE Marc, CONDEMINE Cyril
Energy Autonomous Micro and Nano Systems

CLAVERIE Alain
Transmission Electron Microscopy in Micro-nanoelectronics

LACAZE Pierre Camille
Nanotechnologies: Concepts, Processing and Applications

2010

BALESTRA Francis
Nanoscale CMOS / Innovative Materials, Modeling and Characterization

QUERLIOZ Damien, DOLLFUS Philippe
The Wigner Monte Carlo Method for Nanoelectronic Devices: Particle Description of Quantum Transport and Decoherence

TAKADOUM Jamal
Nanomaterials and Surface Engineering

VODA Alina
Micro, Nanosystems and Systems on Chips: Modeling, Control and Estimation

2008

DAVIM J. Paulo, JACKSON J.
Nano and Micromachining

NOUAILLHAT Alain
An Introduction to Nanoscience and Nanotechnology

OUISSE Thierry
Electron Transport in Nanostructures and Mesoscopic Devices

2006

RIGNEAULT Hervé *et al.*
Nanophotonics

Printed and bound by CPI Group (UK) Ltd, Croydon, CR0 4YY
16/09/2026
14963083-0007